OXFORD SURVEYS
PLANT MOLECULAR AND C[...]

Volume 4

1987

OXFORD SURVEYS OF
PLANT MOLECULAR
AND
CELL BIOLOGY

EDITED BY
B. J. MIFLIN

ASSISTANT EDITOR
H. F. MIFLIN

VOLUME 4

1987

OXFORD UNIVERSITY PRESS

Published in co-operation with the
International Society for Plant Molecular Biology

Oxford University Press, Walton Street, Oxford OX2 6DP

Oxford New York Toronto
Delhi Bombay Calcutta Madras Karachi
Petaling Jaya Singapore Hong Kong Tokyo
Nairobi Dar es Salaam Cape Town
Melbourne Auckland

and associated companies in
Beirut Berlin Ibadan Nicosia

OXFORD is a trade mark of Oxford University Press

Published in the United States
by Oxford University Press, New York

British Library Cataloguing in Publication Data

Oxford surveys of plant molecular and
cell biology. — Vol. 4: (1987)–
1. Plants. Cells
581.87
ISBN 0-19-854233-X

Printed in Great Britain
at the University Printing House, Oxford
by David Stanford
Printer to the University

CONTENTS

Oxford Surveys of Plant Molecular & Cell Biology, Vol.4 (1987) 1-45

RESISTANCE TO PLANT VIRUSES

R.S.S. Fraser

*Biochemistry Section, AFRC Institute of Horticultural Research
Wellesbourne, Warwick, CV35 9EF, U.K.*

*Present Address: AFRC Institute of Horticultural Research,
Worthing Road, Littlehampton, West Sussex, BN17 6LP, U.K.*

INTRODUCTION

Plant viruses can cause serious losses of yield and quality in most crops, although the extent of loss can be difficult to quantify. Some infections are not diagnosed, or are caused by 'cryptic' viruses. A recent survey of 24 apparently healthy species showed six containing double-stranded RNAs, which are thought to be a marker for virus infection (Valverde et al., 1986). Losses have mostly been assessed in artificially inoculated experimental plots, and typically can range from 5-90% (reviewed by Fraser, 1987). It is difficult to extrapolate these values to field conditions where the timing and incidence of infection is an uncontrolled variable. However, it seems likely that most crops suffer losses of between 1 and 10%, with occasional and localized, catastrophic loss (Carr, 1984).

There are fewer methods for control of plant virus diseases than for microbial pathogens or insect pests. Chemical controls are restricted to a few specialised applications, such as assisting in virus elimination from breeding or propagation lines in tissue culture. There is as yet nothing analogous to a fungicide for field use. Host plant resistance is the most important strategy for virus control.

Resistance to viruses is worth studying for various reasons. The full potential for crop protection can only be realized if the mechanisms are understood at the genetical and biochemical levels. There are shortcomings in current resistance strategies: these gaps can only be filled if the underlying mechanisms become accessible to manipulation. At a more academic level, resistance is one of the most intriguing aspects of plant-virus interactions. How do the two genomes interact? What is the nature of the host-pathogen recognition event which determines the pattern of development of the association? Plant viruses – amongst the simplest pathogens at the molecular level – offer particular opportunities to study recognition.

Definitions of Types of Resistance.

Visual inspection suggests that plants have many types of resistance to viruses. Classification is difficult because the mechanisms are not fully understood. However, it is useful to propose three basic mechanisms, operating at different levels of complexity of host population. At each level, the response of the plant can be of two types: these are described by a pair of antonyms. Similarly, the corresponding aspects of viral behaviour can generally be described by a matching pair of antonyms.

Non–host immunity is a species concept. A plant is a **non–host** for a particular virus if there are no detectable signs of multiplication or pathogenesis after inoculation. The virus is **non–pathogenic** to the species.

The other two types of resistance occur within a species which is a **host** for a virus which is **pathogenic** for that species

Genetically controlled resistance occurs in cultivars which have a gene or genes conferring **resistance** to a virus normally pathogenic on that species. Plants without the resistance gene are **susceptible**. The virus interaction with the resistance gene can vary: a strain which is inhibited by the gene is **avirulent**, while a strain which overcomes the inhibitory effects is **virulent** with respect to that gene. The virus thus may have a gene (or genes) for virulence. But as viral genomes are very small, it is unlikely that control of virulence/avirulence is the sole function of that gene product; it may have some other, primary role in pathogenesis.
Host response to a virus may also be described in terms of the severity of disease symptoms, and virus isolates may cause symptoms of varying severity. The terms 'tolerance' and 'aggressiveness' are used to describe these responses; they are separate from the concepts of resistance and virulence, although the mechanisms might overlap. For the virus, 'virulence' and 'aggressiveness' are qualifying attributes of the wider concept of pathogenicity.

Induced resistance operates at the single plant level. A susceptible individual has a form of resistance conferred by a previous infection or other treatment. Normally, this type of resistance is not heritable, although ultimately it must have some genetic basis.

In this article, I will review knowledge of the genetics of these types of resistance, and discuss what is known of the biochemical mechanisms. Where appropriate, the matching aspects of virus behaviour will be considered. At some stages, theoretical models will be described, predictions will be made from them, and compared with observations. Finally, I will discuss the opportunities for further exploitation of resistance mechanisms by recombinant DNA and plant transformation techniques.

NON-HOST IMMUNITY

This is one of the most intriguing and least understood problems in plant virology. The mechanisms involved are, by definition, extremely effective; they would be useful in crop protection. But the mechanisms cannot yet be transferred to host species. The matching question from the virus side is what determines host range. How is it that some viruses are pathogenic on hundreds of species, while others are restricted to one or a few hosts?

Possible Genetic and Biochemical Models

There are three possibilities: two involve action of genes which directly determine host range or immunity, while the third involves indirect effects of genes controlling other features.

In the **positive** model, plants contain a gene or genes conferring completely effective resistance to all known isolates of a particular virus. This is a limit case of genetically controlled (cultivar) resistance, against which the virus has not yet evolved virulence. Holmes (1955) suggested that the durability of this model might reflect additive effects of many resistance genes, perhaps 20 to 40. Evidence for such complex polygenic resistance systems in plants would be hard to obtain by present methods.

In the **negative** model, plants are non-hosts because they lack particular genes required by the virus for pathogenesis. These host 'susceptibility' factors have to match a corresponding series of 'pathogenicity' factors specified by the virus (Bald and Tinsley, 1967). One attraction of this model is that it can explain two phenomena of virus host ranges: congruence and containment. In the former, two viruses have host ranges with species in common, while in the latter, the host range of one virus is completely included in the wider host range of the second. It is suggested that the viruses have common host-range determinants, and that the shared hosts have common susceptibility determinants.

There is little direct experimental evidence for this model. Attempts to show virus-specific attachment sites, or virus-specific uncoating and translation, have generally been unsuccessful or have given equivocal results (reviewed by Atabekov, 1975; Fraser, 1985a).

However, recent work showing that a viral RNA replicase may contain host- and virus-encoded subunits (Mouches et al., 1984) suggest one way in which this type of mechanism could operate. Evans (1985) isolated a mutant of cowpea mosaic virus (CPMV) unable to grow on cowpea varieties which are hosts of the parent strain, but able to grow in other species. The implication is that a pathogenicity function specifically necessary for growth on cowpea was altered by mutation.

In the **passive** model, plants are non-hosts because they have indirect physical or chemical barriers to establishment of infection, such as unsuitable intracellular pH or ionic conditions, or cell wall barriers to infection. There is some experimental evidence for the last mechanism: some viruses infect and multiply well in protoplasts prepared from non-host species (e.g. Furusawa and Okuno, 1978; Huber et al., 1981). However, other evidence suggested that non-host immunity could also be determined at least partly at the intracellular level (Maekawa et al., 1981). Very few combinations of plants and viruses have been examined in this light, so it is premature to suggest that many non-host cells are able to replicate essentially any virus (Van Loon, 1987). But the cell wall or other physical barriers in the intact plant could nonetheless be important in some examples of non-host immunity.

GENETICALLY-CONTROLLED RESISTANCE:
THE GENETIC BASIS OF PLANT-VIRUS INTERACTIONS

The genetics of host-pathogen interactions have been extensively studied for fungi and bacteria. Theoretical models have been developed (reviewed by Crute, 1985) and have served as guides for further work, including the isolation of genes directly controlling host-pathogen interactions (e.g. Staskawicz et al., 1984). For plant-virus interactions, however, genetic models are less well developed.

The Genetics of Resistance

Table 1 summarizes some features of the genetics of resistance, in a random sample of 63 crops and viruses. Mostly, resistance is controlled at a single locus. In some cases, there is evidence for other genes which modify resistance gene action. Most of these are indirect or non-specific, such as through an effect on growth rate. In a few examples, resistance is oligogenic.

The overwhelming simplicity of resistance in genetic terms may reflect the fact that the sample is composed of cultivated species which have been in the hands of plant breeders for decades. Selection

4

TABLE 1. Genetics of resistance to viruses in crop species, and some features of resistance gene action and virulence. Derived from data in Fraser (1986; 1987).

GENETIC BASIS:	Number of host-virus combinations
Single dominant gene	29
Incompletely dominant (gene-dosage dependent)	10
Apparently recessive	11
Sub-total: monogenic	50
Possibly oligogenic	5
Monogenic, with possible modifier genes or effects of host genetic background	8
Sub-total: oligogenic (?)	13
Total number of host-virus combinations in sample	63

LOCALIZATION:	Immune	Yes	Partial	No	Not known	Total
Dominant alleles	0	19	0	2	8	29
Incompletely dominant	0	0	4	8	0	12
Apparently recessive	5	1(?)	1	2	4	13

'Immune' = no virus detectable. 'Yes' = normally involving lesion formation. 'No' = resistance permitting some systemic spread. 'Not known' = not tested, or not reported in the literature.

TEMPERATURE RESPONSE:	ts	tr	Not known	Total
Dominant alleles	7	2	20	29
Incompletely dominant	1	1(?)	10	12
Apparently recessive	1(?)	3	9	13

'ts' = temperature sensitive. 'tr' = temperature resistant.

VIRULENT ISOLATES REPORTED:	Yes	No	Not known	Total
Dominant alleles	16	1	12	29
Incompletely dominant	8	3	1	12
Apparently recessive	4	1	8	13

has been for simply-inherited resistance. The genetics of resistance in wild species might be more complex, but has not been examined. In a few crops, such as tomato and potato, there have been attempts to breed for oligogenic resistance, by introduction of individual resistance genes from several related species.

For the monogenically-inherited resistances in Table 1, most are phenotypically dominant, with smaller proportions being incompletely dominant (i.e. gene–dosage dependent) or recessive. In fact, both dominant and recessive categories are probably overestimates, as some examples of both have turned out to show gene–dosage dependence when symptom expression and virus multiplication were compared in plants with heterozygous or homozygous resistance alleles (e.g. Fraser and Loughlin, 1980; 1984).

Three Types of Genetic Model for Plant Virus Interactions

In **positive** models (Fig. 1a), the resistant plant produces a factor which directly or indirectly inhibits pathogenesis; the susceptible plant does not. Virulent isolates of the virus fail to interact with the factor and so are not inhibited.

In **negative** models (Fig. 1b), the resistant plants lacks some function required by the virus for normal pathogenesis; this function is present in the susceptible plant. Virulent isolates of the virus are able to establish pathogenesis without the assistance of the host function.

Note that both models involve a recognition event which controls a 'go/no-go' decision. In the positive model, resistance is switched on, while in the negative model, susceptibility is switched on.

The third model (Fig. 1c) is not an 'all-or nothing' response, but involves quantitative interactions between host- and virus-specific functions. These interact to form a product: the outcome is determined by the concentration and nature of the product. These in turn depend on the nature and on the concentrations of the host and virus products, and among other things will be strongly influenced by the dosage of resistance or susceptibility alleles. Note that quantitative interaction models could be either positive or negative in action.

Table 2 shows some predictions made from these models, about resistance phenotypes and the nature of virulence. Mutation to virulence in the negative models may be a difficult step for the virus, as it means becoming able to multiply without a host helper function, or with a defective one. In contrast, in positive models, it should be easier for the virus to retain the ability to multiply while losing the interaction with the host inhibitor.

Comparison of the Models and Predictions with Observation

Table 1 summarizes some further properties of resistance and virulence in the sample, as tests of these models and their predictions. In many cases, the literature is unclear about the property tested: these are listed under 'not known'.

Phenotypically, resistance to viruses in plants may be expressed in one of three forms: as a localization response around the site of the initial infection, usually with formation of a necrotic lesion; as apparent complete immunity with no detectable symptoms or multiplication after inoculation; or as non-necrotic resistance which allows some limited virus multiplication and spread through the plant. It is clear from Table 1 that localizing resistance is strongly associated with phenotypically dominant genes. For recessive alleles there is only 1 case of localization and this has been shown to be anomalous, as the alternative response is complete immunity (Thompson et al., 1962). Complete localization is also absent for incompletely dominant genes. The association of localization with dominant alleles strongly suggests a positive model involving a go/no-go switch. Such mechanisms may also be induced rather than constitutive: localization and necrosis generally do not occur until 48-72 h after inoculation (reviewed in Fraser, 1985b). In the few cases examined, the rates of virus multiplication during the early stages of infection of resistant (localizing) and susceptible plants were identical (Takahashi, 1973).

Incompletely dominant genes tend to permit some systemic spread of virus, but to inhibit multiplication. This is consistent with a quantitative interaction model, which also explains gene-dosage dependence. In at least one case, the resistance mechanism appears to be constitutive rather than induced (Fraser and Loghlin, 1980), in that virus multiplication is strongly inhibited from the time of inoculation.

Recessive resistance alleles, leaving aside the anomalous case and those where the response is not know, tend to give complete immunity. These examples are consistent with the negative model, which should involve complete loss of susceptibility and be completely recessive. Resistance in this model is, by definition, constitutive rather than induced.

The temperature-response of resistance has not been examined in most cases, but the limited data suggest that temperature sensitivity is most associated with dominant, localizing alleles, which are thought to involve production of a virus inhibitor. Why this inhibitor or related metabolism should tend to be temperature-sensitive under conditions where plant growth is still fully possible is not clear. Perhaps the localization mechanism of necessity involves some step which is intrinsically sensitive to heat.

The single temperature-sensitive recessive example involves apparent complete immunity, which would be difficult to reconcile with

1. Positive model: resistance is switched on

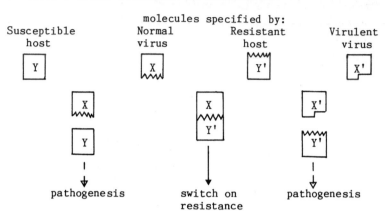

molecules specified by:

Susceptible host · Normal virus · Resistant host · Virulent virus

Y · X · Y' · X'

X / Y → pathogenesis

X / Y' → switch on resistance

X' / Y' → pathogenesis

2. Negative model: susceptibility is not switched on

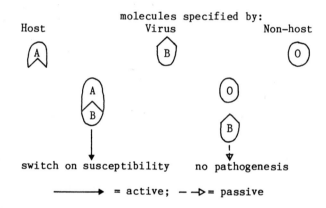

molecules specified by:

Host · Virus · Non-host

A · B · 0

A / B → switch on susceptibility

0 / B ⇢ no pathogenesis

⟶ = active; ⇢ = passive

3. The quantitative interaction model

molecules specified by:

Normal virus	Virulent virus	Susceptible host	Resistant host
V	V'	H	H'

$$[V] + [H] \rightleftharpoons [VH] \longrightarrow \text{susceptibility}$$

$$[V] + [H'] \rightleftharpoons [VH'] \longrightarrow \text{resistance}$$

$$[V'] + [H] \rightleftharpoons [V'H] \longrightarrow \text{susceptibility}$$

$$[V'] + [H'] \rightleftharpoons [V'H'] \longrightarrow \text{susceptibility}$$

8

TABLE 2. Predictions from the resistance models

POSITIVE MODELS (Resistance = inhibition)	**Resistance is:** ■ dominant if recognition is a go/no-go event ■ gene-dosage dependent for quantitative interactions ■ never fully recessive ■ possibly temperature sensitive **Virulence is:** ■ a virus function has an altered interaction with the host resistance gene function, or fails to interact
NEGATIVE MODELS (Resistance = reduced or lack of susceptibility)	**Resistance is:** ■ probably recessive for a go/no-go recognition event ■ gene-dosage dependent for qualitative interactions ■ never fully dominant ■ unlikely to be temperature sensitive **Virulence is:** ■ ability of the virus to multiply without the host susceptibility factor

Figure 1. (opposite) Models for interactions between host- and virus- specified molecules which may determine susceptibility or resistance. (Reproduced from Fraser (1987) by permission of Research Studies Press/John Wiley and Sons, Chichester.)

the predictions, as there should be no resistance gene products, and therefore no temperature-sensitivity. However, it appears that this example involved selection of resistance-breaking 'thermal' strains of virus, and not simple temperature sensitivity of resistance to normal strains.

Fig. 2 summarizes these properties of resistance genetics and proposed mechanisms. The 'subliminal' infections occur in a few hosts, and have been shown to involve active virus multiplication in the directly-inoculated cells only (Sulzinski and Zaitlin, 1982). Resistance appears to operate against cell-to-cell spread of the virus, with the exception of the subliminal infection of cotton with cauliflower mosiac virus, where some limited spread occurs (Hussain et al., 1987). Such mechanisms might be regarded as examples of localization and arguably, placed on the left in Fig. 2. However, they are probably constitutive rather than induced; the resistance might be a lack of host function facilitating cell-to-cell transport. As yet there appears to be no information on the inheritance of the response.

SOME PROPERTIES OF MECHANISMS OF RESISTANCE TO VIRUSES

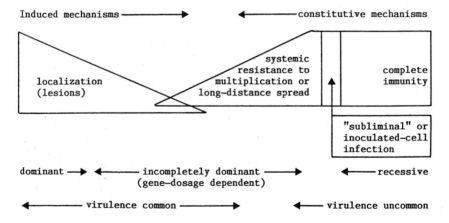

Figure 2. Some general conclusions about mechanisms of resistance to viruses and the nature of genetic controls. (Reproduced from Fraser (1988) by permission of the Ciba Foundation/John Wiley and Sons, Chichester).

Overall, the results from this genetic survey suggest that positive (inhibitor) resistance mechanisms predominate, but that a minority of cases may involve negative (lack of susceptibility) mechanisms.

The Frequency of Virulence

Table 1 shows that virulent isolates have been reported for about 60% of resistance genes. Many of the remaining genes do not appear to have been tested with a number of virus isolates, so their durability is at least questionable. Less than 10% of the resistance genes in the sample have not been overcome after protracted challenge by numerous isolates. Thus for virus resistance, durability appears to be the exception, and resistance-breaking isolates are to be expected for most genes.

Although the data are limited by the large proportion of 'not known' cases, it would appear that virulence is commoner against dominant and gene-dosage dependent alleles than against recessive alleles. This agrees with the predictions of the models – that virulence against a negative mechanism should be a more difficult evolutionary step for the virus – but further observations would be useful.

The Genetics of Virulence

With microbial pathogens, virulence has been analysed by crossing and study of segregation; this is of course impossible with viruses. However, viruses are easier than microbial pathogens to analyse at the molecular level. The genetics of virulence will ultimately be explained at the level of nucleic acid sequences.

In the meantime, some genetic analysis has been achieved by indirect routes. Many plant viruses have divided genomes: the genetic information is distributed over a number (2 to 16) of RNAs, all of which are required for full pathogenesis. Individual genetic segments can be purified from isolates differing in biological properties, including virulence, and infectious pseudorecombinants can be prepared. This allows 'mapping' of genes controlling particular biological properties to individual segments of RNA.

Pseudorecombinants have been prepared between isolates differing in virulence or host range for at least six viruses with di- or tri-partite genomes (reviewed by Fraser and Gerwitz, 1986). Interpretation is hindered because the other functions specified by each particular segment are incompletely understood. However, in several examples the virulence determinant did not map on the segment containing the coat protein gene, which eliminates it as an element controlling virulence in these cases. In cucumber mosaic virus (CMV), two determinants were required for systemic infection of *Lactuca saligna* (Edwards et al., 1983), suggesting that virulence could be oligogenic.

For viruses with monopartite genomes, pseudorecombinant analysis requires different approach. They can be constructed using DNA clones by cutting at appropriate restriction sites and religating. This has already been achieved for partial genomes of different tobacco mosaic virus (TMV) strains, to produce chimaeric progeny (M. Bevan, A. Goldsborough, C. Thomas and R. Fraser, unpublished results). As plants can be transformed or infected with cloned DNA copies of plant viruses or viroids, and are able to express these as functional pathogens (e.g. Meshi et al., 1984; Grimsley et al., 1987), the route to fine-scale mapping of virulence determinants should be open. These analyses can then be combined with site- directed mutagenesis and sequencing. In vitro mutagenesis of CaMV DNA clones has already been used to modify symptom expression and infectivity (Melcher et al., 1986).

Mapping of a virulence determinant on the monopartite genome of TMV has been achieved by mutagenesis. In some Nicotiana species and tobacco varieties, the *N'* gene localizes some isolates of TMV, while others are virulent and spread systemically. Kado and Knight (1966) mutagenized virulent (systemic) isolates to avirulence (local lesion) by treatment with nitrous acid. The frequency of mutation of naked TMV RNA is much higher than when it is encapsidated in coat protein. The TMV rod can be sequentially stripped from the 5'-end by treatment with sodium dodecyl sulphate. Kado and Knight mutagenized populations of rods with different degrees of stripping, and found an increase in the number of avirulent mutants formed when a region of RNA about 30% of the distance from the 3'-end became exposed. This area contains the gene for the 30 kDa protein, which is thought to be involved in cell-to-cell spread of the virus (Leonard and Zaitlin, 1982). Zimmern and Hunter (1983) compared a systemic strain and a complex local lesion mutant derived from it, and found a base-substitution in this gene, and a consequent alteration in the cryptic peptide map of the protein product. These changes are consistent with a role of the 30 kDa protein in determining localization or virulence, but do not exclude a determinant elsewhere in the genome, as the TMV RNAs were not completely sequenced.

Host-Virus Interactions: The Gene-for-Gene Relationship

Breeding for resistance genes in the host, and the resulting selection pressure on the pathogen for matching virulence genes, has led to the co-evolution of 'gene-for-gene' interactions. These are most highly developed and complex for fungal pathogens and can involve combinations of scores of resistance and virulence genes (Flor, 1956; Person, 1959). For viruses, there have been two schools of thought. Robinson (1976) considered that plant viruses have too great a capacity for rapid evolution to permit the relative stability required for gene-for-gene interactions. In contrast, Russell (1978) pointed to some examples of extreme durability of resistance to plant viruses as evidence against gene–for–gene systems. In fact, as Table 1 illustrates, extreme durability is infrequent, and most resistance genes have been matched by corresponding virulence. However, evolution of virulence by the virus is sufficiently slow and geographically separated to allow the co–evolution of comparatively stable gene–for–gene interactions.

Fig. 3 shows one well–established example, resistance to TMV in tomato (Pelham, 1972). The virus has evolved virulence genes to overcome all three host resistance genes individually, although different virulence genes have occurred with varying frequency. TMV–1 isolates are very common, TMV–2 fairly rare, and TMV–2² extremely rare. TMV isolates have also been found which contain combinations of virulence genes: 1 plus 2, or 1 plus 2². However, no isolate yet has 2 plus 2². This might be because these two virulence 'genes' are allelic and mutually exclusive. More trivially, the infrequency of

Host genotypes	Virus genotypes							
	0	1	2	2^2	1.2	1.2^2	2.2^2	$1.2.2^2$
+	M	M	M	M	M	M	m	m
Tm–1	R	M	R	R	M	M	r	m
Tm–2 [a]	R	R	M	R	M	R	m	m
Tm–2² [a]	R	R	R	M	R	M	m	m
Tm–1/Tm–2	R	R	R	R	M	R	r	m
Tm–1/Tm–2²	R	R	R	R	R	M	r	m
Tm–2/Tm–2²	R	R	R	R	R	R	m	m
Tm–1/Tm–2/Tm–2²	R	R	R	R	R	R	r	m

Figure 3. Genetic interactions between TMV–resistant tomato plants and strains of the virus. 'M' indicates systemic mosaic, 'R' indicates resistance. 'm' and 'r' represent predicted responses with virulence genes which have not yet been demonstrated in virus isolates. ᵃPlants with these genotypes may show local and variable systemic necrosis when inoculated with a virulent strains. (Reproduced from Fraser (1985b) by permission of Martinus Nijhoff/Dr W Junk Publishers, Dordrecht)

2^2 virulence may make the double virulence highly improbable.

The most complex gene–for–gene system so far characterized for plant–virus interactions is in resistance to bean common mosaic virus (BCMV) in *Phaseolus vulgaris* (Drijfhout, 1978; Day, 1984). This involves host genes at four loci, allelic series at two of these, and numerous resistance–breaking isolates of the virus. All but one of the host resistance genes have been overcome. One interesting point is that virulent isolates able to overcome resistance gene *bc-1* but not the allelic *bc-1²* have been found, whereas all isolates able to overcome *bc-1²* so far have also been able to overcome *bc-1*. It may be that the 1^2 virulence gene automatically encompasses the property of virulence against *bc-1*, although the converse is not true.

The examples of apparent mutual exclusion of different virulence 'alleles' in TMV, and of containment of virulence against other resistance genes in BCMV, may mean that there is a limit to the number of virulence 'genes' that can be accommodated in the limited genetic information of a virus. Breeding for complex host resistance systems might place an impossible evolutionary barrier in front of the virus. In contrast, fungi and bacteria appear to be able to accumulate large numbers of virulence genes; this may stem in part from an ability to accumulate extra copies of genes, which is not open to viruses.

GENETICALLY–CONTROLLED RESISTANCE AND VIRULENCE: BIOCHEMICAL MECHANISM

Resistance mechanisms may operate before the establishment of infection, or after replication has commenced. The former mechanisms are largely indirect: resistance to vectors, vector non–preference and resistance to seed transmission (reviewed by Fraser, 1986). A special case concerns the resistance of the tobacco variety T.I. 245 to establishment of infection by a number of different viruses (Holmes, 1961). This appears to be correlated with structural properties of the epidermis which disfavour infection (Thomas and Fulton, 1968). None of these mechanisms involve specific interactions between plant and virus, and they are therefore not considered further.

Resistances which operate within the plant, where there is specific interaction, will be considered under two headings: localization mechanisms normally involving lesion formation, and other types. Local lesion mechanisms probably share many biochemical features in different host–virus combinations. Those which do not cause local lesions, and frequently permit some spread of virus, are probably more diverse.

Local Lesion Mechanisms

A lesion forms around each point of infection, and in the majority of cases becomes necrotic. This response is also frequently called hypersensitivity. A few localization mechanisms are not necrotic, but involve chlorotic lesions, or cryptic lesions which only become visible when the leaf is stained to show local accumulation of starch.

In localization, the early phase of virus multiplication is typically uninhibited. Prevention of virus spread occurs only after several hundreds of cells have become infected, and necrosis appears 2-4 days after inoculation. However, most evidence suggests that necrosis is not the primary factor preventing virus spread. Virus particles can frequently be detected by electron microscopy in green tissue around the necrotic area (e.g. Konate et al., 1983), and the size of the necrotic area can be reduced by various treatments, without reducing virus spread or multiplication (e.g. Takusari and Takahasi, 1979). This leaves a question: why, if necrosis is not directly responsible for inhibition of virus spread is it found so frequently in localizing resistance? One factor is that necrosis may serve to reduce the infectivity of the virus already formed, by formation of toxic quinones and other antiviral compounds (reviewed by Ireland and Pierpoint, 1980). This may inhibit secondary infections. Another possibility is that necrogenesis is an almost inevitable secondary consequence of the processes primarily involved in localization. The early changes in membrane permeability (discussed below) and subsequent increases in ethylene biosynthesis and lignification, suggest a possible pathway.

Two experimental approaches have been used to study the biochemistry of localization. Some workers have described changes in various aspects of metabolism, and have attempted to relate these to the development of resistance. These experimentas have often involved the use of inhibitors, to establish causal relationships, and temperature-shift experiments, since localization is frequently temperature-sensitive. A major problem in all investigations has been that necrogenesis causes secondary changes in metabolism which are not directly involved in the primary localization of virus. The second approach, therefore, has been to attempt to isolate molecules with a demonstrable antiviral activity, or which will induce an antiviral state in test plants.

Metabolic Changes During Localization

Most studies have been with TMV infection of tobacco containing the *N* gene for hypersensitivity, although CMV, tomato bushy stunt (TBSV) and tobacco necrosis virus (TNV) have also been used in tobacco, and in various other hosts. Fig. 4 shows the relationship of a number of changes. Although these are derived from various systems, the Figure gives a picture of the relative timings.

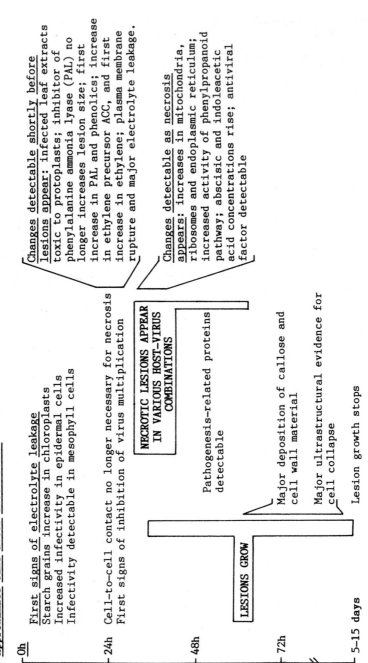

Figure 4. (opposite). Relative timing of changes of host metabolism during the localisation of viruses in (normally) necrotic lesions. The data are drawn from a number of different combinations of resistant hosts and viruses, therefore the timings are relative rather than absolute. References are given in the text. Reproduced from Fraser (1987) with permission of Research Studies Press/John Wiley and Sons, Chichester.

Membrane and permeability changes

Electrolyte leakage, an indicator of increased membrane permeability, is one of the earliest changes detectable, and precedes the onset of necrosis (Weststeijn, 1978; Pennazio and Sapetti, 1982). The later, major increase in permeability is probably associated with necrosis rather than with localization, and is likely to be an early manifestation of the wounding effect causing the increased rate of ethylene systhesis at about the time of lesion appearance. Ruzicska et al. (1983) found an earlier increase in saturation of the phospholipid-bound fatty acids, but the relationship of this to localization is not clear.

Plant growth regulators.

Many workers have applied exogenous plant growth regulators and have shown numerous effects on lesion size and number. The results (reviewed by Fraser and Whenham, 1982) have given little insight into the processes of virus localization. Studies of endogenous growth regulators have shown increases in auxin and abscisic acid (reviewed by Fraser, 1987), without suggesting any role in localization. Changes in ethylene synthesis, which have been comprehensively investigated by De Laat and Van Loon (1982; 1983), are too late to be involved in the early stages of localization, but undoubtedly have a role in controlling secondary metabolic changes, and especially the induction of the pathogenesis-related proteins. These are considered later.

Phenolic compounds and oxidation

A general increase in respiration accompanies localization, and is indicative of enhanced metabolic activity. Various phenolic compounds such as chlorogenic acid and scopoletin accumulate during necrogenesis: although the major changes occur after necrosis appears, rates of synthesis are enhanced before necrosis (Fritig et al., 1972). Peroxidase and polyphenoloxidase activities increase, and toxic quinones accumulate. The phenolics are thought unlikely to be involved in primary localization (Weststeijn, 1976).

Phenylpropanoid metabolism.

One of the most detailed studies of metabolic changes associated with necrosis has been that of the phenylpropanoid pathway by Fritig and his collaborators. Increased activity of the first enzyme, phenylalanine ammonia lyase (PAL), was detectable before the appearance of necrosis (Fritig et al., 1973), and was shown by density labelling to be due to de novo synthesis rather than activation (Duchesne et al., 1977). Subsequent enzymes in the pathway, the 0-methyl transferases, became activated in sequence (Collendavelloo et al., 1983). Two competitive inhibitors of PAL, α-amino-oxyacetate (AOA) and L-α-aminooxy-β-phenylpropionic acid (AOPP) caused increased lesion size, which was interpreted as a weakened resistance response (Massala et al., 1980; 1987). AOPP was a more potent inhibitor of PAL activity, and AOA of accumulation of insoluble lignin polymers. AOA weakened resistance more than AOPP. This led Massala et al.(1987) to suggest that insoluble lignins are likely to be the products of the phenylpropanoid pathway which are involved in virus resistance. Ultrastructural and histochemical studies have shown that lignification and many other wall modifications accompany lesion formation. However, the major changes occur rather late, and several authors have concluded that they are probably not involved in the early processes of virus localization (e.g. Russo et al., 1981).

Changes in transcription and protein synthesis.

There are two ultimate objectives: to study the synthesis of the RNA product of the single gene which controls the inheritance of the localization response, and to characterize and study the function of its (presumed) protein product. It is also of interest to study any other proteins which may be involved in the primary resistance reaction. Making the reasonable assumption that transcription of the gene for localization is induced during the early part of virus multiplication, one would expect to find its mRNA in total mRNA from infected plants, but not from healthy plants. mRNA preparations of the two types can be used to screen bacterial colonies containing clones of cDNAs to host polyadenylated mRNAs, to select those which 'light up' with probes from infected plant mRNA but not with those from healthy plants (Smart et al., 1987). There seem to be two problems. Firstly, the changes in gene activity after infection involve many genes, such as those coding for the pathogenesis related proteins (Hooft van Huijsduijnen et al. (1986b), and it is difficult to select the single product of the localization gene against this background. Secondly, as the nature of the gene product is not known, there are no simple criteria for recognizing it other than activation.

At the protein level, changes in numerous enzymes have been found. Apart from those discussed above, others are reviewed by Fraser (1987). There is little evidence for direct involvement of any of these in localization.

The most striking protein change accompanying lesion formation is the induction of the 'pathogenesis-related' (PR) proteins. Demonstration of enzyme activity is recent, but they have long been considered to be associated with an induced resistance mechanism known as 'acquired systemic resistance'. Their properties are considered later in that context. But are they involved in the primary localization and lesion formation? Surprisingly, this question has only been addressed recently, even though the proteins have been known since the early 1970s. One problem was the difficulty in assaying the small amounts of PR proteins present during the early stages of infection. Antoniw and White (1986) have now developed an ELISA method which can measure low concentrations of PR protein in different parts of single lesions. Fig. 5 shows changes in the concentration and distribution of TMV and the commonest PR protein, PR1a, during lesion formation on tobacco leaves. In later stages, the highest PR concentrations were in a ring around the necrotic centre, while the highest TMV concentrations were in the middle of the lesion. Antoniw and White (1986) argued that this would be consistent with a role of PR1a in preventing virus spread, although there is no direct evidence of an antiviral effect. Also, PR1a was only detected by day 5 after inoculation, whereas necrosis and localization were

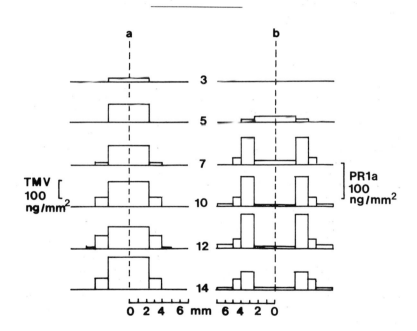

Figure 5. Changes with time in (a) tobacco mosaic virus (TMV) and in (b) pathogenesis-related protein (PR1a) content in different parts of single local necrotic lesions. The data are presented as cross-sections through lesion centres, shown by the broken line. (Reproduced from Antoniw and White (1986) by permission of the authors and Martinus Nijhoff/Dr W Junk, Dordrecht)

established much earlier. This suggests that PR proteins are unlikely to be involved in the earliest stages of localization, although they could be involved in later restriction of the virus.

Antiviral Factors

Many plants with or without a virus infection, contain substances which will inhibit virus infection or multiplication on a second, test species. The biological basis for this apparent altruism is far from clear, but the inhibitors probably have some separate function, unrelated to viruses, in the donor species. The donor species do not appear to be less susceptible to viruses.

The substances include proteins, frequently glycoproteins, RNA, and several low molecular weight compounds. The glycoproteins from *Phytolacca americana* and *Dianthus caryophyllus* are well characterized (e.g. Stirpe et al., 1981), and are potent inhibitors of translation. Other compounds appear to induce a systemically–effective resistance, and the formation of secondary antiviral compounds, in the test species (e.g. Verma and Dwivedi, 1984). Mostly, the resistance induced is not specific for any particular virus and the mode of action remains to be demonstrated. These phenomena are of interest, but one step removed from specific resistance mechanisms operating with the plant. They are therefore not considered further here. Recent reviews include those by Chessin (1983) and Fraser (1987).

Two main types of antiviral factor have been reported in genetically resistant hosts. Both concern resistance to TMV in *Nicotiana* controlled by the *N* gene, although there are a very few reports of similar compounds from other combinations of host and virus (e.g. in TNV-infected *Chenopodium*; Facciolo and Capponi, 1983).

Sela and his co-workers have partially purified and characterized an 'Antiviral Factor' (AVF). This appears to be a phosphorylated glycoprotein (Mozes et al., 1978; Antignus et al., 1977). Most tests of the antiviral activity of AVF have involved mixing it with TMV and showing that it inhibited lesion formation on *N*–gene tobacco. This may inhibit establishment of infection, rather than TMV multiplication as such. However, Sela (1981) has also proposed complex schemes for the metabolism and mode of action of AVF, including a role in the induction of some further antiviral state in uninfected tissue. Most proposed modes of action have still to be supported by experimental evidence. The considerable literature on this topic is more extensively reviewed elsewhere (Sela, 1981; Fraser, 1985b; 1987).

Throughout the work on AVF, parallels have been drawn with the purification, properties, and mode of action of interferon in mammalian cells. These parallels became even more striking with reports by Orchansky et al. (1982) and Reichmann et al. (1983) that human interferon was a potent inhibitor of TMV multiplication in tobacco, and indeed was considerably more effective than tobacco AVF.

Others, however, have been unable to demonstrate effects of mammalian interferons against TMV and other plant viruses (Antoniw et al., 1984; Huisman et al., 1985; Loesch-Fries et al., 1985).

Two proteins, together named 'Inhibitor of Virus Replication' (IVR) have been purified and characterized by Loebenstein and co-workers. IVR is released from TMV-infected protoplasts prepared from *N*-gene tobacco leaves. IVR was shown to inhibit TMV multiplication in protoplasts from resistant or susceptible tobacco (Loebenstein and Gera, 1981), and also to inhibit CMV and potato virus X (PVX) multiplication in other species (Gera and Loebenstein, 1983). IVR is therefore neither virus- nor host-specific. As many of the experiments showed incomplete inhibition of virus multiplication, it would be worth checking whether IVR also inhibits aspects of host metabolism, i.e. whether the antiviral effect is merely an expression of general phytotoxicity.

Hooley and McCarthy (1980) concentrated a low molecular weight protein from TMV-infected leaves of *N*-gene tobacco, and have shown that it is detrimental to the survival of protoplasts, with or without the *N*-gene. They suggested that this factor may be involved in necrogenesis in whole infected leaves. Presumably, infected protoplasts from *N*-gene plants do not normally turn necrotic because the factor does not reach high enough concentrations in culture.

Resistance Mechanisms which do not form Lesions

This diverse group contains mechanisms which are systemically effective: i.e. which may permit spread of the virus throughout the plant, but with a reduced level of multiplication and symptom development. Other examples prevent long distance spread of virus, especially from the inoculated leaf, but do not prevent local (cell-to-cell) spread from the point of infection. Yet others contain the virus within a single infected tissue, such as the phloem, while the 'subliminal infection' mechanisms appear to restrict the virus to the initially infected cell, or to those close by. It is probable, though not fully established, that many resistance mechanisms in this group are constitutive. Mechanisms in this group may have various targets: virus multiplication, processes of virus spread over short or long distances, or from tissue to tissue, and symptom formation as such.

'Tolerance' is a rather loose term describing resistance which allows some virus multiplication and spread, but inhibits development of visible symptoms. An implication is that the resistance activity is directed against the expression of pathogenic effects, rather than against virus multiplication as such. However, 'tolerant' plants can show inhibition of virus multiplication (e.g. Fraser and Loghlin, 1980) compared to susceptible plants, and symptomless 'tolerant' plants can still show reduced yield compared with healthy plants (e.g. Kooistra, 1968). There do not appear to be any substantiated examples of resistance mechanisms which can operate solely against

21

symptom formation, without interacting with other components in the host–pathogen relationship (see Fraser et al., 1986).

Experiments with isolated protoplasts have given some insight into the differences between local lesion resistance and the other types. But only a few examples of each have been studied in protoplasts, so the following generalizations are preliminary. Localization mechanisms tend not to give effective resistance in protoplasts (e.g. the *N* gene in tobacco; Otsuki et al., 1972, and the *Tm–2²* gene in tomato; Motoyoshi and Oshima, 1975). However, the resistance may still influence the amount of virus made in protoplasts, and the interaction with factors such as growth regulators in the culture medium (Loebenstein et al., 1980). In contrast, plants with resistance which does not cause lesion formation tend to yield protoplasts where resistance is still effective (e.g. the *Tm–1* gene in tomato; Motoyoshi and Oshima, 1979, and the genetically undefined resistances to CMV in cucumber; Maule et al., 1980). The situation is particularly complex in cowpeas, where protoplasts from the majority of 'immune' host lines are susceptible to cowpea mosiac virus, but those from one line (Arlington) are partly resistant, accumulating only one-tenth as much virus as protoplasts from susceptible cultivars (Beier et al., 1979). Interestingly, the authors regarded 'immunity' in Arlington as not representing an exaggerated form of hypersensitivity. It is perhaps better to consider it as a limit case of resistance to multiplication, which prevents replication so effectively in the intact leaf that restriction of spread must follow. Some specific examples of these mechanisms will now be considered in detail.

Resistance to cowpea mosaic virus (CPMV) in cowpea.

Kiefer et al. (1984) showed that CPMV infected protoplasts from resistant 'Arlington' cowpeas produced less capsid protein per unit of (+) strand RNA than susceptible protoplasts. Translation of CPMV RNAs produces polyproteins, which are cleaved by a virus–encoded protease to the mature, functional proteins. The protease activity is highly specific, and will not process the polyproteins of related comoviruses (Goldbach and Krijt, 1982). Sanderson et al. (1985) showed that Arlington protoplasts contain an inhibitor of the proteinase activity of the common (avirulent) isolate of CPMV, and suggested that this inhibitor is the basis of the resistance. A severe isolate, CPSMV-DG, which is not inhibited in Arlington protoplasts, was not affected by the protease inhibitor. Some of these results require expansion, for example, more tests of inhibitor specificity are required. But if confirmed, this system would provide one of the most complete explanations of the biochemistry of resistance and virulence.

Resistance to long-distance transport.

Resistance may operate by preventing spread from the inoculated leaf to the remainder of the plant in at least three hosts, but in all might involve further components.
Thus in resistance of cowpea to cowpea chlorotic mottle virus

(CCMV), Wyatt and Kuhn (1979) showed that the virus failed to move from the inoculated leaf, but that its multiplication in the inoculated leaf was also inhibited. The virus which did accumulate in the inoculated leaf contained comparatively little genomic RNA-3. This contains two cistrons. One codes for the coat protein, and is expressed via a monocistronic mRNA, RNA-4. This was present in normal proportions in resistant plants, so coat protein synthesis was unlikely to have been the target of resistance. The product of the other cistron is a 35,000 M_r protein which may be involved in replication. Experiments with pseudorecombinants between avirulent and virulent isolates of CCMV have suggested that both RNA-1 and RNA-3 influence replication in resistant cowpeas, whereas the ability to invade systemically is controlled by RNA-1 (Wyatt and Kuhn, 1980). One possibility is that RNA-1 plus the resistance factor regulate the amount of RNA-3 made, and this controls the resistance phenotype. Perhaps surprisingly, protoplasts from resistant plants were unable to inhibit virus multiplication (Wyatt and Wilkinson, 1984). Bijaisoradat and Kuhn (1985) found a similar resistance to movement of CCMV from inoculated leaves in certain lines of soybean.

In *Phaseolus vulgaris* plants containing the *bc* gene system for resistance to BCMV, the *bc-1* and *bc-2* genes appear to operate by preventing virus spread from the inoculated leaf, where multiplication was as high as in susceptible plants (Day, 1984). Virus was undetectable even in the petioles, suggesting a failure of loading into the phloem, which is the pathway of long-distance transport (Ekpo and Saettler, 1975).

Full expression of *bc-1* or *bc-2* resistance is also dependent on the presence of the *bc-u* gene: from experiments where the genes were tested separately, it appears that *bc-u* alone has no effect; *bc-1* alone allows systemic spread but inhibits formation of systemic symptoms, and prevention of systemic spread and symptoms is only obtained when both resistance factors are present (Day, 1984).

Resistance involving tissue limitation.

Luteoviruses, which are transmitted by aphids feeding only from the phloem, are restricted to the phloem in susceptible plants. The mechanisms which prevent access to other tissues are of interest in terms of how virus movement is controlled, and may also be enhanced in resistant plants (Barker and Harrison, 1986).

There appears to be nothing intrinsic to the non-phloem tissues which causes resistance. Kubo and Takanami (1979) showed that leaf mesophyll protoplasts could be infected with tobacco necrotic dwarf virus (TNDV), which is phloem-limited in intact plants. Barker (1987) showed that invasion of non-phloem tissue by potato leaf roll virus (PLRV) was enhanced by a simultaneous infection with the unrelated potato virus Y (PVY), although the nature of the complementation was not clear.

Resistance allowing full systemic spread.

There are two main examples. In tomato, the *Tm*-1 gene inhibits TMV multiplication in a gene–dosage–dependent manner, but is dominant for suppresion of symptom formation (Fraser and Louglin, 1980). The gene is temperature-sensitive for inhibition of multiplication, but temperature-resistant for prevention of symptoms (Fraser and Loughlin, 1982). These examples of apparent independence of the two antiviral functions might suggest that resistance is actually controlled by separate, though closely-linked genes, or that expression involves a branched pathway with two functional end-products. However, the observations could still be consistent with a single functional end product, if the kinetics of reactions leading to the two effects are different. The mathematical analysis of genotype–phenotype relationships by Kacser and Burns (1981) may prove useful in this respect, as well as in other analyses of host–pathogen interactions.

Evans et al. (1985) suggested that the *Tm*-1 gene might operate by preventing synthesis or modification of the enzyme activity required to form a functional TMV RNA replicase, but could not exclude more trivial explanations of the results.

In cucumbers, there are two sources of systemically-effective resistance to CMV (Coutts and Wood, 1977). In plants, resistance suppresses mosiac formation and inhibits virus multiplication by 90% (Wood and Barbara, 1971; Amemiya and Misawa, 1977). Resistance only became effective 36h after inoculation, and could be prevented by treatment with inhibitors of transcription or translation (Barbara and Wood, 1974; Levy et al., 1974), suggesting that an inhibitor of virus multiplication is synthesized soon after infection. However, the effectiveness of resistance in protoplasts (Maule et al., 1980) suggests that the mechanism must operate within the single cell: in protoplasts there was no initial phase of unrestricted virus multiplication. Furthermore, treatment with actinomycin-D or ultraviolet light did not prevent inhibition (Boulton et al., 1985). These results are somewhat at odds with the whole-leaf experiments, leading Boulton et al. (1985) to suggest that there might be two components of resistance: one constitutive, and operating to reduce virus multiplication within the individual cell, and an induced mechanism, not operating in protoplasts, which requires co-operation between cells.

Mechanisms of Virulence

With the exception of the limited information from quasi-genetical studies, and the example of the CCMV proteinase inhibitor, we are not close to a biochemical explanation of the nature of virulence in most plant-virus interactions. Genetic studies suggest that virulence is likely to involve different mechanisms, and different virus functions, in different host-virus combinations.

In resistance to TMV conferred by the N' gene in tobacco, the evidence discussed above suggests that the 30,000 M_r viral protein, thought to be involved in cell-to-cell transport, contains the virulence function. However, there is also evidence that the virus coat protein may be involved: Tsugita (1962) showed that mutants with an altered pattern of interaction with the N' gene had a high frequency of altered coat proteins, while Fraser (1983) found an associated between the temperature-sensitivity of the coat protein, and the size of lesions. The coat protein might at least be able to modify the pattern of interaction, even if it is not the major determinant of virulence.

Coat protein properties have also been implicated in the interaction between TMV and the Tm-1 gene in tomato. Dawson et al., (1979) suggested that there was no correlation between ability to overcome Tm-1 and the amino acid composition of the coat protein, but their data suggested that all the resistance-breaking isolates could have had altered coat proteins. Fraser et al. (1983) found that 20 resistance-breaking isolates, derived from an avirulent isolate under conditions favouring single-hit mutations, had altered coat proteins.

De Jager and Wesseling (1981) studied two natural isolates of CPMV which overcome the local lesion reaction on cowpeas. Both spread systemically, but one induced necrosis, whereas the other caused mosaic. They suggested two components of virulence: the ability to evade localization, and the ability to induce or not to induce necrosis. Analysis of pseudorecombinants showed that both mapped on the same genomic segment.

INDUCED RESISTANCE

This topic is very diverse. Some mechanisms are induced by a range of pathogens, chemicals and environmental or developmental factors, while others are induced only by specific agents. Some types are effective against various types of pathogen and even insect pests (McIntyre et al., 1981), while others are only effective against certain isolates of a particular virus. Induced resistance can be localized, effective only in the proximity of the inducer, or it can be systemic. Plants in which a single agent is used to induce resistance may show different mechanisms in different parts. The mechanisms may be divided into three main types.

Green Islands

Levels with a systemic virus infection often have islands of darker green tissue. This has a higher chlorophyll concentration than healthy leaf, is almost or completely virus-free, and can be at least partly resistant to infection in a second (challenge) inoculation. This induced resistance can be specific to the original virus (Loebenstein et al., 1977). Aspects of the biology of green island formation were reviewed by Fraser (1987).

Sziraki and Balazs (1975) suggested that dark green areas contained higher concentrations of cytokinins than healthy or light green (infected) tissue in TMV-infected tobacco. But the differences were small, and based on bioassays. We have shown by gas chromatographic assay that the concentration of zeatin is decreased in dark green tissue, while that of the less active metabolite zeatin-0-glucoside is increased (Whenham and Fraser, 1987). Thus while infection seems to reduce the concentration of the active cytokinin, it increases the rate of metabolism, and must increase the average rate of cytokinin synthesis.

In a parallel study of abscisic acid (ABA) metabolism, we have shown that ABA concentration is higher in dark green areas (Whenham et al., 1986), and that this increase may cause the increased chlorophyll concentration of dark green areas, and the resistance to challenge infection, at least in part (Fraser, 1982).

PR Proteins and 'Acquired' Resistance

In plants reacting to a virus by formation of local lesions, it has frequently been found that the lesions formed after a second inoculation are smaller, and less numerous, than those formed on previously inoculated control plants. Thus formation of the first lesions, in itself a genetically-controlled resistance response, in some way enhances resistance to the second inoculation. This 'acquired' resistance may be localized, effective only in the vicinity of the primary lesions, or it may be systemically effective. The latter has been studied most.

Acquired systemic resistance is now known to be induced by various pathogens (viruses, bacteria, fungi); chemicals and environmental or developmental factors (reviewed by Van Loon, 1983; Fraser, 1985c). The resistance tends to be non-specific, affecting various pathogens (Kuc, 1985) and insects (McIntyre et al., 1981). Induced resistance to challenging viruses causing necrotic lesion infections has been examined more than effects on those causing systemic infection.

The discovery of PR proteins (Van Loon and Van Kammen, 1970; Gianinazzi et al., 1970), and suggestions that they might be involved in induced resistance (Kassanis et al., 1974), gave a further stimulus to research. PR proteins have now been identified in at least 16 species (Van Loon, 1985). The prediction is that they are likely to be very common; perhaps universal, in the dicotyledons. Far fewer monocotyledonous and lower plants have been examined, and searches for PR proteins in these groups would be interesting.

26

PR proteins from several sources have been characterized to
varying extents (reviewed by Van Loon, 1985; Fraser, 1985c and
Pierpoint, 1986). In the absence of defined functions, the
characterization has necessarily been somewhat diffuse, but most tend
to show many of the following properties: solubility at low pH;
induced by a number of biotic and abiotic factors, molecular weight
within the range of approximately 10,000 – 25,000 M_r; resistance to
digestion by trypsin, and apoplastic location.

Van Loon (1983) has suggested a central role for ethylene in the
induction of PR proteins. The model is that factors such as necrotic
virus infection cause an increase in rate of ethylene synthesis,
which leads to synthesis of a benzoic acid–type compound, which then
induces PRs. Benzoic acid–derivatives such as acetylsalicylic acid
(White, 1979) and others (Van Loon and Antoniw, 1982) are powerful
inducers of PR proteins and resistance. Hooft van Huijsduijnen et al.
(1986a) showed that in cowpea protoplasts, salicylic acid inhibited
the replication of alfalfa mosaic virus (A1MV) by up to 99%, but had
no effect on host protein synthesis or protoplast viability. The data
suggested that salicylic acid must inhibit a very early stage of
virus replication. In whole leaves, salicylic acid inhibited A1MV
multiplication, and induced PR proteins. The implication is that a PR
or other protein induced by salicylic acid was responsible for the
observed resistance. However, comparatively high concentrations of
salicylic acid were required for inhibition; about 1 mM. Other
compounds can inhibit virus multiplication at much lower
concentrations. For example, Fraser and Gerwitz (1984) recorded
complete inhibition of TMV multiplication in tobacco leaf discs with
only 3 µM of 2–α–hydroxybenzylbenzimidazole. The possibility remains
that salicylic acid inhibited A1MV multiplication directly, rather
than by inducing some antiviral protein.

A stimulus to the study of PR protein induction came with the
report by Carr et al. (1982) suggesting that plants contained stable
PR mRNA, and that induction of PR protein synthesis operated as a
translational control. Others, however, found no evidence for
significant levels of PR mRNAs in healthy plants (Hooft van
Huijsduijnen et al., 1985); it now appears that the original
suggestion of translational control was based on results with an
antibody of poor specificity (Carr et al., 1985). There now seems
little doubt that PR proteins are induced by transcription of PR
mRNAs.

PR proteins are interesting in their own right, because of their
somewhat peculiar properties, some unique structural features (e.g.
Lucas et al., 1985), and because they seem to be a widespread
response to stress conditions. Three questions can be asked:

1. Are they involved in resistance to viruses, and especially in
   acquired systemic resistance?
2. If they are not involved in resistance, what do they do?
3. If PR proteins are not involved in acquired systemic
   resistance, then what are the mechanisms?

27

The arguments on whether PR proteins are involved in acquired systemic resistance or not have been rehearsed in detail elsewhere, and need not be repeated at length here. The evidence in favour of an involvement is circumstantial, but based on an impressive number of cases where PR proteins and resistance are induced together (see, for example, Van Loon (1983) and Antoniw and White (1986) for comprehensive treatments). The contrasting view is based on evidence of poor temporal and quantitative correlations between the amounts of PR proteins and induced resistance, and evidence that PR proteins and acquired resistance can, in some cases, be induced independently. The arguments are summarized by Fraser (1982), Fraser and Clay (1983) and Dumas and Gianinazzi (1986). It is logically impossible to prove the negative statement, that PR proteins are **not** involved in acquired systemic resistance.

Proof that PR proteins **are** involved in acquired resistance will require demonstration that added exogenous PR proteins confer resistance, which is a direct effect on some aspect of viral replication, and not a trivial, secondary effect of a wider action on other aspects of cell metabolism. Unfortunately, attempts to show effects of exogenous PR proteins on virus replication seem to have failed. Kassanis and White (1978) found no effects of PR proteins on TMV multiplication in tobacco protoplasts, and Hooft van Huijsduijnen et al. (1986a and unpublished evidence cited therein) found that intercellular fluid from PR-containing cowpea leaves did not cause any resistance to A1MV in cowpea protoplasts. Furthermore, although in this system salicylic acid caused strong resistance to A1MV replication in protoplasts, and induced PR proteins in whole leaves, there was no evidence for the presence of PR proteins in protoplasts: PR proteins are normally extracellular (Parent and Asselin, 1984) and undetectable in protoplasts (e.g. Wagih and Coutts, 1981).

Perhaps the answer to the question of whether PR proteins are involved in resistance will come when plants are transformed with, and express, cloned PR protein genes. Hooft van Huijsduijnen et al. (1986b) have already cloned cDNAs for six of the mRNAs induced by necrotic TMV infection of tobacco, and have evidence that some of these represent known PR proteins. Consequent isolation of PR protein genes from a tobacco genomic library, and transformation of tobacco plants with them, should be comparatively simple procedures, since all the techniques are well established.

What might be the function of PRs if they are not directly involved in resistance? The continual discovery of more PR proteins, and an increase in the apparent biochemical diversity of the group, suggest that there may be several functions. The apoplastic location could mean a structural role in the cell wall, or an enzyme activity there. As yet, there is little consistent evidence for an association with ultrastructural changes (Fraser and Clay, 1983), and early work only on enzyme activity. Metraux and Boller (1986) have shown the induction of cucumber chitinase by necrotic viral infection, and that this enzyme can be important in induced resistance to fungi. Chitinase shares some features with PRs, especially induction by ethylene. Pierpoint (1983) showed that three tobacco PR proteins could bind to chitin, and recently, Fritig and co-workers (personal communication) have shown that various tobacco PRs have chitinase and β-1-3 glucanase activities. The possibility is therefore that these PR proteins at least are involved in defence mechanisms against fungi and bacteria.

Lucas et al. (1985) determined the complete amino acid sequence of the tomato PR protein p14, and failed to find any sequence homology with over 3000 other published protein sequences. This suggested that p14 is a previously unknown type of protein structure. The amino acid sequence reveals five hydrophobic domains, suggesting an association with membranes. Similarities in amino acid composition, in partial sequences, and in antigenic properties, were then found between tomato p14 and some tobacco PR proteins (Nassuth and Sänger, 1986), suggesting some common function. A further and intruiguing sequence homology has been found between a protein specified by one of the mRNAs induced after necrotic TMV infection of tobacco, and thaumatin, a sweet-tasting protein from the monocotylendonous shrub *Thaumatococcus danielli* (Cornelissen et al., 1986).

If PR proteins are not related to acquired resistance, what mechanisms are involved? It is necessary first to ask whether virus multiplication is actually inhibited, or whether the reported resistance is merely a reduction in the expression of visible necrotic symptoms. The evidence for reduction in virus multiplication in the smaller lesions formed on leaves with acquired resistance is mixed: approximately equal number of papers report less virus in the smaller lesions, and unchanged amounts (reviewed by Fraser, 1987). Multiplication of a challenging virus causing a systemic infection of leaves with acquired resistance appears, in general, to be uninhibited by the induced resistance (e.g. Fraser, 1979). Other explanations for the reduced number of lesions formed after challenge inoculation of leaves with acquired resistance have been based on changed susceptibility to infection as a result of altered water status. Alterations in lesion size have been ascribed to factors controlling necrosis rather than virus spread, and mechanisms based on virus-induced changes in plant growth regulators have been advanced. As none of these mechanisms involves a direct antiviral effect, they are not considered in detail here.

Cross Protection

There are important differences between cross protection and acquired systemic resistance: cross protection tends to be highly specific, giving resistance only against the protecting virus, or even particular strains of it (Fulton, 1978); the cross-protecting virus is systemic, or nearly so (Cassells and Herrick, 1977); and cross protection has been used successfully in crop protection (reviewed by Sequiera, 1984).

Practical crop protection, for obvious reasons, has normally involved inoculation with attenuated strains. Aggressive strains will also inhibit infection by a challenging strain, but the effects are less clear because of the severe symptoms caused by the first infection. The property of an isolate causing attenuation might be related to its ability to cross protect. Another question which has to be answered by any complete theory of the mechanism of cross protection is why different isolates of a virus can differ in ability to cross protect; why some are not protected against, and why some viruses show no cross protection at all.

The literature on cross protection is large, and contains a diverse terminology and several arguments about the mechanisms. Some of these arguments have arisen because different cases of 'cross protection' undoubtedly involve different types of mechanism. Sometimes, the resistance of dark green, virus-free islands has been confused with the separate type of resistance in the virus-infected tissues, and sometimes, results from one system or host-virus combination have been applied to other cases without adequate qualification. The evidence for the more likely mechanisms which have been suggested is reviewed here.

Mechanisms operating at the site of infection.

These are normally assayed using a local lesion host for the challenge strain, and reduced lesion number is taken to represent interference with establishment of infection. Sherwood and Fulton (1983) suggested that a TMV isolate causing a systemic infection of *Nicotiana sylvestris* reduced lesion formation by a necrotic strain, by blocking infection sites. But they also suggested a later level of interference, which inhibited multiplication of the challenging strain.

Mechanisms involving coat protein.

De Zoeten and Fulton (1975) originally suggested for TMV infection of *N. sylvestris*, that the RNA of the challenging strain (local lesion) was encapsidated in the coat protein of the protecting (systemic) strain, and was thus unable to initiate the replicative cycle. Hence fewer lesions were formed. They argued that when multiplication of the protecting strain was well-established, the cellular environment was in favour of encapsidation, and likely to contain adequate amounts of coat protein. This theory was criticized by Zaitlin (1976), on the grounds that a TMV mutant with a coat protein defective in assembly could cross-protect. Later, supporting evidence came from experiments in other systems, showing that a coat protein-free mutant of TMV could protect (Sarkar and Smitamana, 1981); from the demonstration of cross protection between viroids, which have no proteins (Niblett et al., 1978), and from experiments with pseudorecombinants of multicomponent viruses, showing that the specific ability of a strain to cross-protect can map on a genomic segment other than that coding for the coat protein gene (reviewed by Fulton, 1980).

However, all these lines of evidence show only that cross-protection effects can operate by mechanisms other than those involving coat protein; they do not exclude the possibility that coat protein mechanisms could also be involved. Recently, more direct evidence has been obtained for these (e.g. Sherwood and Fulton, 1982; Zinnen and Fulton, 1986). Perhaps the best demonstration that coat protein may be involved comes from experiments in which susceptible plants are transformed with and express the coat protein gene, and have been shown to be at least partly protected against challenge inoculation (e.g. Abel et al., 1986). These experiments are considered in more detail later.

Autoregulation.

The tomato strain of TMV designated at $L_{11}A$ has been used to cross-protect tomato crops, and shown to cause very mild symptoms and multiply to low concentrations (Kiho and Nishiguchi, 1984). The authors suggested that the early multiplication of $L_{11}A$ was at the same rate as wild type, but that by four days after inoculation, the rate of accumulation suddenly slowed: an effect they named 'autoregulation'. They suggested that the virus-coded 183,000 M_r protein, which is probably involved in replication, is over-produced by $L_{11}A$, and that this may have caused the decline in rate of synthesis, and by implication, the ability to cross-protect. In a study of another attenuated, cross-protecting strain, MII-16 (Rast, 1975), we found that the rate of accumulation was lower than that of wild type for 20 days after inoculation, and did not alter during this time (R. Fraser and A. Gerwitz, unpublished results). This suggests that there may be more than one mechanism leading to attenuation and ability to cross-protect.

The $L_{11}A$ nucleic acid has been cloned and sequenced (Nishiguchi et al., 1985), and compared with its ancestral strain L which multiplies normally. Of ten base substitutions in $L_{11}A$, seven were in the third bases of in-phase codons and did not alter the amino acid specified. The three others did cause amino acid changes, and were located in the common reading frame of the 126,000 and 183,000 M_r proteins. By comparison with a partial sequence for a related attenuated isolate L, Nishiguchi et al. (1985) suggested that the exchange at position 1117 controlled attenuation. Watanabe et al. (1987) examined the performance of $L_{11}A$ in protoplasts: multiplication was not reduced, but that there was reduced synthesis of the 30,000 M_r virus-coded protein which is thought to be involved in cell-to-cell movement. The implication is that the 126,000 or 183,000 M_r proteins are involved in synthesis of the subgenomic mRNA for the 30,000 M_r protein, and that the altered high molecular weight proteins of $L_{11}A$ are responsible for the low production of 30,000 M_r protein. How this might contribute to cross protection is not immediately clear.

Mechanisms based on minus-sense RNA.

In a model proposed by Palukaitis and Zaitlin (1984), the minus-sense RNA of the challenging virus is sequestered by the excess plus–sense RNA of the protecting strain. This model would explain the ability of viroids and coat protein–less or defective mutants of viruses to cross–protect. It also offers some explanation for why less–related strains may fail to cross protect, as the mechanism must depend on the degree of sequence complementarity between the two RNAs. There are, however, two aspects of the model which require further explanation. Firstly, the majority of the plus–sense RNA of the protecting strain is encapsidated in coat protein, and not necessarily available to sequester the minus–sense RNA of the challenge strain. Secondly, why does molecular hybridization between homologous plus and minus strains of any virus not severely inhibit multiplication in single infections? Palukaitis and Zaitlin argued that such feedback inhibition does occur in singly–infected plants, and full–length double stranded RNAs do accumulate late in TMV infection (Jackson et al., 1971). But there is no evidence yet that this limits virus multiplication in singly–infected plants.

PROSPECTS FOR EXPANDED CONTROL OF PLANT VIRUS DISEASES BY GENETIC MANIPULATION

The genetic base for control of plant virus diseases by resistance genes needs to be expanded. Sometimes there is no useable resistance. For vegetables alone in the United Kingdom, J.A. Tomlinson (personal communication) has listed 22 crops and 25 viruses for which no resistance is available. Furthermore, in many crops, resistance depends on one or a few genes only, and the frequency of development of resistance–breaking strains of the virus makes the protection somewhat precarious. Thus there is a need to develop new genes for virus resistance, and to expand the utilization of existing genes.

There are two possible approaches: to isolate natural resistance genes by cloning and gene identification techniques, and to 'invent' new types of resistance gene.

Isolation of Natural Resistance Genes

With a few possible exceptions, the protein products of genes which are involved in specific resistance to viruses are not known. This means that isolation of the gene from a genomic library from resistant plants must be by indirect means, for example by transposon tagging. The indirect methods generally lead to large experiments and low probabilities of detecting the desired genes: some of the approaches and problems are considered by Hepburn et al. (1985). In several methods, there is no simple way to recognize the resistance gene when it is isolated. Differences in genes, or in genes expressed, may be found between resistant and susceptible varieties, but these are likely to involve 'background noise' variation between

32

the genotypes, or secondary differences in gene expression which are not primary products of the resistance genes.

However, the problem is of such interest that much effort will be devoted to it, and resistance genes will probably be isolated in the near future. Genes conferring resistance to viruses such as TMV, which affect a number of crop species, will be available for protection in these species if transformation and regeneration systems are available. However, genetically-engineered resistance genes will arguably be as likely as normally-bred genes to be overcome by virulent isolates of the virus, and this should be considered when they are deployed. Genetic engineering, and the consequent ability to transfer genes between genera or families, may allow the non-conventional breeder to develop more robust, oligogenic systems.

Then there are two wider possibilities: to alter the nature of the cloned resistance gene by site-specific mutagenesis, and to use the knowledge of host-virus interactions which will flow from sequencing of resistance and virulence genes, to develop completely new types of control. These could have either a genetic or chemical basis.

Development of Novel Types of Resistance Gene

The molecular biology of virus replication, and of how different viruses may interact with each other or with satellites, is now beginning to be quite well understood. It is becoming possible to design ways of interfering with replication using parts of the virus itself or related elements. There are several possible strategies; in comparison with the problems of isolating natural resistance genes, all involve accessible genetic elements. Furthermore, the genomic size of viruses is so limited, and transformation systems for some plant hosts are now so efficient, that a 'shotgun' approach to selecting useful antiviral clones becomes feasible. In most cases, the strategy involves insertion of the cloned viral sequence into the host DNA using an *Agrobacterium* Ti-plasmid or other vector, and expressing either plus- or minus-sense RNA by an appropriately placed promoter sequence. Some mechanisms operate at the RNA level, while others can require expression of the protein specified by the viral sequence.

Antimessengers and minus-sense RNAs.

Expression of viral RNAs of minus-polarity may interfere with translation of viral mRNAs. Coleman et al. (1985) prepared a plasmid which would produce large amounts of RNA complementary to certain bacteriophage mRNAs, and showed that these were able to inhibit phage multiplication in the bacterium. Transgenic tobacco plants, expressing a minus-sense RNA complementary to the 5'-end of TMV RNA seemed to be required. Other possible target areas include replicase-binding sites (on either plus- or minus-sense viral RNAs)

33

and sites for the initiation of assembly of virus particles. Targets early in the viral replication cycle, such as at the initial uncoating and translation event, are likely to be more economical, in terms of the level of expression required, than targets at a later stage such as assembly, or cell-to-cell spread. Another possible problem is the range of virus isolates which will display sufficient sequence-complementarity to form a stable hybrid with the minus-sense resistance RNA. Some viruses, such as TMV, can show great diversity of sequence in at least some regions of their genomes. Two possibilities would be to concentrate on regions which are likely to have been more highly conserved, or to transform plants with genomic elements producing a number of related minus-strand sequences.

Satellites.

Some plant viruses have satellites: sequences unrelated to the genomic RNAs, which require the presence of replicating genomic RNAs for their own replication, and which are normally transmitted with the virus particles. Satellites can increase or inhibit the disease expression of the supporting virus, and can inhibit its multiplication (Waterworth et al., 1979). The satellite of CMV has now been cloned and inserted into the nuclear genome of *N. tabacum* (Baulcombe et al., 1986). The satellite is expressed as an RNA, and CMV inoculated to transgenic plants has been shown to acquire the satellite, and then to cause less severe symptoms on *N. clevelandii* than satellite-free virus. Unfortunately, CMV causes very mild symptoms on *N. tabacum* in any case, so it was not possible to test whether the transgenic plants themselves were protected by the expressed satellite sequence.

Before cloned satellite sequences can be deployed as useful resistance genes, it will be necessary to study their interaction with many isolates of virus, to ensure that they do not cause more severe disease symptoms with some, or under different environmental conditions, and that they do not affect plant growth. The problem of mutation to disease-enhancing forms also needs to be assessed.

Coat protein mechanisms.

Abel et al. (1986) have transformed tobacco plants with the coat protein gene of TMV, and have shown that this is expressed. Transformed plants, when inoculated with TMV, developed symptoms at a slower rate than control plants: there is therefore the promise of a useful degree of resistance. Subsequent experiments (Nelson et al., 1987) showed that expression of the coat protein gene in transgenic plants inhibited both the establishment of infection (measured as a reduction in number of lesions formed on challenge inoculation of transgenic *N*-gene tobacco) and multiplication. Inoculation with TMV RNA largely overcame the inhibition of infection. This suggests that the expressed coat protein gene may have interfered with an early stage of infection specific to TMV particles, such as uptake or uncoating. Wilson and Watkins (1986) have shown that homologous coat protein can inhibit the combined translation and disassembly (Wilson,

1985) of TMV particles in an *in vitro* protein synthesizing system. While this result may not be a complete model of the interaction *in vivo*, it does offer one level of explanation of how cross protection might work. It does not exclude other possible ways in which the coat protein of the protecting strain, or indeed its mRNA, might interfere.

There are already indications from various laboratories that a coat protein mechanism of protection may operate for other viruses such as alfalfa mosaic. It may be possible to interfere with the multiplication of many plant viruses by transforming plants with coat protein genes, which are readily accessible. However, much remains to be established about the mechanisms involved.

The objective for the longer term must be to eliminate some of the more important virus diseases of plants. But plant viruses are very diverse, and their capacity for rapid mutation may enable them to overcome novel mechanisms of resistance derived from molecular biology, as readily as they have overcome many of those evolved in host plants. It will probably be necessary to employ many strategies in an attempt to construct robust barriers to the evolution of virulence.

ACKNOWLEDGEMENTS

I thank colleagues at Wellesbourne who have worked with me on resistance to viruses: Robert Whenham, Sheila Thompson, Tony Gerwitz and Su Loughlin. I am grateful for useful discussion, and for allowing me to read manuscripts and cite results in advance of publication to Kees van Loon, Milt Zaitlin, John van Emmelo, Wolfgang Mundry, Michael Wilson, Ulrich Melcher, Masamichi Nishiguchi, Hugh Barker, Bernard Fritig, Roger Beachy and John Bol.

REFERENCES

ABEL, P.P., NELSON, R.S., DE, B., HOFFMANN, N., ROGERS, S.G. FRALEY, R.T. and BEACHY, R.N. (1986) Delay of disease development in transgenic plants that express the tobacco mosaic virus coat protein gene. Science 232: 738-743.

AMEMIYA, Y. and MISAWA, Y. (1977) Studies on the resistance of cucumber to cucumber mosaic virus. II. Induction of resistance by infection. Tokohu J. Agric. Res. 28: 18-25.

ANTIGNUS, Y., SELA, I. and HARPAZ, I. (1977) Further studies of the biology of an antiviral factor from virus infected plants and its association with the *N*-gene of *Nicotiana* species. J. Gen. Virol. 35: 107-116.

ANTONIW, J.F. and WHITE, R.F. (1986) Changes with time in the distribution of virus and PR protein around single local lesions of TMV infected tobacco. Plant Mol. Biol. 6: 145–150.

ANTONIW, J.F., WHITE, R.F. and CARR, J.P. (1984) An examination of the effect of human α–interferons on the infection and multiplication of tobacco mosaic virus in tobacco. Phytopathol. Z. 109: 367–371.

ATABEKOV, J.G. (1975) Host specificity of plant viruses. Annu. Rev. Phytopathol. 13: 127–145.

BALD, J.G. and TINSLEY, T.W. (1967) A quasi-genetic model for plant virus host ranges. III. Congruence and relatedness. Virology 32: 328–336.

BARBARA, D.A. and WOOD, K.R. (1974) The influence of actinomycin D on cucumber mosaic virus (strain W) multiplication in cucumber cultivars. Physiol. Plant Pathol. 4: 45–50.

BARKER, H. (1987) Invasion of non-phloem tissue in *Nicotiana clevelandii* by potato leafroll luteovirus is enhanced in plants also infected by potato Y potyvirus. J. Gen. Virol., in press.

BARKER, H. and HARRISON, B.D. (1986) Restricted distribution of potato leafroll virus antigen in resistant potato genotypes and its effect on transmission of the virus by aphids. Ann. Appl. Biol. 109: 595–604.

BAULCOMBE, D.C., SAUNDERS, G.R., BEVAN, M.W., MAYO, M.A. and HARRISON, B.D. (1986) Expression of biologically active virus satellite RNA from the nuclear genome of transformed plants. Nature 321: 446–449.

BEIER, H., BRUENING, G., RUSSELL, M.L. and TUCKER, C.L. (1979). Replication of cowpea mosaic virus in protoplasts isolated from immune lines of cowpeas. Virology 95: 165–175.

BIJAISORADAT, M. and KUHN, C.W. (1985) Nature of resistance in soybean to cowpea chlorotic mottle virus. Phytopathology 75: 351–355.

BOULTON, M.I., MAULE, A.J. and WOOD, K.R. (1985) Effect of actinomycin D and UV irradiation on the replication of cucumber mosaic virus in protoplasts isolated from resistant and susceptible cucumber cultivars. Physiol. Plant Pathol. 26: 279–287.

CARR, A.J.H. (1984) Infection by viruses and subsequent host damage. In: Plant Diseases. Infection, Damage and Loss, Wood, R.K.S. and Jellis, J.G., eds. Blackwells, Oxford. pp. 199–209.

CARR, J.P., ANTONIW, J.F., WHITE, R.F. and WILSON, T.M.A. (1982) Latent messenger RNA in tobacco (*Nicotiana tabacum*) Biochem. Soc. Trans. 10: 353–354.

CARR, J.P., DIXON, D.C. and KLESSIG, D.F. (1985) Synthesis of pathenogenesis-related proteins in tobacco is regulated at the level of mRNA accumulation and occurs on membrane-bound polysomes. Proc. Natl. Acad. Sci. USA 82: 7999–8003.

CASSELLS, A.C. and HERRICK, C.C. (1977) Cross protection between mild and severe strains of tobacco mosaic virus in doubly inoculated tomato plants. Virology 78: 253–260.

CHESSIN, M. (1983) Is there a plant interferon? Bot. Rev. 49: 1–27.

COLEMAN, J., HIRASHIMA, A., INOKUCHI, Y., GREEN, P.J. and INOUYE, M. (1985) A novel immune system against bacteriophage infection using complementary RNA (micRNA) Nature 315: 601–603.

COLLENDAVELLOO, J., LEGRAND, M. and FRITIG, B. (1983) Plant disease and the regulation of enzymes involved in lignification. Increased rate of de novo synthesis of the three tobacco 0-methyl transferases during the hypersensitive response to infection by tobacco mosaic virus. Plant Physiol. 73: 550–554.

CORNELISSEN, B.J.C., HOOFT VAN HUIJSDUIJNEN, R.A.M. and BOL, J.F. (1986) A tobacco mosaic virus–induced tobacco protein is homologous to the sweet-tasting protein thaumatin. Nature 321: 531–532.

COUTTS, R.H.A. and WOOD, K.R. (1977) Inoculation of leaf mesophyll protoplasts from a resistant and a susceptible cucumber cultivar with cucumber mosaic virus. FEMS Microbiol. Lett. 1: 121–124.

CRUTE, I.R. (1985) The genetic bases of relationships between microbial parasites and their hosts. In: Mechanisms of Resistance to Plant Diseases, Fraser, R.S.S., ed. Martinus Nijhoff/Dr W. Junk, Dordrecht. pp. 80–142.

DAWSON, J.R.O., REES, M.W. and SHORT, M.N. (1979) Lack of correlation between the coat protein composition of tobacco mosaic virus isolates and their ability to infect resistant tomato plants. Ann. Appl. Biol. 91: 353–358.

DAY, K.L. (1984) Resistance to bean common mosaic virus in *Phaseolus vulgaris*. PhD Thesis, University of Birmingham, U.K.

DE JAGER, C.P. and WESSELING, J.B.M. (1981) Spontaneous mutations in cowpea mosaic virus overcoming resistance due to hypersensitivity in cowpea. Physiol. Plant Pathol. 19: 347–358.

DE LAAT, A.M.M. and VAN LOON, L.C. (1982) Regulation of ethylene biosynthesis in virus-infected tobacco leaves. II. Time course of levels of intermediates and in vivo conversion rates. Plant Physiol. 69: 240–245.

DE LAAT, A.M.M. and VAN LOON, L.C. (1983) The relationship between stimulated ethylene production and symptom expression in virus-infected tobacco leaves. Physiol. Plant Pathol. 22: 261–273.

DE ZOETEN, G.A. and FULTON, R.W. (1975) Understanding generates possibilities. Phytopathology 65: 221–222.

DRIJFHOUT, E. (1978) Genetic interaction between *Phaseolus vulgaris* and bean common mosaic virus with implications for strain identification and breeding for resistance. Agric. Res. Rep. (Neth.) 872: 1–98.

DUCHESNE, M., FRITIG, B. and HIRTH, L. (1977) Phenylalanine ammonia lyase in tobacco mosaic virus-infected hypersensitive tobacco. Density-labelling evidence of de novo synthesis. Biochim. Biophys Acta 485: 465–481.

DUMAS, E. and GIANINAZZI, S. (1986) Pathogenesis-related (b) proteins do not play a central role in TMV localization in *Nicotiana rustica*. Physiol. Plant Pathol. 28: 243–250.

EDWARDS, M.C. GONSALVES, D. and PROVVIDENTI, R. (1983) Genetic analysis of cucumber mosaic virus in relation to host resistance: location and determinants for pathogenicity to certain legumes and *Lactuca saligna*. Phytopathology 73: 269–273.

EKPO, E.J.A. and SAETTLER, A.W. (1975) Multiplication and distribution of bean common mosaic virus in *Phaseolus vulgaris*. Plant Dis. Repr. 59: 939–943.

EVANS, D. (1985) Isolation of a mutant of cowpea mosaic virus which is unable to grow in cowpeas. J. Gen. Vriol. 66: 339–343.

EVANS, D.M.A., FRASER, R.S.S. and BRYANT, J.A. (1985) RNA–dependent RNA polymerase activities in tomato plants susceptible or resistant to tobacco mosaic virus. Ann. Bot. 55: 587–591.

FACCIOLI, G. and CAPPONI, R. (1983) An antiviral factor present in plants of *Chenopodium amaranticolor* locally infected by tobacco necrosis virus: 1. Extraction, partial purification, biological and chemical properties. Phytopathol. Z. 106: 289–301.

FLOR, H.H. (1956) The complementary genetic systems in flax and flax rust. Adv. Genet. 8: 29–54.

FRASER, R.S.S. (1979) Systemic consequences of the local lesion reaction to tobacco mosaic virus in a tobacco variety lacking the *N* gene for hypersensitivity. Physiol. Plant Pathol. 14: 383–394.

FRASER, R.S.S. (1982) Are 'pathogenesis–related' proteins involved in acquired systemic resistance of tobacco plants to tobacco mosaic virus. J. Gen. Virol. 58: 305–313.

FRASER, R.S.S. (1983) Varying effectiveness of the *N'* gene for resistance to plant tobacco mosaic virus in tobacco infected with virus strains differing in coat protein properties. Physiol. Pathol. 22: 109–119.

FRASER, R.S.S. (1985a) Host range control and non–host immunity to viruses. In: Mechanisms of Resistance to Plant Diseases, Fraser, R.S.S., ed. Martinus Nijhoff/Dr W. Junk, Dordrecht. pp. 13–28.

FRASER, R.S.S. (1985b) Mechanisms involved in genetically controlled resistance and virulence: virus diseases. In: Mechanisms of Resistance to Plant Diseases, Fraser, R.S.S., ed. Martinus Nijhoff/Dr W. Junk, Dordrecht. pp. 143–196.

FRASER, R.S.S. (1985c) Mechanisms of induced resistance to virus disease. In: Mechanisms of Resistance to Plant Diseases, Fraser, R.S.S., ed. Martinus Nijhoff/Dr W. Junk, Dordrecht. pp. 373–404.

FRASER, R.S.S. (1986) Genes for resistance to plant viruses. CRC Critical Reviews in Plant Sciences 3: 257–294.

FRASER, R.S.S. (1987) Biochemistry of Virus–infected Plants. Research Studies Press, Letchworth/John Wiley and Sons, New York. 259 pp.

FRASER, R.S.S. (1988) Genetics of plant resistance to viruses. Ciba Foundation Symposia 133. John Wiley and Sons, Chichester. In press.

FRASER, R.S.S. and CLAY, C.M. (1983) Pathogenesis–related proteins and acquired systemic resistance: causal relationship or separate effects? Neth. J. Plant Pathol. 89: 283–292.

FRASER, R.S.S. and GERWITZ, A. (1984) Effects of 2–α–hydroxy-benzylbenzimidazole on tobacco mosaic virus and host RNA synthesis in tobacco leaf disks and plants. Plant Sci. Lett. 34: 111–117.

FRASER, R.S.S. and GERWITZ, A. (1986) The genetics of resistance and virulence in plant virus disease. In: Genetics and Plant Pathogenesis, Day, P.R. and Jellis, G.J., eds. Blackwells, Oxford. pp. 33–44.

FRASER, R.S.S., GERWITZ, A., LOUGHLIN, S.A.R. and LEARY, J.A. (1983) Resistance to tobacco mosaic virus in tomato plants. Rep. Natl. Veg. Res. Stn. for 1982, pp. 18–19.

FRASER, R.S.S., GERWITZ, A. and MORRIS, G.E.L. (1986) Multiple regression analysis of the relationships between tobacco mosaic virus multiplication, the severity of mosaic symptoms, and the growth of tobacco and tomato. Physiol. Molec. Plant Pathol. 29: 239–249.

FRASER, R.S.S. and LOUGHLIN, S.A.R. (1980) Resistance to tobacco mosaic virus in tomato: effects of the *Tm-1* gene on virus multiplication. J. Gen. Virol. 48: 87–96.

FRASER, R.S.S. and LOUGHLIN, S.A.R. (1982) Effects of temperature on the *Tm-1* gene for resistance to tobacco mosaic virus in tomato. Physiol. Plant Pathol. 20: 109–117.

FRASER, R.S.S. and WHENHAM, R.J. (1982) Plant growth regulators and virus infection: a critical review. Plant Growth Regul. 1: 37–59.

FRITIG, B., GOSSE, J., LEGRAND, M. and HIRTH, L. (1973) Changes in phenylalanine ammonia lyase during the hypersensitive reaction of tobacco to TMV. Virology 55: 371–379.

FRITIG, B., LEGRAND, M. and HIRTH, L. (1972) Changes in the metabolism of phenolic compounds during the hypersensitive reaction of tobacco to TMV. Virology 47: 845–848.

FULTON, R.W. (1978) Superinfection by strains of tobacco streak virus. Virology 85: 1–8.

FULTON, R.W. (1980) Biological significance of multicomponent viruses. Ann. Rev. Phytopathol. 18: 131–146.

FURUSAWA, I. and OKUNO, T. (1978) Infection with BMV of mesophyll protoplasts isolated from five plant species. J. Gen. Virol. 40: 489–491.

GERA, A. and LOEBENSTEIN, G. (1983) Further studies of an inhibitor of virus replication from tobacco mosaic virus-infected protoplasts of a local lesion-responding tobacco cultivar. Phytopathology 73: 111–115.

GIANINAZZI, S., MARTIN, C. and VALEE, J.-C. (1970) Hypersensitivite aux virus, temperature et proteines solubles chez le *Nicotiana Xanthi* n.c., Apparition de nouvelles macromolecules lors de la repression de la synthesis virale. Compt. Rend. Acad. Sci. Paris 270D: 2383–2386.

GRIMSLEY, N., HOHN, T., DAVIES, J.W. and HOHN, B. (1987) *Agrobacterium*-mediated delivery of infectious maize streak virus into maize plants. Nature 325: 177–179.

GOLDBACH, R. and KRIJT, J. (1982) Cowpea mosaic virus-encoded protease does not recognize primary translation products of mRNAs from other comoviruses. J. Virol. 43: 1151–1154.

HEPBURN, A.G., WADE, M. and FRASER, R.S.S. (1985) Present and future prospects for exploitation of resistance in crop protection by novel means. In: Mechanisms of Resistance to Plant Diseases, Fraser, R.S.S. ed. Martinus Nijhoff/Dr W. Junk, Dordrecht. pp. 425–452.

HOLMES, F.O. (1955) Additive resistance to specific viral diseases in plants. Ann. Appl. Biol. 42: 129–139.

HOLMES, F.O. (1961) Concomitant inheritance of resistance to several virus diseases in tobacco. Virology 13: 409–413.

HOOFT VAN HUIJSDUIJNEN, R.A.M., ALBLAS, S.W., DE RIJK, R.H. and BOL, J. (1986) Induction by salicylic acid of pathogenesis-related proteins and resistance to alfalfa mosaic virus infection in various plant species. J. Gen. Virol. 67: 2135–2143.

HOOFT VAN HUIJSDUIJNEN, R.A.M., CORNELISSEN, B.J.C., VAN LOON, L.C., VAN BOOM, J.H., TROMP, M, and BOL, J.F. (1985) Virus-induced synthesis of messenger RNAs for precursors of pathogensis-related proteins in tobacco. EMBO J. 4: 2167–2171.

HOOFT VAN HUIJSDUIJNEN, R.A.M., VAN LOON, L.C. and BOL, J.F. (1986) cDNA cloning of six mRNAs induced by TMV infection of tobacco and a characterisation of their translation products. EMBO J. 5: 2057–2061.

HOOLEY, R. and McCARTHY, D. (1980) Extracts from virus infected hypersensitive tobacco leaves are detrimental to protoplast survival. Physiol. Plant Pathol. 16: 25–38.

HUBER, R., HONTILEZ, J. and VAN KAMMEN, A. (1981) Infection of cowpea protoplasts with both the common strain and the cowpea strain of TMV. J. Gen. Virol. 55: 241–245.

HUISMAN, M.J., BROXTERMAN, H.J.G., SCHELLEKENS, H. and VAN VLOTEN-DOTING, L. (1985) Human interferon does not protect cowpea plant cell protoplasts against infection with alfalfa mosaic virus.

Virology 143: 622–625.

HUSSAIN, M.M., MELCHER, U., WHITTLE, T., WILLIAMS, A., BRANNAN, C.M. and MITCHELL, E.D. (1987) Replication of cauliflower mosaic virus DNA in leaves and suspension culture protoplasts of cotton. Plant Physiol., in press.

IRELAND, R.J. and PIERPOINT, W.S. (1980) Reaction of strains of potato virus X with chlorogenoquinone in vitro, and the detection of a naturally modified form of the virus. Physiol. Plant Pathol. 16: 81–92.

JACKSON, A.O., MITCHELL, D.M. and SIEGEL, A. (1971) Replication of tobacco mosaic virus. I. Isolation and characterization of double-stranded forms of ribonucleic acid. Virology 45: 182–191.

KACSER, H. and BURNS, J.A. (1981) The molecular basis of dominance. Genetics 97: 639–666.

KADO, C.I. and KNIGHT, C.A. (1966) Location of a local lesion gene in tobacco mosaic virus RNA. Proc. Natl. Acad. Sci. USA 55: 1276–1283.

KASSANIS, B., GIANINAZZI, S. and WHITE, R.F. (1974) A possible explanation of the resistance of virus-infected tobacco plants to second infection. J. Gen. Virol. 23: 11–16.

KASSANIS, B. and WHITE, R.F. (1978) Effect of polyacrylic acid and b proteins on TMV multiplication in tobacco protoplasts. Phytopathol. Z. 91: 269–272.

KIEFER, M.C., BRUENING, G. and RUSSELL, M.L. (1984) RNA and capsid accumulation in cowpea protoplasts that are resistant to cowpea mosaic virus strain SB. Virology 137: 371–381.

KIHO, Y. and NISHIGUCHI, M. (1984) Unique nature of an attenuated strain of tobacco mosaic virus: autoregulation. Microbiol. Immunol. 28: 589–599.

KONATE, G., KOPP, M. and FRITIG, B. (1983) Studies on TMV multiplication in systemically and hypersensitively reacting tobacco varieties by means of radiochemical and immunoenzymatic methods. Agronomie 3: 95.

KOOISTRA, E. (1968) Significance of the non-appearance of visible disease symptoms in cucumber (Cucumis sativus L.) after infection with Cucumis virus 2. Euphytica 17: 136–140.

KUBO, S. and TAKANAMI, Y. (1979) Infection of tobacco mesophyll protoplasts with tobacco necrotic dwarf virus, a phloem-limited virus. J. Gen. Virol. 42: 387–398.

KUC, J. (1985) Expression of latent genetic information for disease resistance in plants. In: Cellular and Molecular Biology of Plant Stress, Key, J.L. and Kosuge, T., eds. Alan R. Liss, New York. pp. 303–318.

LEONARD, D.A. and ZAITLIN, M. (1982) A temperature-sensitive strain of tobacco mosaic virus defective in cell-to-cell movement generates an altered viral-coded protein. Virology 117: 416–424.

LOEBENSTEIN, G., COHEN, J., SHABTAI, S., COUTTS, R.H.A. and WOOD, K.R. (1977) Distribution of cucumber mosaic virus in systemically infected tobacco leaves. Virology 81: 117–125.

LOEBENSTEIN, G. and GERA, A. (1981) Inhibitor of virus replication released from tobacco mosaic virus infected protoplasts of a local lesion responding tobacco cultivar. Virology 114: 132–139.

LOEBENSTEIN, G., GERA, A., BARNETT, A., SHABTAI, S. and COHEN, J. (1980) Effect of 2,4-dichlorophenoxyacetic acid on multiplication of TMV in protoplasts from local-lesion and systemic responding hosts. Virology 100: 110–115.

LOESCH-FRIES, L.S., HALK, E.L., NELSON, S.E. and KRAHN, K.J. (1985) Human leukocyte interferon does not inhibit alfalfa mosaic virus in protoplasts or tobacco tissue. Virology 143: 626-629.

LEVY, A., LOEBENSTEIN, G., SMOOKLER, M. and DROVI, T. (1974) Partial suppression by UV irradiation of the mechanism of resistance to cucumber mosaic virus in a resistant cucumber cultivar. Virology 60: 37-44.

LUCAS, J., HENRIQUEZ, C.A., LOTTSPEICH, F., HENSCHEN, A. and SÄNGER, H.L. (1985) Amino acid sequence of the "pathogenesis-related" leaf protein p14 from viroid-infected tomato reveals a new type of structurally unfamiliar proteins. EMBO J. 4: 2745-2749.

McINTYRE, J.L., DODDS, J.A. and HARE, J.D. (1981) Effects of localized infections of *Nicotiana tabacum* by tobacco mosaic virus on systemic resistance against diverse pathogens and an insect. Phytopathology 71: 297-301.

MAEKAWA, K., FURUSAWA, I. and OKUNO, T. (1981) Effects of actinomycin D and ultraviolet irradiation on multiplication of brome mosaic virus in host and non-host cells. J. Gen. Virol. 53: 353-356.

MASSALA, R., LEGRAND, M. and FRITIG, B. (1980) Effect of α-aminooxyacetate, a competitive inhibitor of phenylalanine ammonia lyase, on the hypersensitive resistance of tobacco to tobacco mosaic virus. Physiol. Plant Pathol. 16: 213-226.

MASSALA, R., LEGRAND, M. and FRITIG, B. (1987) Comparative effects of two competitive inhibitors of phenylalanine ammonia-lyase on the hypersensitive resistance of tobacco to tobacco mosaic virus. Plant Biochem. Physiol. 25: in press.

MAULE, A.J., BOULTON, M.I. and WOOD, K.R. (1980) Resistance of cucumber protoplasts to cucumber mosaic virus: a comparative study. J. Gen. Virol. 51: 271-279.

MELCHER, U., STEFFENS, D.L., LYTTLE, D.J., LEBEURIER, G., LIN, H., CHOE, I.S. and ESSENBERG, R.C. (1986) Infectious and non-infectious mutants of cauliflower mosaic virus DNA. J. Gen. Virol. 67: 1491-1498.

MESHI, T., ISHIKAWA, M., OHNO, T., OKADA, Y., SANO, T., VEDA, I. and SHIKATE, E. (1984) Double-stranded cDNAs of hop stunt viroids are infectious. J. Biochem. Tokyo 95: 1521-1524.

METRAUX, J.P. and BOLLER, T. (1986) Local and systemic induction of chitinase in cucumber plants in reponse to viral, bacterial and fungal infections. Physiol. Molec. Plant Pathol. 28: 161-170.

MOTOYOSHI, F. and OSHIMA, Y. (1975) Infection with tobacco mosaic virus of leaf mesophyll protoplasts from susceptible and resistance lines of tomato. J. Gen. Virol. 29: 81-91.

MOTOYOSHI, F. and OSHIMA, N. (1979) Standardization in inoculation procedure and effect of a resistance gene on infection of tomato protoplasts with tobacco mosaic virus RNA. J. Gen. Virol. 44: 801-806.

MOUCHES, C., CANDRESSE, T. and BOVE, J.M. (1984) Turnip yellow mosaic virus RNA-replicase contains host and virus-encoded subunits. Virology 134: 78-90.

MOZES, R., ANTIGNUS, Y., SELA, I. and HARPAZ, I. (1978) The chemical nature of an antiviral factor (AVF) from virus-infected plants. J. Gen. Virol. 38: 241-249.

NASSUTH, A. and SÄNGER, H.L. (1986) Immunological relationship between pathogenesis-related proteins from tomato, tobacco and cowpea. Virus Res., in press.

NELSON, R.S., ABEL, P.P. and BEACHY, R.N. (1987) Lesions and virus accumulation in inoculated transgenic tobacco plants expressing the coat protein gene of tobacco mosaic virus. Virology, in press.

NIBLETT, C.L., DICKSON, E., FERNOW, K.H., HORST, R.K. and ZAITLIN, M. (1978) Cross protection among four viroids. Virology 91: 198–203.

NISHIGUCHI, M., KIKUCHI, S., KIHO, Y., OHNO, T., MESHI, T. and OKADA, Y. (1985) Molecular basis of plant viral virulence; the complete nucleotide sequence of an attenuated strain of tobacco mosaic virus. Nucleic Acids Res. 13: 5585–5590.

ORCHANSKY, P., RUBINSTEIN, M. and SELA, I. (1982) Human interferons protect plants from virus infection. Proc. Natl. Acad. Sci. USA 79: 2278–2280.

OTSUKI, Y., SHIMOMURA, T. and TAKEBE, I. (1972) Tobacco mosaic virus multiplication and expression of the N gene in necrotic responding tobacco varieties. Virology 50: 45–50.

PALUKAITIS, P. and ZAITLIN, M. (1984) A model to explain the "cross–protection" phenomenon shown by plant viruses and viroids. In: Plant–Microbe Interactions, Molecular and Genetic Perspectives, vol. 1., Kosage, T. and Nester, E.W., eds. Macmillan, New York and London. pp. 420–429.

PARENT, J.-G. and ASSELIN, A. (1984) Detection of pathogenesis-related (PR or b) and of other proteins in the intercellular fluid of hypersensitive plants infected with tobacco mosaic virus. Can. J. Bot. 62: 564–569.

PIERPOINT, W.S. (1983) The major proteins in extracts of tobacco leaves that are responding hypersensitively to virus–infection. Phytochemistry 22: 2691–2697.

PIERPOINT, W.S. (1986) The pathogenesis–related proteins of tobacco leaves. Phytochemistry 25: 1595–1601.

PELHAM, J. (1972) Strain–genotype interaction of tobacco mosaic virus in tomato. Ann. Appl. Biol. 71: 219–228.

PENNAZIO, S. and SAPETTI, C. (1982) Electrolyte leakage in relation to viral and abiotic stresses inducing necrosis in cowpea leaves. Biol. Plant. 24: 218–225.

PERSON, C. (1959) Gene–for–gene relationships in host : parasite systems. Can. J. Bot. 37: 1101–1130.

RAST, A.T.B. (1975) Variability of tobacco mosaic virus in relation to control of tomato mosaic in glasshouse crops by resistance breeding and cross protection. Agric. Res. Rep. (Neth.) 834: 1–76.

REICHMAN, M., DEVASH, Y., SUHADOLNIK, R.J. and SELA, I. (1983) Human leukocyte interferon and the antiviral factor (AVF) from virus-infected plants stimulate plant tissues to produce nucleotides with antiviral activity. Virology 128: 563–570.

ROBINSON, R.A. (1976) Plant Pathosystems. Springer Verlag, Berlin. pp. 1–184.

RUSSELL, G.E. (1978) Plant Breeding for Pest and Disease Resistance. Butterworths, London. 485 pp.

RUSSO, M., MARTELLI, G.P. and FRANCO, A. DI. (1981) The fine structure of local lesions of beet necrotic yellow vein virus in Chenopodium amaranticolor. Physiol. Plant Pathol. 19: 237–242.

RUZICSKA, P., GOMBOS, Z. and FARKAS, G.L. (1983) Modification of the fatty acid composition of phospholipids during the hypersensitive reaction in tobacco. Virology 128: 60–64.

SANDERSON, J.L., BRUENING, G. and RUSSELL, M.L. (1985) Possible molecular basis of immunity of cowpeas to cowpea mosaic virus. In:

Cellular and Molecular Biology of Plant Stress. Key, J.L. and Kosuge, T., eds. Alan R. Liss, New York. pp. 401–412.
SARKAR, S. and SMITAMANA, P. (1981) A proteinless mutant of tobacco mosaic virus: evidence against the role of a coat protein for interference. Mol. Gen. Genet. 184: 158.
SELA, I. (1981) Plant-virus interactions related to resistance and localization of viral infections. Adv. Virus Research 26: 201–238.
SEQUIERA, L. (1984) Cross protection and induced resistance: their potential for plant disease control. Trends Biotechnol. 2: 25–29.
SHERWOOD, J.L. and FULTON, R.W. (1982) The specific involvement of coat protein in tobacco mosaic virus cross protection. Virology 119: 150–158.
SHERWOOD, J.L. and FULTON, R.W. (1983) Competition for infection sites and multiplication of the competing strain in plant viral interference. Phytopathology 73: 1363–1365.
SMART, T.E., DUNIGAN, D.D. and ZAITLIN, M. (1987) In vitro translation products of mRNAs derived from TMV-infected tobacco exhibiting a hypersensitive response. Virology, in press.
STASKAWICZ, B.J., DAHLBECK, D. and KEEN, N.T. (1984) Cloned avirulence gene of Pseudomonas syringae pv. glycineae determines race-specific incompatibility on Glycine max (L.) mem. Proc. Natl. Acad. Sci. USA 81: 6024–6028.
STIRPE, F., WILLIAMS, D.G., ONYON, L.J., LEGG, R.F. and STEVENS, W.A. (1981) Dianthins, ribosome-damaging proteins with antiviral properties from Dianthus caryophyllus L. Biochem. J. 195: 399–405.
SULZINSKI, M.A. and ZAITLIN, M. (1982) Tobacco mosaic virus replication in resistant and susceptible plants: in some resistant species virus is confined to a small number of initially infected cells. Virology 121: 12–19.
SZIRAKI, I. and BALAZS, E. (1975) The effect of infection by TMV on cytokinin level of tobacco plants, and cytokinins in TMV RNA. In: Current Topics in Plant Pathology, Kiraly, Z., ed. Akademiai Kiado, Budapest. pp. 345–352.
TAKAHASHI, T. (1973) Studies on viral pathogenesis in plant hosts. IV. Comparison of early processes of tobacco mosaic virus infection in the leaves of 'Samsun NN' and 'Samsun' tobacco plants. Phytopathol. Z. 77: 157–168.
TAKUSARI, H. and TAKAHASHI, T. (1979) Studies on viral pathogenesis in plant hosts IX. Effect of citrinin on the formation of necrotic lesions and virus localization in the leaves of 'Samsun NN' plants after tobacco mosaic virus infection. Phytopathol. Z. 96: 324–329.
THOMAS, P. and FULTON, R.W. (1968) Correlation of the ectodesmata number with nonspecific resistance to initial virus infection. Virology 34: 459–469.
THOMPSON, A.E., LOWER, R.L. and THORNBERRY, H.H. (1962) Inheritance in beans of the necrotic reaction to tobacco mosaic virus. J. Hered. 53: 89–91.
TSUGITA, A. (1962) The proteins of mutants of TMV: composition and structure of chemically evoked mutants of TMV RNA. J. Mol. Biol. 5: 284–292.
VALVERDE, R.A., DODDS, J.A. and HEICK, J.A. (1986) Double-stranded ribonucleic acid from plants infected with viruses having elongated particles and undivided genomes. Phytopathology 76: 459–465.
VAN LOON, L.C. (1983) The induction of pathogenesis related proteins by pathogens and specific chemicals. Neth. J. Plant Pathol. 89:

265–273.

VAN LOON, L.C. (1985) Occurrence and properties of pathogenesis-related proteins. Plant Molec. Biol. 4: 111–116.

VAN LOON, L.C. (1987) Disease induction by plant viruses. Adv. Virus Res. 32, in press.

VAN LOON, L.C. and ANTONIW, J.F. (1982) Comparison of the effects of salicylic acid and ethephon with virus-induced hypersensitivity and acquired resistance in tobacco. Neth. J. Plant Pathol. 88: 237–256.

VAN LOON, L.C. and VAN KAMMEN, A. (1970) Polyacrylamide disc electrophoresis of the soluble leaf proteins from N. tabacum var. Samsun and Samsun NN. II. changes in protein constitution after infection with TMV. Virology 40: 199–211.

VERMA, H.N. and DWIVEDI, S.D. (1984) Properties of a virus inhibiting agent, isolated from plants which have been treated with leaf extracts from *Bougainvillea spectabilis*. Physiol. Plant Pathol. 25: 93–101.

WAGIH, E.E. and COUTTS, R.H.A. (1981) Similarities in the soluble protein profiles of leaf tissue following either a hypersensitive reaction to virus infection or plasmolysis. Plant Sci. Lett. 21: 61–69.

WATANABE, Y., MORITA, N., NISHIGUCHI, M. and OKADA, Y. (1987) Attenuated strains of tobacco mosaic virus: reduced synthesis of a viral protein with a cell-to-cell movement function. J. Mol. Biol., in press.

WATERWORTH, H.E., KAPER, J.M. and TOUSIGNANT, M.E. (1979) CARNA 5, the small cucumber mosaic virus-dependent replicating RNA, regulates disease expression. Science 204: 845–847.

WESTSTEIJN, E.A. (1976) Peroxidase activity in leaves of *Nicotiana tabacum* var. Xanthi nc. before and after infection with tobacco mosaic virus. Physiol. Plant Pathol. 8: 63–71.

WESTSTEIJN, E.A. (1978) Permeability changes in the hypersensitive reaction of *Nicotiana tabacum* cv. Xanthi-nc. after infection with tobacco mosaic virus. Physiol. Plant Pathol. 13: 253–258.

WHENHAM, R.J. and FRASER, R.S.S. (1987) Cytokinins in control of resistance and symptom development. Rep. Natl. Veg. Res. Stn. for 1986. In press.

WHENHAM, R.J., FRASER, R.S.S., BROWN, L.P. and PAYNE, J.A. (1986) Tobacco mosaic virus-induced increase in abscisic acid concentration in tobacco leaves: intracellular location in light and dark green areas, and relationship to symptom development. Planta 168: 592–598.

WHITE, R.F. (1979) Acetylsalicylic acid (aspirin) induces resistance to tobacco mosaic virus in tobacco. Virology 99: 410–412.

WILSON, T.M.A. (1985) Nucleocapsid disassembly and early gene expression by positive strand RNA viruses. J. Gen. Virol. 66: 1201–1207.

WILSON, T.M.A. and WATKINS, P.A.C. (1986) Influence of exogenous virus coat protein on the cotranslational disassembly of tobacco mosaic virus (TMV) particles in vitro. Virology 149: 132–135.

WOOD, K.R. and BARBARA, D.J. (1971) Virus multiplication and peroxidase activity in leaves of cucumber (*Cucumis sativus L.*) cultivars systemically infected with the W strain of cucumber mosaic virus. Physiol. Plant Pathol. 1: 73–81.

WYATT, S.D. and KUHN, C.W. (1979) Replication and properties of cowpea chlorotic mottle virus in resistant cowpeas. Phytopathology 69: 125–129.

WYATT, S.D. and KUHN, C.W. (1980) Derivation of a new strain of cowpea chlorotic mottle virus from resistant cowpeas. J. Gen. Virol. 49: 289–296.

WYATT, S.D. and WILKINSON, T.C. (1984) Increase and spread of cowpea chlorotic mottle virus in resistant and fully susceptible cowpeas. Physiol. Plant Pathol. 24: 339–345.

ZAITLIN, M. (1976) Virus cross protection: more understanding is needed. Phytopathology 66: 382–383.

ZIMMERN, D. and HUNTER, T. (1983) Point mutation in the 30-K open reading frame of TMV implicated in temperature-sensitive assembly and local lesion spreading of mutant *Ni 2519*. EMBO J. 1893–1900.

ZINNEN, T.M. and FULTON, R.W. (1986) Cross-protection between sunn-hemp mosaic and tobacco mosaic viruses. J. Gen. Virol. 67: 1679–1687.

Oxford Surveys of Plant Molecular and Cell Biology Vol.4 (1987) 47-70

PHYSIOLOGY AND MOLECULAR BIOLOGY
OF MEMBRANE ATPASES

Michael R. Sussman and Terry K. Surowy

Department of Horticulture
University of Wisconsin
Madison, WI 53706

INTRODUCTION

Several recent comprehensive review articles and books have appeared concerning the proton translocating ATPases found in the tonoplast and plasma membrane of plant and fungal cells (Goffeau and Slayman, 1981; Leonard, 1904; Marin, 1984; Marre and Ballarin-Denti, 1985; Serrano, 1984a, 1985; Sze, 1985; Bowman and Bowman, 1986; Pederson and Carafoli, 1987a,b; Slayman, 1985, 1987). This chapter will focus on bringing this information up-to-date by incorporating recent results obtained using molecular biology techniques. Emphasis will also be placed on studies concerned with regulation of the plasma membrane H^+-ATPase of higher plants.

Membrane-bound transport proteins are divided into three broad categories:- pumps, carriers and channels. This classification is based mainly on:

(i) the speed of transport (i.e., turnover number) and
(ii) the type of energy coupling mechanism.

Pump proteins are the slowest of the three, with typical turnover numbers of 10-1,000 (the unit here is solute molecules transported per second per protein molecule). Pump proteins perform energy transduction, e.g., conversion of chemical to electrical energy, and the H^+-ATPases found in tonoplast and plasma membrane are two examples. Carriers usually have intermediate turnover numbers, in the order of 1,000-100,000. Carriers utilize the electrochemical energy gradients established by pumps and catalyze symport and antiport reactions, i.e., they perform co-transport functions. Example of carriers include a fungal plasma membrane $K^+ \cdot 2H^+$ carrier (Rodriguez-Navarro et al., 1986) and a $Ca^{2+} \cdot H^+$ carrier in plant tonoplasts (Schumaker and Sze, 1985, 1986; Bush and Sze, 1986). Channel proteins are the fastest transporters, with turnover numbers typically 10^6 or greater. These proteins perform uniport functions, i.e., they simply act as 'pores' in the membrane that allow a particular solute to move down its electrochemical gradient without coupling to the chemical gradient of another solute. In so doing, channels utilize an electrochemical gradient established by carriers and pumps. An example of a channel protein is a K^+ channel recently observed in the plasma membrane of plants and fungi (Takeda et al., 1985; Gustin et al., 1986). The term 'channel' has also been used to describe portions of pumps or carriers that are specifically involved in binding and moving a solute across the membrane.

Because of their speed, carriers and channels can be present at very low concentrations; only a few per cell may be sufficient to regulate solute flow. However, because of their greater abundance and their ease of detection with a simple but sensitive biochemical assay, the H^+-ATPases (proton pumps) were the first plant and fungal transport proteins to be purified and biochemically characterized. There are three main membrane–bound transport ATPases found in all plant and fungal cells and all are proton pumps: the mitochondrial H^+-ATPase, the tonoplast H^+-ATPase, and the plasma membrane H^+-ATPase. The mitochondrial H^+-ATPase, also known as the F_oF_1 H^+-ATPase, is the most abundant of the three and functions *in vivo* as an ATP synthase. A similar type of enzyme, with extensive amino acid sequence homology, exists in chloroplasts and bacteria. Because this enzyme is not unique to plants and fungi, and is so well characterized, its structure and function will not be discussed in this chapter. The reader is referred to Pederson and Carofoli (1987a,b) and to Bowman and Bowman (1986) for further information on this enzyme. A brief discussion of recent results concerning the structure of the tonoplast H^+-ATPase and a more extensive description of the plasma membrane H^+-ATPase, is presented below.

TONOPLAST H^+-ATPASE

Most of the volume of plant cells is occupied by a large central vacuole that functions as a storage organ for solutes, and as a 'lysosome' that contains proteases and other hydrolytic enzymes. Fungal cells contain similar vacuoles but these are generally smaller and more numerous. A protonmotive force exists across the vacuolar membrane which, by analogy to the mitochondrial membrane and the plasma membrane, provides the driving force for solute accumulation in this organelle. A H^+-ATPase has been observed in intact vacuoles and with purified vacuolar membranes, and this enzyme has a pH optimum, substrate specificity and inhibitor sensitivity that is distinct from that of the mitochondrial and plasma membrane H^+-ATPases (Mandala et al., 1982; Lew and Spanswick, 1985; Mandala and Taiz, 1985a,b; Sze, 1985; Wang and Sze, 1985; Giannini et al., 1987). Key distinguishing characteristics of this activity are that it is insensitive to azide and vanadate, potent inhibitors of the mitochondrial and plasma membrane H^+-ATPase activities respectively, and that it is uniquely sensitive to nitrate ions. A study of the putative tonoplast H^+-ATPase in *N. crassa* mutants with altered mitochondria has shown conclusively that the purified activity is not due to mitochondrial contamination (Bowman, 1983).

Detergent solubilization and glycerol or sucrose density gradient centrifugation techniques have been used to purify the H^+-ATPase from the tonoplast of several plant and fungal species (Mandala and Taiz, 1985a,b; Randall and Sze, 1986). In all cases, the most purified enzyme preparation contains two prominent polypeptides of ca. 60,000 and 70,000 M_r . In addition, a third subunit of ca.15-20,000 M_r has been identified. Although this latter polypeptide stains poorly with Coomassie Blue on sodium dodecyl sulphate polyacrylamide gel electrophoresis, it is readily identified by its reactivity with $[^{14}C]$-N,N'-dicyclohexylcarbodiimide (DCCD) a hydrophobic carboxyl-modifying reagent that irreversibly inactivates the H^+-ATPase. This observation is noteworthy because with the mitochondrial and plasma membrane H^+-ATPases, there is genetic and electrophysiological data suggesting that DCCD identifies a portion of proton pumps directly involved in proton translocation (Pederson and Carafoli, 1987a,b). Further studies are needed in which the low molecular weight DCCD-reactive vacuolar peptide is purified and then reconstituted into purified liposomes. Its reactivity with DCCD predicts that this peptide may contain protonophoric activity that could be directly assayed in a reconstituted system. Results with covalent modifying reagents indicate that the 70,000 M_r polypeptide contains amino acids involved in nucleotide binding and that the 60,000 M_r polypeptide may contain a regulatory nucleotide binding site (Mandala and Taiz, 1986).

Polyclonal antibodies have been raised to the two larger subunits, and immunological cross-reactivity between plant and fungal species has been demonstrated on Western Blots for the 70,000 M_r, but not the 60,000 M_r polypeptide (Bowman et al., 1986). Immunological cross-reactivity was not observed between the vacuolar membrane H^+-ATPase and subunits of the mitochondrial and plasma membrane H^+-ATPases. Successful production of an antibody against the smaller DCCD-binding subunit has not yet been reported.

Efforts in several laboratories are underway to clone the structural gene for the tonoplast H^+-ATPase. Progress has been reported in isolating tomato cDNA clones from an expression vector library using antibodies against the two larger polypeptides of red beet vacuolar H^+-ATPase (Rosichan et al., 1987).

PLASMA MEMBRANE H^+-ATPASE

In animal cells, the transport carriers are sodium-coupled, i.e., they perform sodium co-transport functions while transporting glucose, amino acids and other essential solutes across the plasma membrane. A Na^+,K^+-ATPase, ubiquitous to all animal plasma membranes, utilizes cytoplasmic ATP to create and maintain the sodium gradient. Plant and fungal plasma membranes contain a similar set of transporting enzymes which are coupled to a proton gradient maintained by an H^+-ATPase. By inserting microelectrodes into the cytoplasm of cells, it is possible to measure the activity of plasma membrane carriers and pumps with intact cells. This electrophysiological analysis has proven particularly useful in the characterization of plasma membrane transport proteins in plant and fungal cells because of considerable problems in extracting these proteins without loss of activity. A reasonable explanation for the lability of these extracted proteins centres on the release of deleterious hydrolytic enzymes (proteases, lipases) from the vacuole during cell breakage (Gallagher and Leonard, 1987).

With large cells (e.g., hyphal cells of *Neurospora crassa* and root hair cells of plants), it is possible to insert several microelectrodes and perform a detailed 'current-voltage analysis', the electrophysiologist's equivalent of a biochemist measuring an enzyme's catalytic parameters. There is a striking similarity in results obtained from the current-voltage analysis of large hyphal cells of *N. crassa* (Slayman, 1987) and root hair cells of a higher plant (Felle, 1982). This analysis has revealed that the plasma membrane of these cells contains an electrogenic proton pump (H^+-ATPase) with the following two characteristics:

(1) there is a 1:1 stoichiometry between protons pumped and ATP hydrolyzed and
(2) the pump has a K_m of ca. 1 mM for ATP.

These measurements also indicate that this single enzyme consumes 25-50% of the cytoplasmic ATP, i.e., is probably the major energy consumer in these actively transporting cells.

Biochemical investigations with cell extracts over the past decade have borne out these predictions derived from work with intact cells. An H^+-ATPase has been purified from fungal and plant plasma membrane vesicles, and this enzyme has the expected stoichiometry, substrate affinity and inhibitor specificity (Bowman et al., 1981; Perlin et al. 1984; Serrano, 1984b; Anthon and Spanswick, 1986; Kasamo, 1986; Perlin et al., 1986). The purified plasma membrane H^+-ATPase, in all species studied, contains a single catalytic polypeptide of 100,000 M_r, although the native enzyme most likely exists as a multimer of this subunit (Dufour and Goffeau, 1980; Bowman et al., 1985; Briskin et al., 1985; Goormaghtigh et al., 1986; Chadwick et al., 1987). This enzyme constitutes ca. 1-5% of the protein in purified *N. crassa* plasma membrane vesicles (Sussman, 1984) and ca. 0.1-0.5% of the protein in the most highly enriched vesicles derived from *Avena sativa* (oat) roots (Schaller and Sussman, submitted). Antibody generated against the enzyme of one fungal species cross-reacts strongly with other fungi (Vai et al., 1986) and that generated against the oat enzyme cross-reacts well with both monocot and dicot higher plant species (Surowy and Sussman, 1986). There is weak but detectable immunological cross-reactivity between the fungal and plant enzymes (Clement, 1986).

The plant/fungal plasma membrane H^+-ATPase belongs to a class of cation-pumping ATPases that show similar reaction mechanisms, inhibitor-sensitivities and contain a similar 100,000 M_r polypeptide with several stretches of amino acid sequence homology. Four other known members of this group are the Na^+,K^+-ATPase isolated from animal plasma membranes, a Ca^{2+}-ATPase isolated from muscle sarcoplasmic reticulum, an H^+,K^+-ATPase isolated from stomach gastric mucosa and a K^+-ATPase found in the plasma membrane of *E. coli*. The H^+-, Ca^{2+}- and H^+,K^+-ATPases contain only this single 100,000 M_r polypeptide whereas the Na^+,K^+- and K^+-ATPases contain one and two additional lower molecular weight polypeptides, respectively. Although there is no direct evidence apart from co-purification for an essential function of the lower molecular weight polypeptide in the Na^+,K^+-ATPase, genetic evidence has proven that the two lower molecular polypeptides in the *E. coli* K^+-ATPase are essential for catalytic activity (Epstein and Laimins, 1980).

Covalent amino acid modifying reagents that inactivate the H^+-ATPase have been used to identify amino acids essential for catalytic activity. Although more extensive studies have been performed with the fungal enzyme, similar experiments with the plant enzyme have demonstrated that it too contains N-ethylmaleimide-reactive essential cysteine and fluorescein isothiocyanate-reactive lysine residues in its ATP binding site (Katz and Sussman, 1987; Gildensoph et al., 1987). Of particular interest in these covalent modification studies is the observation that the plant (Cid et al., 1987; Oleski and Bennett, 1987) and fungal (Sussman and Slayman, 1983) plasma membrane H^+-ATPase is highly sensitive to DCCD, a hydrophobic carboxyl-modifying reagent that reacts with glutamate and aspartate residues. The plasma membrane proton pump reacts with, and is inhibited by, DCCD at a rate ca. 10^4 times that observed with free glutamate or aspartate. Hydrophilic carbodiimides do not react with the enzyme. These results, together with electrophysiological studies with intact cells (Kishimoto et al., 1984) and by analogy to more extensive studies with the F_0F_1 H^+-ATPase, suggest that DCCD is reacting with an essential glutamate or aspartate residue in a portion of the 100,000 M_r polypeptide that is directly involved in proton translocation. The location of the DCCD-reactive residue in the fungal plasma membrane H^+-ATPase has recently been identified as glutamate-129, which is located in the middle of the first trans-membrane segment of the 100,000 M_r polypeptide (Sussman et al., 1987).

Structural genes corresponding to five types of 100,000 M_r cation-pumping ATPases have been isolated and sequenced (Brandl et al, 1986; Hager et al., 1986; Hesse et al., 1984; Kawakami et al., 1985, 1986; MacLennan et al., 1985; Ovchinnikov et al., 1986; Schneider et al., 1985; Serrano et al., 1986a; Shull and Lingrel, 1986, 1987; Shull et al., 1985, 1986). The deduced amino acid sequences show a remarkably similar overall structure (Fig. 1). Using a hydropathy plot to identify hydrophobic regions, all of the predicted transmembrane segments lie within the N-terminal third, or the C-terminal third, with 4 segments in the former section and a controversial number (4, 5 or 6) in the latter. The middle one-third is an uninterrupted hydrophilic sequence that extends into the cytoplasm.

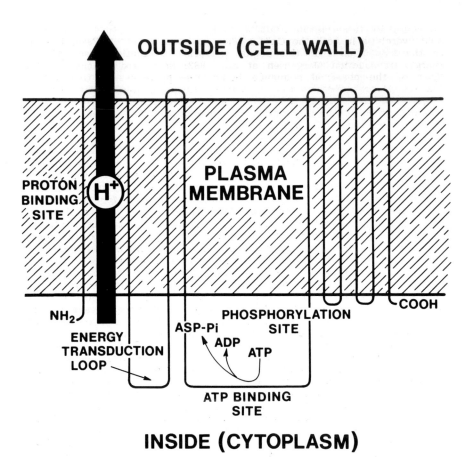

Figure. 1. Structure of the plasma membrane H⁺-ATPase, as deduced from a hydropathy plot to determine transmembrane segments, and from studies with covalent amino acid modifying agents used to identify an ATP binding site, a phosphorylated aspartyl residue, and a proton binding site. The putative energy transduction loop, as identified by tryptic cleavage studies of the Ca^{2+} -ATPase and Na^+,K^+ -ATPase, is also indicated.

This water–soluble portion contains an ATP binding site (Ohta et al. 1986) and a phosphorylated aspartate residue (Walderhaug et al., 1985). There is an additional hydrophilic loop extending into the cytoplasm that resides between the second and third transmembrane segments. Although covalent amino acid modifying reagents have not yet identified an essential amino acid in this loop, a proteolytic cleavage in this region inhibits cation transport without affecting ATP hydrolysis in both the Ca^{2+}-ATPase and the Na^+,K^+-ATPase, leading to the suggestion that this region performs a function essential for energy transduction (Jorgensen et al., 1982; Scott and Shamoo, 1982). Most of the observed sequence homologies between the five cation translocating ATPases occurs within the two hydrophilic 'loops' (Figure 2). In addition, practically all of the protein sequence data obtained from Edman degradation of isolated peptides has been derived from the two hydrophilic sections.

Although the plasma membrane H^+-ATPase gene of both *N. crassa* and *S. cerevisiae* has been cloned and from this the complete amino acid sequence determined, this work is still in progress with the higher plant gene. A four amino acid long stretch surrounding the phosphorylated aspartyl residue has been determined for the plasma membrane H^+-ATPase of corn roots, and this sequence was found to be 100% homologous to that surrounding the phosphorylated aspartate of other cation-transporting ATPases. More recently, a method has been developed by which large amounts of the plasma membrane H^+-ATPase can be purified from oat roots, and using this method, extensive protein sequence information has recently been obtained for the higher plant plasma membrane H^+-ATPase. For unambiguous protein sequence determinations, it is helpful to start with at least several nanomoles of the 100,000 M_r polypeptide. Following cleavage to completion with trypsin (1-2 µg trypsin/100 µg of ATPase), the cleaved peptides are resolved by reverse-phase high pressure liquid chromatography. Tryptic peptides are detected by OD_{220}, where the peptide bond absorbs and by OD_{280}, where only tryptophan and tyrosine absorb. In a typical experiment with tryptic peptides isolated from 3-5 nanomoles of the oat root plasma membrane H^+-ATPase, 30-40 peaks of 220 nm absorbing material are observed when elution was performed with a 0- 60% v/v acetonitrile gradient (constant 0.1% v/v trifluoroacetic acid); less than 5 of these peaks showed prominent absorbance at 280 nm. Eight peptides were picked at random, further purified by re-chromatographing in a shallower acetonitrile gradient and then sequenced in an Applied Biosystems gas-phase sequenator equipped with an on-line micro-bore HPLC for high- sensitivity PTH-amino acid analysis. The average length of these tryptic peptides was 10-15 amino acids, and yields of PTH-amino acid observed in the first cycle ranged from 0.3 to 3 nanomoles. The sequence of these eight peptides was then compared to the complete known sequences of four cation-transporting ATPases (H^+-, Na^+,K^+-, Ca^{2+}-, and K^+-). This analysis revealed that six of the eight random tryptic peptides isolated from the oat root plasma membrane H^+-ATPase showed strong homology to hydrophilic regions containing energy- transduction, phosphorylation and nucleotide binding sites (Schaller and Sussman, submitted; Figure 2). Overall, the oat-derived peptides showed slightly greater similarity to the fungal H^+-ATPase than to the other cation transporting ATPases.

It is not surprising to note sequence similarities in the water-soluble regions of the polypeptides responsible for energy-transduction and nucleotide hydrolysis since the five cation ATPases must all perform these functions. The major difference between these enzymes is their cation-specificity, i.e., which

Figure.2.(opposite and following two pages). Comparison of amino acid sequence (with one letter codes) for the plasma membrane H⁺-ATPase ("H") of Neurospora crassa *(Hager et al., 1986), the sheep kidney plasma membrane Na⁺,K⁺-ATPase ("NaK") (Shull et al., 1985), the rabbit muscle sarcoplasmic reticulum Ca²⁺-ATPase ("Ca") (MacLennan et al., 1985), the* Escherichia coli *plasma membrane K⁺-ATPase ("K") (Hesse et al., 1984) and the rat stomach plasma membrane H⁺,K⁺-ATPase ("HK") (Shull and Lingrel, 1986). The sequence comparison program called PROFILE was used to introduce gaps and align all five sequences (Gribskov et al., 1987). The "Consensus" sequence shown in capital letters throughout, represents the amino-acid most mutationally similar to all of the aligned residues. Sequences covered with a solid line represent transmembrane regions of the H⁺- ATPase predicted from a hydropathy plot. Arrows denote the location of essential amino acids: at position 136, a DCCD-reactive glutamate (E) involved in proton binding/transport; at position 437, a phosphorylated aspartyl residue (D) involved in ATP hydrolysis; at position 606, a FITC-reactive lysine (K) essential for nucleotide binding; and at positions 769 and 832, amino acids labeled with FSBA, an ATP affinity reagent. Sequences shown in italics above the fungal H⁼-ATPase represent amino acids known to occur in the higher plant plasma membrane H⁺-ATPase. The -CSDK- sequence starting at position 426 was determined by Walderhaug et al. (1985) with ATPase from corn roots, while all others were obtained by automated Edman degradation of randomly selected tryptic peptides isolated from the plasma membrane H⁺-ATPase of oat roots (Schaller and Sussman, submitted).*

---

particular ion gets transported. Progress in identifying amino acids essential in the ion binding and transport function has been limited by the lack of inhibitors specific for this function, and in technical difficulties associated with handling the hydrophobic, transmembrane peptides. One approach that has proven useful for studying the cation transport function involves the hydrophobic carboxyl modifying reagent, DCCD. As mentioned earlier, this hydrophobic carbodiimide has been found to react with glutamate-129 in the fungal plasma membrane H⁺-ATPase, causing irreversible inactivation of ATPase activity. These results provide the first demonstration of a catalytic function for a transmembrane section of the 100,000 M_r cation-translocating ATPase (Sussman et al., 1987).

If DCCD specifically blocks an amino acid essential only for proton binding or movement, without affecting residues involved in ATP binding or hydrolysis, why is the overall ATPase activity reduced? The answer is that with this enzyme, as with the F_0F_1 H⁺-ATPase, the proton transport and ATP hydrolytic activities are obligately coupled, i.e. blocking one completely prevents the enzyme from turning over. It is interesting to note two instances in which this strict coupling breaks down:

54

```
           1                                                       50
H          .madhsasga paLstniesg kfdekAaeaa a...yqpkpK veddEd..eD
NaK        .......... ..Mgkgaase kyqpaAtsen aknskksksK ttdlDelkkE
Ca         .......... .......... .......... .......... ..........
K          .......... .......... .......... .......... ..........
HK         mgkenyelys veLgtgpggd m...aAkmsk kkaggggkK keklEnmkkE
Consensus  ---------- --L------- -----A---- --------K ----E----E


           51                                                      100
H          idaLieDles hdghdaeeee eeaTPGggrv vpeDmLqtDt rvgLtseEvv
NaK        v..sldDHkl nldElhqkYg tdlTqGLtpa raKEiLarDg PnALtPpptt
Ca         ...MenaHtk tveEvlghFg vnestGLsle qvK.kLKerw gsnelPaEeg
K          .......... ...msrkqLa l.fePtLvvq alKEavKkln PqAqw....r
HK         m..emnDHql svsElenkYn tsaTkGLkas laaElLlrDg PnALrPprgt
Consensus  ------DH-- ---E----Y- --TPGL--- --KE-LK-D- P-AL-P-E--


           101                                            150
H          qrRrKYglnQ mkEekenhFL kFlgffvgpi qFv.MeGaAV LAAgleDWvd
NaK        peWiKFcrql fggFs...iL LWtgaIlcfL AYg.iqvatV dnpa.nDnLY
Ca         ktlLeLvieQ feDLlVRIL. LLaaCISfvL AWfeegeetI tAfve.pFvi
K          npvM...... ...FiVWIgs LLttCISiaM AsgaMpGnAl FsAaisgWLW
HK         peYvKFarql aggL...qcL MWvaaaicli AFa.iqaseg .dlttdDnLY
Consensus  --R-KF---Q -----V-I-L LL--CIS--L AF--M-G-AV -AA---DWLY


           151                                            200
H          F.gvIcglll lNavvGFv.Q EfqAgs.... ...IVDeLKK .....tLalk
NaK        Lgvvlstvvl itgcfsY.YQ Eakssk.... ...ImDsFKn .....mvPqq
Ca         LlilVanaIV .....Gv.WQ ErnAEn.... ...aIEaLKe ye.....PEm
K          M.....tVlf aN......Fa EalAEgrska qanslkgvKK tafarkLrDa
HK         LalaliaVVV vtgcfGY.YQ Efkstn.... ...IIasFKn l.....vPqq
Consensus  L------VVV -N----GY-YQ E--AE----- ---IID-LKK ------LP--


           201                                                     250
                      wgegeAsilV PGDIVsiklG DIVPADaR          iD
H          AvVlRDG..t lkeIeApeVV PGDIlqVEeG tIIPADGRIV t.ddafLqVD
NaK        AlViRDGEKs ..sInAEQVV vGD1VEVkgG DrIPAD1RII S.acsc.kVD
Ca         gkVyRqdrKs vqrIkAkdIV PGDIVEIavG DkVPAD1Rlt SiksttLrVD
K          kygaa.aDK. ...VpADQlr kGDIV1VEaG DIIFcDGeVI eggas...VD
HK         AtViRDGDKf ..qInADQlV vGD1VEmkgG DrVPAD1Ril Saqgc..kVD
Consensus  A-V-RDGDK- ---I-ADQVV PGDIVEVE-G DIIPADGRII S-----L-VD


           251                                                     300
           QSgLTGESlp VtK                 npgd      eVFsgSt  ck
H          QSALTGESla VDK........ ......hKgd ...qVFaSsa VkrGeAfvVI
NaK        nSsLTGES.. .Epqsrspey ssenplEtkN ia...FfSTn cveGtArGIV
Ca         QSiLTGESvs ViKhidpvpd pravnqDKkN m...lFsgTn IaaGkAmGVV
K          eSAiTGESap Viresg...g df........ ...asVtggTr Ilsdwlviec
HK         nSsLTGES.. .Epqtrspec thesplEtrN ia...FfSTm cleGtAqGlV
Consensus  QSALTGES-- VEK------- ------EK-N ----VF-ST- I--G-A-GVV


           301                                            350
H          tATGDNTFVG RaAALVnaAs ggsghFtevL ngiGtILlIl VIFtlliivWV
NaK        IniGDhTvmG RIAtLasGlE vgqTPiaaei EhFihIitgV aVFlgvsFFI
Ca         VATGvNTeIG kIrdemvatE qerTPLqqkL DeFGeqLskV IsLiciavWI
K          svnpgeTFld RmiAMVeGAq rrkTPneiaL tiLliaLtIV flLatatLWp
HK         VsTGDrTiIG RIAsLasGvE nekTPiaiei EhFvdIiagl aILfgatFFV
Consensus  VATGDNTFIG RIAALV-GAE ---TP----L E-FG-IL-IV -IL----FW-


           351                                            400
H          sSFYrsNpIV q......... .iLeFtlAIt IigVPvGLPA VVTTtMAvGA
NaK        lSLilGytwl e......... .aviFLIgIi VAnVPEGLlA tVTvCLtLtA
Ca         inighfNdpV hggswirgai yyFkiaVAla VAaIPEGLPA VITTCLALGc
K          fSaWgGNaV. .......... .svtvLVAll VclIPttigg llsasavaGm
HK         vaMCiGytfl r......... .aMvFFmAIv VAyVPEGLlA tVTvCLsLtA
Consensus  -SF--GN--V ---------- ----FLVAI- VA-VPEGLPA VVTTCLALGA
```

```
                                        ↓
        401                         CSDKT GTLTlNk                    450
H       ayLAKKkAIV qkLsAIEsLa gveIlCSDKT GTLTkNkLS1 hDpYtvaGVD
NaK     kRMArKNclV knLeAVETLG sTStICSDKT GTLTqNrMtV ahMWfdnqIh
Ca      rRMAKKNAIV rSLpsVETLG cTSVICSDnT GTLTtNqMSV crMFildkVD
K       sRMlgaNvIa tSgrAgEaaG dvdVlllDKT GTiTlvnrqa sEFipaqGVD
HK      kRLAsKNcVV knLeAVETLG sTSVICSDKT GTLTqNrMtV shLWfdnhIh
Consensus -RMAKKNAIV -SL-AVETLG -TSVICSDKT GTLT-N-MSV --MW---GVD

        451                                                        500
H       pE........ ..Dlmlt... .......... .......... .........A
NaK     ea........ ..DtTenqsG isfdktsl.. .....swnaL srIaaLCNrA
Ca      gDtcslneft itgsTyapiG evhkddkpvk chqydglveL atIcaLCNds
K       e......... .k........ .......... .......... ..........
HK      ta........ ..DtTedqsG qtfdqss... ....etwraL crVltLCNrA
Consensus ---------- --D-T----G ---------- ---------L --I--LCN-A

        501                                                        550
H       cLAasrkKkg I.......DA iDkAFLKsLk yYpraksvL. skykvl....
NaK     vFqagqdsvp IlKrsvAgDA sEsALLKcie lccgsvsqMR DRnpkiveI.
Ca      aLdyneaKgv yeKvgeAtEt alTcLvekMn VFdtelkgLs .......kIe
K       tLAdaaq... .....lAslA dETpegrmiv ILakqrfnLR ERdv......
HK      aFksgqdavp VpKrivigDA sETALLKfse ltlgnamgyR DRfpkvceI.
Consensus -LA----K-- I-K---A-DA -ETALLK--- --------LR DR------I-

        551                                                        600
H       .......... ........qf hPFDpvSKkV vaVVe..... ..Spqg.Er.
NaK     .......... .......... .PFnstnKYq lsIhe..... ...ndkaDsR
Ca      ranacnsvik qlmkkeftle fsrDrkSmsV yctpn..... ...kpsrtsm
K       .......... .......... ......qsLh atfVpftaqs rmSginiDnR
HK      .......... .......... .PFnstnKFq lsI....... htledprDpR
Consensus ---------- ---------- -PFD--SKLV --IV------ --S----D-R

        601    ↓                                                   650
H       .itcVKGAP. .lfvLktv.. ........ee DhPiPEEVDQ AYknkVaEF.
NaK     ylLVmKGAPE rI..LDRC.. ..stillnge DkPLnEEmke AFqnaylEL.
Ca      skMfVKGAPE gV..iDRCth i....rVgst kvPMtagVkQ kiisvIrEWg
K       ..sIaKGsvD aIr....... ...rhVean gghFPtDVDQ kvdq......
HK      hlLVmKGAPE rV..LERCss il....Ikgq ElPLdEqwre AF..qtayLs
Consensus --LVVKGAPE ----LDRC-- ------V--- D-PLPEEVDQ AF-----EL-

        651                                                        700
H       .atrGFRsLG vArkrgegs. .......... ........we iLGImpcMDP
NaK     .GglGeRvLG Fchlklstsk fpegypfDve epnfpitdLc FvGlmsMiDP
Ca      sGsdtLRcLa LA.thdnplr reemhlkDsa nfikyetnLt FvGcVgMLDP
K       varqGatpLv vv........ .......... .....egsr vLGVIaLkDi
HK      lGglGeRvLG Fcqlylnekd yppgytfDve amnfpssgLc FaGlVsMiDP
Consensus -G--G-R-LG FA-------- -------D-- --------L- FLG-V-M-DP

        701                                                        750
H       PRhDtyktVc eaktlGlsIk MlTGDavGiA retsRqlGl. ....gTniyn
NaK     PRaaVpDAVg kCRsAGIkVI MVTGDhpiTA kAIakgVGIi segnETvedi
Ca      PRiEVassVk lCRqAGIrVI MITGDnkGTA vAIcRrIGIf gqeeDv....
K       vkggIkEAfa qlRkmGIktV MITGD..... .......... ..........
HK      PRatVpDAVl kCRtAGIrVI MVTGDhpiTA kAIaasVGIi segsETvedi
Consensus PR--V-DAV- -CR-AGI-VI MITGD--GTA -AI-R-VGI- ----ET----

        751                    ↓                                   800
H       AeRLglgggg dmpGsEvYDF VeaAd..... .......... ....gFAeVf
NaK     AaRLnipvnq vnprdakacv Vhgtdlkdls henlddilhy hteivFARts
Ca      .......tak AftGrEfdEL npsAq..... ......rdac lnarcFARVe
K       .nRLtaaaia AeaGvDdF.. .......... .......... .....LAeat
HK      AaRLrmpvdq vnkkdaracv Ingmqlkdmd pselvealrt hpemvFARts
Consensus A-RL------ A--G-E---- V--A------ ---------- -----FARV-
```

```
            801              gIVgM TGDGVNDAPA  LK                    850
H           PQHKynVVEi  lQqrvylVAM  TGDGVNDAPs  LKKADtGIAv  e.GsSDAArs
NaK         PQqKLiIVEg  CQrqGaIVAv  TGDGVNDsPA  LKKADIGVAM  GiagSDvsKq
Ca          PsHKskIVEf  lQsfdeItAM  TGDGVNDsPA  LKKAEIGIAM  GSGta.vAKt
K           PeaKLalIrq  YQaeGr1VAM  TGDGtNDAPA  LaqADVaVAM  nSG.tqAAKe
HK          PQqKLvIVEs  CQr1GaIVAv  TGDGVNDsPA  LKKADIGVAM  GiagSDAAKn
Consensus   PQHKL-IVE-  CQ--G-IVAM  TGDGVNDAPA  LKKADIGVAM  GSG-SDAAK-

            851                                                     900
H           AADiVFLapg  LgaIIDAlkt  sRQIFhrMya  YVvYrIAlsI  .hleiFLgLW
NaK         AADMILLDDN  FasIVtgVEe  GR1lFdNLKk  sIaYtltsNI  pEitpFLvFi
Ca          AsEMVLaDDN  FstIVaAVEe  GRaIYnNMKq  FIrYlIssNV  gEvvciFlta
K           AgnMVdLDsN  ptklIEvVhi  GkQmLmtrvs  LttFsIAndV  ak......YF
HK          AADMILLDDN  FasIVtgVEq  GR1lFdNLKk  sIaYtltkNI  pEltpYLiYi
Consensus   AADMVLLDDN  F--IV-AVE-  GRQIF-NMK-  -I-Y-IA-NI  -E----FL-Y-

            901                                                     950
H           iailnrsLni  elVvFIaiFa  D.vatlAiAY  DnA.......  ..........
NaK         ianVPlpLgt  VtILcIdLgT  DmvPaisLAY  ErAESDIM..  ..........
Ca          AlgfPeALip  VqlLWViLvT  DglPatALgF  nppD1DIM..  ..........
K           Ai.IPaAFaa  t.......Yp  qlnalniMcL  hspDSaILsa  vifnaliivf
HK          tvsVPlpLgc  ItILpIeLCT  DifPsvsLAY  EkAESDIM..  ..........
Consensus   A--VP-AL--  V-IL-I-L-T  D--P--ALAY  E-ADSDIM--  ----------

            951                                                    1000
H           ..........  ..........  .......pYa  QtpvkWnLpk  Lwgm.sV...
NaK         .......krq  PRNPktDkLV  nerLismaYg  QIGmIqaLgG  Ffsyf.V...
Ca          .......nkp  PRNPk.EPLI  sgwLF.frYl  aIGcyvGaat  vga...aa..
K           lipl.....a  lkgvsykPLt  asaMLrrnl.  ...wIYGrgG  L.lvpfIgik
HK          ...hl....r  PRNPrrDrLV  nepLaaysYf  QIGaIqsFaG  Fadvfta
Consensus   ........... PRNP--DPLV  ---L----Y-  QIG-I-GL-G  L-----V---

            1001                                                   1050
H           1L.....vvV  lAVGtWitvt  tMYaQgengg  ivQnFgnmDe  VlFlqis...
NaK         iLaengFLpI  dlIGiRekWd  eLWtQdleds  YgQqWt.YEq  rkiveYtcht
Ca          ......WWfI  aAdG..gpRv  sFYqlshflq  CkednpdFEg  VdCaiFes..
K           vidl..LLtV  eglv......  ..........  ..........  ..........
HK          .MaqegWFpl  lcVGlRpqWw  dhhlQdlqds  YgQeWt.Fgq  rlYqqYtc..
Consensus   -L----WL-V  -AVG-R--W-  --Y-Q-----  Y-Q-W--FE-  V----Y----

            1051                                                   1100
H           .Ltenwl...  ifITran..g  pf......Ws  Si..pswqLs  gaIFlvdI.L
NaK         sFfv.siviv  qwad...l..  ..iicktrRn  SiFqqg..mk  NkILIfglfe
Ca          pYpmtmalsv  l.VTiemc.n  a..lnslsen  qsLlrmppWe  NiwLVgsIcL
K           ..........  ..........  ..........  ..........  ..........
HK          .....ytvff  isIemcqi.a  dvlirktrRl  SaFqqgf.Fr  NrILViaIvF
Consensus   ----------  --IT------  --------R-  S-F------   N-ILV--I-L

            1101                                                   1150
H           atcFtiWgwF  ehsdtsIvaV  vri.......  .WiFSFgIFc  imggvyYIlq
NaK         etaLaaFlsY  tpgtdiaLrm  yPlkpswWFc  aFpYSLiIFl  yDEarrFIlR
Ca          smsLhfLilY  veplplIFqI  tPrnvtqWLm  vLkiSLpVil  mDEtlkFVaR
K           ..........  ..........  ..........  ..........  ..........
HK          qvcigcFlcY  cpgmpnIFnf  mPirfqwWLv  pmpFgLlIFv  yDEirkLgvR
Consensus   ---L--F--Y  ------IF--  -P-----WL-  ---FSL-IF-  -DE---FI-R

            1151                              1193
H           dsvG..fdnl  mhgkspkgnq  kqrsledfvv  slqrvstqhe  ksq
NaK         rnpG....gw  veqetyy...  ..........  ..........  ...
Ca          nylepaile.  ..........  ..........  ..........  ...
K           ..........  ..........  ..........  ..........  ...
HK          ccpG....sw  wdqelyy...  ..........  ..........  ...
Consensus   ---G------  ----------  ----------  ----------  ---
```

57

(1) when the F_1 is physically separated from the F_0 (membranous portion) and
(2) in the Ca^{2+}-, and Na^+,K^+-ATPases, when the energy transduction loop is proteolytically cleaved. From these results one may predict that a proteolytic cleavage in the energy transduction domain of a DCCD-inactivated plasma membrane H^+-ATPase might actually reverse the inhibition elicited with DCCD-modified glutamate-129. The experiment to test this prediction has not yet been performed.

DCCD has also been reported to inactivate the Ca^{2+}- and Na^+,K^+-ATPases (Schoner, 1971; Robinson, 1974; Pick and Racker, 1979; Murphy, 1981; Gorga, 1985; Scofano et al., 1985; Pedemonte and Kaplan, 1986). The observation that this inhibition is prevented by the presence of Ca^{2+} and Na^+ ions, respectively, suggests that like the H^+-ATPase, DCCD is reacting with an essential glutamate or asparate residue located within a portion of the enzyme that forms a Ca^{2+} or Na^+ 'channel'. Further conclusions must wait until the location of these residues is determined. For studies with [^{14}C]-DCCD it is important to optimize DCCD specificity by performing the inactivation at low temperature and high pH. Under these conditions reactivity with the proton channel is maintained while that with other glutamate or aspartate residues is greatly diminished (Pougeois et al., 1980; Sussman, unpublished observations).

GENETIC STUDIES WITH THE PLASMA MEMBRANE H^+-ATPASE

Several laboratories are engaged in efforts to clone the structural gene(s) encoding the plasma membrane H^+-ATPase of higher plants, and progress has been reported (Surowy and Sussman, 1987). Polyclonal antibody directed against the 100,000 M_r subunit of the H^+-ATPase was used to isolate plant DNA fragments which express immunoreactive protein in a lambda gt11 expression vector system. This approach also was used recently to clone the plasma membrane H^+-ATPase structural gene of the mycelial fungus, *Neurospora crassa* (Hager et al., 1986) and the yeast, *Saccharomyces cerevisiae* (Serrano et al., 1986a). The cloned yeast DNA maps to a locus that was previously suggested to be the ATPase gene on the basis of altered catalytic activity in inhibitor-resistant mutants (Ulaszewski et al., 1986, 1987). Gene disruption experiments performed with *S. cerevisiae* demonstrate that the single cloned plasma membrane H^+-ATPase gene is essential for growth (Serrano et al., 1986b). Two mutants were produced, one in which the promoter is disrupted, and a second in which the structural gene itself is disrupted. Haploid mutants containing the first genetic construction contained ca. 1/3 the normal level of ATPase polypeptide, and these cells grew at only 1/3 their normal growth rate. Haploid spores containing the second construction do not germinate, demonstrating that complete disruption of the ATPase is a lethal mutation. These results further suggest that there is a direct correlation between the growth rate and ATPase activity in *S. cerevisiae*. These mutants may be useful in physiological studies concerned with ATPase function. For example, if the proton-motive force generated by the ATPase is only necessary to power solute uptake systems and is not directly involved in other essential eukaryotic cellular processes such as cell wall biosynthesis or maintenance of cell polarity (Kropf et al., 1983), it may be possible to devise a highly concentrated nutrient medium that supports cell division in the absence of a protonmotive force.

In a more recent study, the *S. cerevisiae* chromosomal ATPase gene was placed under the control of an inducible galactose promoter, permitting the use of site-directed mutagenesis to obtain a temperature-sensitive ATPase gene located in an extra-chromosomal yeast expression plasmid (Serrano et al., 1986b). With this temperature-sensitive 'mutant', it was possible to demonstrate that solute movement into the cell was dependent on the galactose concentration and temperature, thereby proving through genetic means the conclusion suggested from electrophysiological and biochemical studies that the ATPase-dependent protonmotive force plays an obligate role in solute movement across the fungal plasma membrane.

With the chromosomal ATPase locus under control by galactose, it will now be possible to clone and express via the extrachromosomal plasmid, ATPase genes specifically altered by site-directed mutagenesis as well as with chimeric genes derived by switching various parts of the five different 100,000 M_r ATPases. These 'protein engineering' experiments are essential to confirm conclusions derived from chemical modification and peptide sequencing studies as well as to provide structure-activity insights not readily possible using chemical modification experiments. Genetic manipulation of the melibiose transport carrier in *E. coli* has demonstrated that a single amino acid substitution is sufficient to convert a H^+-coupled carrier protein into one that only performs Li^+-coupled co-transport (Yazyu et al., 1987). Thus, by determining what amino acid substitutions are needed to alter the ATPases' cation specificity, it may be possible to map the location of amino acids specifically involved in cation transport. Finally, these protein engineering experiments will help to resolve the question of whether the plant or fungal plasma membrane H^+-ATPase 100,000 M_r polypeptide also acts as an inwardly directed potassium pump (Villalobo, 1984).

NUMBER OF PLASMA MEMBRANE ATPASE GENES IN HIGHER PLANTS

It is interesting to note that although the yeast studies suggest that there is a single plasma membrane H^+-ATPase gene in this lower eukaryote, the cation-pumping ATPases of animals exist in multiple copies, with each gene showing tissue-specific expression. Thus, for example, there are at least four separate mammalian Na^+,K^+-ATPase genes, three of which are differentially expressed in various tissues of the same animal (Shull and Lingrel, 1987). Higher plants contain tissue with differentiated transport cells not found in fungi. It is therefore reasonable to ask whether higher plants contain multiple plasma membrane H^+-ATPase genes unique to specialized transport cells such as the stomatal guard cells, root hair cells or the phloem-loading cells. If multiple H^+-ATPase genes are found in higher plants, *in situ* hybridization experiments with microscopic thin sections, similar to those recently reported with the Na^+,K^+-ATPase genes (Schneider et al., 1987), will determine whether tissue–specific expression occurs.

Even if higher plants contain a single essential plasma membrane H^+-ATPase gene, there may be other 100,000 M_r-type cation ATPase genes present. Biochemical experiments have suggested that there may exist a Ca^{2+}-pumping ATPase in the fungal and plant plasma membrane (Rasi-Caldogno et al., 1987). However, until this enzyme is purified to homogeneity, free of contamination with the H^+-ATPase, it may be difficult to rule out an alternative mechanism, that is, a Ca^{2+}/H^+ antiporter that is indirectly powered by and therefore dependent on the H^+-ATPase. An alternative to this biochemical approach is to screen a genomic or cDNA library for 100,000 M_r-type ATPase genes that are unlike the H^+-ATPase but which still contain conserved sequences common to this class of enzymes.

REGULATION OF ATPASE GENE EXPRESSION

More is known concerning transcriptional control of the *E. coli* plasma membrane K^+-ATPase gene than any other 100,000 M_r cation-pumping ATPase (Epstein and Laimins, 1980; Laimins et al., 1985). This bacterial gene exists as an operon containing four contiguous open reading frames (A, B, C and D), corresponding to four polypeptides with molecular weights of 47,000, 90,000, 22,000 and 90,000, respectively. Gene B encodes the typical ATPase catalytic polypeptide while A and C are additional lower molecular weight proteins required for catalytic activity, as determined by mutant analysis. Polypeptide D is present in the plasma membrane, and although it is required for proper induction of transcription within the operon, the precise nature of its catalytic activity is unknown. Early experiments demonstrated that K^+-ATPase gene transcription was induced by potassium starvation, and this led to the suggestion that the enzyme was involved in 'scavenging' low concentrations of potassium during conditions of mineral depletion. When the promoter for this operon was cloned and placed next to a 'reporter' gene, β-galactosidase, a more comprehensive study of the mechanism of transcriptional control was possible. These experiments demonstrated that, in fact, the K^+-ATPase promoter was not turned on by the low internal potassium concentrations directly, but rather, by the loss of turgor induced by this condition. Thus, other plasmolysing agents such as high concentrations of permeant sugars (sorbitol, mannitol), were equally effective in 'turning on' the promoter. It was further postulated that polypeptide D was involved in sensing the turgor loss and conveying this information to the chromosome.

Higher plant cells also sense changes in turgor pressure, and adjust the polypeptide composition and transport functions of their plasma membrane accordingly. For example, a decrease in turgor has been correlated with an increase in glucose uptake that may be mediated by changes in the proton-motive force (Reinhold et al., 1984; Wyse et al., 1986). Further experiments are needed to determine whether, as in *E. coli*, this effect is mediated by changes in turgor, and whether transcriptional activity of the plasma membrane H^+-ATPase gene is being affected. Because of the extreme lability of plant plasma membrane transport proteins and the technical difficulties associated with purifying very hydrophobic proteins, it is possible that questions concerning control of gene expression for higher plant plasma membrane proteins may be best answered through measurements of specific mRNA levels via hybridization to cloned gene probes, rather than by measurements of polypeptide levels. Application of molecular biology techniques to answer these questions offers the exciting potential for uncovering new types of transcriptional control (e.g., turgor pressure) unique to genes involved with plasma membrane transport function.

RAPID REGULATION OF THE PLASMA MEMBRANE H^+-ATPASE BY FUSICOCCIN AND AUXIN

Recent electrophysiological experiments with intact cells of higher plants have demonstrated that the plasma membrane H^+-ATPase catalytic activity is specifically increased within seconds after treatment with the growth stimulator, fusicoccin (Felle, 1982; Bertl and Felle, 1985). Slightly slower changes in activity were observed after treatment with the growth-stimulating plant hormone, auxin (Brummer et al., 1985; Felle et al., 1986).

What is the biochemical mechanism for stimulation of the ATPase? The speed of these effects suggest an allosteric regulation by some metabolite or by a covalent modification such as phosphorylation/dephosphorylation mediated with protein kinases and phosphatases. Observations of an activation of the plasma membrane H^+-ATPase prior to the growth rate increases induced by fusicoccin and auxin support the 'acid-secretion' hypothesis for cellular elongation in higher plants (Rayle and Cleland, 1977). According to this theory, a turgor-induced volume increase is prevented by an inelastic cell wall, and when the wall pH is decreased (e.g., by a stimulated H^+-ATPase), the wall becomes more elastic, and increased volume and tissue elongation rates ensue. Although this simple hypothesis seems to adequately explain rapid changes in elongation rate, more sustained long-term changes in tissue growth rates probably also involve an increase in the overall cell wall biosynthetic machinery. A crucial missing link in the acid-secretion theory is the question of how growth regulators affect the plasma membrane H^+-ATPase. A

61

detailed examination of the time course of changes within the first few seconds after fusicoccin and auxin treatment seems to imply that changes in the cytoplasmic pH occur before the observed changes in ATPase activity (Felle et al., 1986). Electrophysiological experiments have demonstrated that the plasma membrane H^+-ATPase is extremely sensitive to changes in the cytoplasmic pH above or below its 'resting' value of ca. 7.2. Thus, when the cytoplasmic pH is decreased by application of high concentrations of butyric acid, the H^+-ATPase responds to the increase in substrate concentration (cytoplasmic protons) by an increase in catalytic activity, and a resultant hyperpolarization of the membrane potential and increase in acid secretion rate is observed (Sanders et al., 1981; Marre et al., 1983). If cytoplasmic pH proves to be the key regulator of ATPase activity the next question becomes how do fusicoccin and auxin alter its value?   A reasonable explanation is that proton cotransport carriers, such as Ca^{2+} / H^+ or Na^+ / H^+ antiporters are involved although changes in oxidative metabolism should not be ruled out (Sanders and Slayman, 1982).

An alternative view is that the electrophysiological measurements cannot resolve cause and effect, and that fusicoccin and auxin are directly modifying the ATPase catalytic activity through some other means. An attractive candidate for this regulatory event is kinase-mediated phosphorylation of the H^+-ATPase. The plant plasma membrane H^+-ATPase has been shown to be a phosphoprotein, and a calcium-dependent protein kinase that uses the ATPase as a major substrate *in vitro*, has been found associated with plasma membrane vesicles (Schaller and Sussman, 1987a,b). The ATPase was found to be phosphorylated at serine and threonine but not tyrosine residues. Kinase-mediated phosphorylation of the ATPase is strongly affected by micromolar concentrations of Ca^{2+} ions, and also, interestingly, by the pH. In fact, as the pH drops from 7.5 to 6.5, there is a substantial and specific increase in phosphorylation of the ATPase. The Ca^{2+}- and pH dependence of this ATPase kinase provides two potentially important links in the signal transduction pathway for higher plants. A critical assessment of whether kinase-mediated phosphorylation of the ATPase actually alters ATPase catalytic activity must await further studies in a better defined system.

There are several reports of *in vitro* effects of fusicoccin (Rasi-Caldogno et al., 1986) and auxin (Scherer, 1984; Gabathuler and Cleland, 1985) on plant plasma membrane H^+-ATPase activity, but these observations have not yet been independently confirmed. There are several ways in which artefactual stimulation of ATPase activity could occur:

(1) A kinase and ATP-dependent change in catalytic activity with membrane vesicles may represent a change in proton 'leakiness' due to phosphorylation of proteins other than the ATPase. The ATPase is known to be strongly affected by changes in the protonmotive force back pressure (Sze, 1980), and these could produce a spurious stimulation.
(2) The plant ATPase in vesicles is not always stable to prolonged treatment at room temperature (heat inactivation, lipases,

proteases?) and changes in this condition could mimic an alteration in catalytic activity.
(3) There is a considerable amount of endogenous lipid kinase associated with plasma membrane vesicles, and it may be difficult to separate effects due to changes in the phosphorylation state of lipids, from that of the ATPase.
(4) Because the ATPase contains so many hydrophobic, transmembrane segments, it tenaciously binds lipid and other hydrophobic molecules. Therefore, it is possible that high concentrations of a lipophilic molecule (e.g., auxin) may bind to the ATPase and create spurious 'pharmacological' effects unrelated to *in vivo* physiological effects.

Complex patterns of phosphorylation and regulation have been found with a number of proteins; glycogen synthase, for example, is regulated by five different protein kinases, which phosphorylate seven separate sites (Picton et al., 1982). Conflicting reports have appeared concerning the role of protein phosphorylation in regulating the H$^+$-ATPase of plant (Zocchi, 1985; Bidwai and Takemoto, 1987; Bidwai et al., 1987) and fungal (Serrano, 1983, 1985; Portillo and Mazon, 1985; Yanagita et al., 1987) plasma membranes. Resolution of this question will probably require identification of the location of the phosphorylated residues and measurement of changes in their phosphorylation state using purified ATPase and protein kinase. Confirmation will be obtained by elimination of the phosphorylated residues through site-directed mutagenesis, and analysis of the effects on expressed protein and *in vivo* response.

Finally, it should be noted that the plant plasma membrane H$^+$-ATPase has been proposed as a key regulatory enzyme in cellular events other than the hormonal response. For example, blue-light induced changes in stomatal aperture (Assmann et al., 1985), adaptation to cold stress (Vemura and Yoshida, 1986; Yoshida et al., 1984; Iswari and Palta, 1987) and developmental changes (MacDonald and Weeks, 1984; Tuduri et al., 1985) have all been attributed to alterations in the plasma membrane H$^+$-ATPase catalytic activity. Regulatory phosphorylation may play one role in such events, but other mechanisms such as rapid changes in lipid composition or other allosteric effectors, have not been ruled out.

CONCLUSION

Historically, research in this field began with electro-physiological investigations, moved on to biochemical studies and more recently, has been probed through genetic means by exploiting advances in recombinant DNA technology. This multi-disciplinary approach has yielded considerable progress in understanding the structure and function of membrane ATPases. Further studies are needed to clarify the molecular mechanisms by which these enzymes are regulated, and their precise role in the life of plants and fungi.

ACKNOWLEDGEMENTS

We wish to thank our colleague Donald B. Katz for assistance in performing the PROFILE sequence comparison analysis and both he and G. Eric Schaller for a critical reading of the manuscript. This review was prepared with assistance from research funds provided to M.R.S. from the Department of Energy (No. DE-AC02-83ER13086), the United States Department of Agriculture (No. 87-CRCR-1-2357) and the College of Agriculture and Life Sciences and the Graduate School of the University of Wisconsin-Madison.

REFERENCES

ANTHON, G.E. and SPANSWICK, R. (1986) Purification and properties of the H^+-translocating ATPase from the plasma membrane of tomato roots. Plant Physiol. 81:1080-1085.

ASSMANN, S.M., SIMONCINI, L. and SCHROEDER, J.I. (1985) Blue light activates electrogenic ion pumping in guard cell protoplasts of *Vicia faba*. Nature 318:285-287.

BERTL, A. and FELLE, H. (1985) Cytoplasmic pH of root hair cells of *Sinapis alba* recorded by a pH-sensitive microelectrode. Does fusicoccin stimulate the proton pump by cytoplasmic acidification? J. Exp. Bot. 36:1142-1149.

BIDWAI, A.P.and TAKEMOTO, J.Y. (1987) Bacterial phytotoxin, syringomycin, stimulates a protein kinase mediated phosphorylation of red beet plasma membrane polypeptides. J. Cell. Biochem. (Supple.) 11B, 93.

BIDWAI, A.P., ZHANG, L., BACHMANN, R.C. and TAKEMOTO, J.Y. (1987) Mechanism of action of *Pseudomonas syringae* phytotoxin, syringomycin: stimulation of red beet plasma membrane ATPase activity. Plant Physiol. 83:39-43.

BOWMAN, E.J. (1983) Comparison of the vacuolar membrane ATPase of *Neurospora crassa* with the mitochondrial and plasma membrane ATPases. J. Biol. Chem. 258:15238-15244.

BOWMAN, B.J., BERENSKI, C.J. and JUNG, C.Y. (1985) Size of the plasma membrane H^+-ATPase from *Neurospora crassa* determined by radiation inactivation and comparison with the sarcoplasmic reticulum Ca^{2+}-ATPase from skeletal muscle. J. Biol. Chem. 260:8726-8730.

BOWMAN, B.J., BLASCO, F. and SLAYMAN, C.W. (1981) Purification and characterization of the plasma membrane ATPase of *Neurospora crassa*. J. Biol. Chem. 256:12343-12349.

BOWMAN, B.J. and BOWMAN, E.J. (1986) H^+-ATPases from mitochondria, plasma membrane and vacuoles of fungal cells. J. Membrane Biology 94:83-97.

BOWMAN, E.J., MANDALA, S., TAIZ, L. and BOWMAN, B.J. (1986) Structural studies of the vacuolar membrane ATPase from *Neurospora crassa* and comparison with the tonoplast membrane ATPase from *Zea mays*. Proc. Natl. Acad. Sci. USA 83:48-52.

BRANDL, C.J., GREEN, N.M., KORCZAK, B. and MACLENNAN, D.H. (1986) Two Ca^{2+}- ATPase genes: homologies and mechanistic implications of deduced amino acid sequences. Cell 44:597–607.

BRISKIN, D.P., THORNLEY, W.R. and ROTI-ROTI, J.L. (1985) Target molecular size of the red beet plasma membrane ATPase. Plant Physiol. 78:642–644.

BRUMMER, B., BERTL, A., POTRYKUS, I., FELLE, H. and PARISH, R.W. (1985) Evidence that fusicoccin and inole-3-acetic acid induce cytosolic acidification of *Zea mays* cells. FEBS Lett. 189:109–114.

BUSH, D.R.and SZE, H. (1986) Calcium transport in tonoplast and endoplasmic reticulum vesicles isolated from cultured carrot cells. Plant Physiol. 80:549–555.

CHADWICK, C.C., GOORMAGHTIGH, E. and SCARBOROUGH, G.A. (1987) A hexameric form of the *Neurospora crassa* plasma membrane H^+-ATPase. Arch. Biochem. Biophys. 252:348–356.

CID, A., VARA, F. and SERRANO, R. (1987) Inhibition of the proton pumping ATPases of yeast and oat root plasma membranes by DCCD. Arch. Biochem. Biophys. 252:496–500.

CLEMENT, J.D., GHISLAIN, M., DUFOUR, J.P. and SCALLA, R. (1986) Immunodetection of a 90,000 M_r polypeptide related to yeast plasma membrane ATPase in plasma membrane from maize shoots. Plant Sci. 5:43–50.

DUFOUR, J.P. and GOFFEAU, A. (1980) Molecular and kinetic properties of the purified plasma membrane ATPase of the yeast *Schizosaccharomyces pombe*. Eur. J. Biochem. 105:145–154.

EPSTEIN, E. and LAIMINS, L. (1980) Potassium transport in *Escherichia coli*: diverse systems with common control by osmotic forces. Trends Bioch. Sci. 5:21–23.

FELLE, H. (1982) Effects of fusicoccin upon membrane potential, resistance and current-voltage characteristics in root hairs of *Sinapis alba*. Pl. Sci. Lett. 25:219–225.

FELLE, H., BRUMMER, B., BERTL, A. and PARISH, R.W. (1986) IAA and fusicoccin cause cytosolic acidification of corn coleoptile cells. Proc. Nat. Acad. Sci. USA 83:8992–8995.

GABATHULER, R. and CLELAND, R. (1985) Auxin regulation of a proton translocating ATPase in pea root plasma membrane vesicles. Plant Physiol. 70:1080–1085.

GALLAGHER, S.R. and LEONARD, R.T. (1987) Electrophoretic characterization of a detergent-treated plasma membrane fraction from corn roots. Plant Physiol. 83:265–271.

GIANNINI, J.L., MILLER, G.W. and BRISKIN, D.P. (1987) Membrane transport in isolated vesicles from sugar beet tap root. III. Effects of fluoride on ATPase activity and transport. Plant Physiol. 83:709–712.

GILDENSOPH, L., GIANNINI, J. and BRISKIN, D.P. (1987) FITC binding to plasma membrane ATPase of red beet (*Beta vulgaris*) storage tissue. Plant Physiol. (Supple.) 83:357.

GOFFEAU, A. and SLAYMAN, C.W. (1981) The proton-translocating ATPase of the fungal plasma membrane. Biochim. Biophys. Acta 639:197–223.

GOORMAGHTIGH, E., CHADWICK, C. and SCARBOROUGH, G.A. (1986) Monomers of the *Neurospora* plasma membrane H^+-ATPase catalyze efficient proton translocation. J. Biol. Chem. 261:7466–7471.

GORGA, F.R. (1985) Inhibition of (Na^+,K^+)-ATPase by DCCD. Evidence for two carboxyl groups that are essential for enzymatic activity. Biochem. 24:6783–6788.

GRIBSKOV, M., MCLACHLIN, A.D. and EISENBERG, D. (1987) Profile analysis: detection of distantly related proteins. Proc. Nat. Acad. Sci. USA 84:4355–4358.

GUSTIN, M.C., MARTINAC, B., SAIMI, Y., CULBERTSON, M. and KUNG, C. (1986) Ion channels in yeast. Science 233:1195–1197.

HAGER, K.W., MANDALA, S.M., DAVENPORT, J.W., SPEICHER, D.W., BENZ, JR., E.J. and SLAYMAN, C.W. (1986) Amino acid sequence of the plasma membrane ATPase of *Neurospora crassa*: deduction from genomic and cDNA sequences. Proc. Nat. Acad. Sci. USA 83:7693–7697.

HESSE, J.E., WIECZOREK, L., ALTENDORF, K., REICIN, A.S., DORUS, E. and EPSTEIN, W. (1984) Sequence homology between a bacterial ATP-driven potassium transport system and a cation transport ATPase of animal cells. Proc. Nat. Acad. Sci. USA 81:4746–4750.

ISWARI, S. and PALTA, J. (1987) Plasma membrane H^+-ATPase as a site of functional alterations during cold acclimation and freezing injury. In: Low temperature stress physiology in crops. P.H. Li, ed., CRC Press, Inc., Boca Raton, FL.

JORGENSEN, P.L., SKRIVER, E., HEBERT, H. and MAUNSBACH, A.B. (1982) Structure of the Na^+,K^+-pump: crystallization of pure membrane-bound Na^+,K^+-ATPase and identification of functional domains of the α-subunit. Ann. N.Y. Acad. Sci. 402:207–255.

KASAMO, K. (1986) Purification and properties of the plasma membrane H^+-translocating ATPase of *Phaseolus mungo* L. roots. Plant Physiol. 80:818– 824.

KATZ, D.B. and SUSSMAN, M.R. (1987) Inhibition and labeling of the plant plasma membrane H^+-ATPase with N-ethylmaleimide. Plant Physiol. 83:977– 981.

KAWAKAMI, R., NOGUCHI, S., NODA, M., TAKAHASHI, H., OHTA, T., KAWAMURA, M., NOJIMA, H., NAGANO, K., HIROSE, T., INAYAMA, S., HAYASHIDA, H., MIYATA, T. and NUMA, S. (1985) Primary structure of the α-subunit of *Torpedo californica* (Na^+,K^+)-ATPase deduced from cDNA sequence. Nature 316:733– 736.

KAWAKAMI, K., OHTA, T., NOJIMA, H. and NAGANO, K. (1986) Primary structure of the α-subunit of human Na^+,K^+-ATPase deduced from cDNA sequence. J. Biochem. 100:389–397.

KISHIMOTO, U., KAMI-IKE, N., TAKEUCHI, Y. and OHKAWA, T. (1984) A kinetic analysis of the electrogenic pump of *Chara corallina*:   I. Inhibition of the pump by DCCD. J. Membr. Biol. 80:175–183.

KROPF, D.L., LUPA, M.D.A., CALDWELL, J.H. and HAROLD, F.M. (1983) Cell polarity: endogenous ion currents precede and predict branching in the water mold *Achyla*. Science 220:1385–1387.

LAIMINS, L.A., RHOADS, D.B. and EPSTEIN, W. (1981) Osmotic control of kdp operon expression in *Escherichia coli*. Proc. Natl. Acad. Sci. USA 78:464– 468.

LEONARD, R.T. (1984) Membrane-associated ATPases and nutrient-absorption by roots. In: Advances in Plant Nutrition, vol. 1, P.B. Tinker and A. Lauchli (eds.), pp. 209–240, Praeger Publishers, N.Y.C., N.Y.

LEW, R.R. and SPANSWICK, R.M. (1985) Characterization of anion effects on the nitrate-sensitive ATP-dependent proton pumping activity of soybean (*Glycine max* L.) seedling root microsomes. Plant Physiol. 77:352–357.

MACLENNAN, D.H., BRANDL, C.J., KORCZAK, B. and GREEN, N.M. (1985) Amino acid sequence of a $Ca^{2+}+Mg^{2+}$-dependent ATPase from rabbit muscle sarcoplasmic reticulum, deduced from its complementary DNA sequence. Nature 316:696– 700.

MACDONALD, J.I.S. and WEEKS, G. (1984) A plasma membrane ATPase in the cellular slime mold *Dictyostelium discoideum*. Arch. Biochem. Biophys. 235:1–7.

MANDALA, S., METTLER, I.J. and TAIZ, L. (1982) Localization of the microsomal proton pump of corn coleoptiles by density gradient centrifugation. Plant Physiol. 70:1743–1747.

MANDALA, S. and TAIZ, L. (1985a) Partial purification of a tonoplast ATPase from corn coleoptiles. Plant Physiol. 78:327–333.

MANDALA, S. and TAIZ, L. (1985b) Proton transport in isolated vacuoles from corn coleoptiles. Plant Physiol. 78:104–109.

MANDALA, S. and TAIZ, L. (1986) Characterization of the subunit structure of the maize tonoplast ATPase. JBC 261(27) 12850–12855.

MARRE, M.T., ROMANI, G. and MARRE, E. (1983) Transmembrane hyperpolarization and increase of K^+ uptake in maize roots treated with permeant weak acids. Plant, Cell and Environment 6:617–623.

MARRE, E. and BALLARIN-DENTI, A. (1985) The proton pumps of the plasmalemma and the tonoplast of higher plants. J. Bioenergetics Biomemb. 17:1–21.

MARIN, B.P. (1984) Biochemistry and function of vacuolar ATPases in fungi and plants. Springer–Verlag, New York 259 pp.

MURPHY, A. (1981) Kinetics of the inactivation of the ATPase of sarcoplasmic reticulum by DCCD. J. Biol. Chem. 256:12046–12050.

OHTA, T., NAGANO, K. and YOSIDA, M. (1986) The active site structure of Na^+/K^+-transporting ATPase: location of the 5'-(p-fluorosulfonyl) benzoyladenosine binding site and soluble peptides released by trypsin. Proc. Nat. Acad. Sci. USA 83:2071–2075.

OLESKI, N.A. and BENNETT, A.B. (1987) H^+-ATPase activity from storage tissue of *Beta vulgaris* IV. DCCD binding and inhibition of the plasma membrane H^+-ATPase. Plant Physiol. 83:569–572.

OVCHINNIKOV, Y.A., MODYANOV, N.N., BROUDE, N.E., PETRUKKIN, K.E., GRISKIN, A.V., ARZAMAZOVA, N.M., ALDANOVA, N.A., MONASTYRSKAYA, G.S. and SVERDLOV, E.D. (1986) Pig kidney Na^+,K^+-ATPase: primary structure and spatial organization. FEBS Letters 201:237–245.

PEDEMONTE, C.H. and KAPLAN, J.H. (1986) Carbodiimide inactivation of Na^+,K^+-ATPase. A consequence of internal cross-linking and not carboxyl group modification. J. Biol. Chem. 261:3632–3639.

PEDERSON, P.L. and CARAFOLI, E.(1987a) Ion motive ATPases. I. Ubiquity, properties and significance to cell function. Trends Biochem. Sci. 12:146–150.

PEDERSON, P.L. and CARAFOLI, E.(1987b) Ion motive ATPases. II. Energy coupling and work output. Trends Bioch. Sci. 12:186–189.

PERLIN, D.S., KASAMO, K., BROOKER, R.J. and SLAYMAN, C.W. (1984) Electrogenic H^+ translocation by the plasma membrane ATPase of *Neurospora*: studies on plasma membrane vesicles and reconstituted enzyme. J. Biol. Chem. 259:7884–7892.

PERLIN, D.S., SAN FRANCISCO, M.J.D., SLAYMAN, C.W. and ROSEN, B.P. (1986) Proton-ATP stoichometry of proton pumps from *Neurospora crassa* and *Escherichia coli*. Arch. Biochem. Biophys. 248:53–61.

PICK, U. and RACKER, E. (1979) Inhibition of the Ca^{2+}-ATPase from sarcoplasmic reticulum by DCCD: evidence for location of the Ca^{2+} binding site in a hydrophobic region. Biochem. 18:108–113.

PICTON, C., AITKEN, A., BILHAM, T. and COHEN, P. (1982) Multiple phosphorylation of glycogen synthase from rabbit skeletal muscle. Eur. J. Biochem. 24:37–45.

PORTILLO, F. and MAZON, M.J. (1985) Activation of yeast plasma membrane ATPase by phorbol ester. FEBS Letters 192:95–98.

POUGEOIS, R., SATRE, M. and VIGNAIS, P.V. (1980) Characterization of DCCD binding site on coupling factor 1 of mitochondrial and bacterial membrane– bound ATPase. FEBS Lett. 117:344–348.

RANDALL, S.K. and SZE, H. (1986) Properties of the partially purified tonoplast H^+-pumping ATPase from oat roots. J. Biol. Chem. 261:1364–1371.

RASI–CALDOGNO, F., DEMICHELIS, M.I., PUGLIARELLO, M.C. and MARRE, E. (1986) H^+-pumping driven by the plasma membrane ATPase in membrane vesicles form radish: stimulation by fusicoccin. Plant Physiol. 82:121–125.

RASI–CALDOGNO, F., PUGLIARELLO, M.C. and DEMICHELIS, M.I. (1987) The Ca^{2+}- transport ATPase of plant plasma membrane catalyzes an H^+/Ca^{2+} exchange. Plant Physiol 83:994–1000.

RAYLE, D.L. and CLELAND, R.E. (1977) Control of plant cell enlargement by hydrogen ions. Curr. Top. Dev. Biol. 11:187–214.

REINHOLD, L., SEIDEN, A. and VOLOKITA, M. (1984) Is modulation of the rate of proton pumping a key event in osmoregulation?   Plant Physiol. 75:846–849.

ROBINSON, J.D. (1974) Affinity of the (Na^+,K^+)–dependent ATPase for Na^+ measured by Na^+-modified enzyme inactivation. FEBS Letters 38:325–328.

RODRIGUEZ–NAVARRO, A., BLATT, M.R. and SLAYMAN, C.L. (1986) A potassium–proton symport in *Neurospora crassa*. J. Gen. Physiol. 87:649–674.

ROSICHAN, J.L., BOCHERTS, K.J., EWING, N.N. and BENNETT, A.B. (1987) Progress toward cDNA cloning of the 72 and 58 kd subunits of the tomato tonoplast H^+-ATPase. Plant Physiol. (Supple.) 83:54.

SANDERS, D., HANSEN, U-P. and SLAYMAN, C.L. (1981) Role of the plasma membrane proton pump in pH regulation in non–animal cells. Proc. Natl. Acad. Sci. USA 78:5903–5907.

SANDERS, D. and SLAYMAN, C.L. (1982) Control of intracellular pH. Predominant role of oxidative metabolism, not proton transport, in the eukaryotic microorganism *Neurospora*. J. Gen. Physiol. 80:377–342.

SCHALLER, G.E. and SUSSMAN, M.R. (1987) Kinase-mediated phosphorylation of the oat plasma membrane H^+-ATPase. In: Plant membranes: structure, function, biogenesis. UCLA symposia on Molecular and Cellular Biology, New Series, Volume 63, Editors, C. Lever and H. Sze. Alan R. Liss, Inc., New York, N.Y.

SCHALLER, G.E. and SUSSMAN, M.R. (1987b) Phosphorylation of the plasma-membrane $H^=$-ATPase of oat roots by a calcium–stimulated protein kinase. Planta. (in press). SCHALLER, G.E. and SUSSMAN, M.R. (submitted) Isolation and sequence of tryptic peptides isolated from the proton pumping ATPase of the oat plasma membrane.

SCHERER, G.F.E. (1984) Stimulation of ATPase activity by auxin is dependent on ATP concentration. Planta 161:394–397.

SCHNEIDER, J.W., MERCER, R.W. and BENZ, JR., E.J. (1987) Molecular cloning and characterization of α–subunit isoforms of the Na^+,K^+-ATPase. 5th International Conference on Na^+,K^+-ATPase Abstracts (Arhus, Denmark; June 14–19).

SCHNEIDER, J.W., MERCER, R.W., CAPLAN, M., EMANUEL, J.R., SWEADNER, K.J., BENZ, JR., E.J. and LEVENSON, R. (1985) Molecular cloning of rat brain Na^+,K^+-ATPase α–subunit cDNA. Proc. Nat. Acad. Sci. USA 82:6357–6361.

SCHONER, W. (1971) Active transport of Na⁺ and K⁺ through animal cell membranes. Angew. Chemie. 10:882–889.
SCHUMAKER, K.S. and SZE, H. (1985) A Ca²⁺/H⁺ antiport system driven by the proton electrochemical gradient of a tonoplast H⁺-ATPase from oat roots. Plant Physiol. 79:1111–1117.
SCHUMAKER, K.S. and SZE, H. (1986) Calcium transport into the vacuole of oat roots. J. Biol. Chem. 261:12172–12178.
SCOFANO, H.M., BARRABIN, H., LEWIS, D. and INESI, G. (1985) Specific DCCD inhibition of the E – P + H₂O E + Pi reaction and ATP/Pi exchange in sarcoplasmic reticulum ATPase. Biochem. 24:1025–1029.
SCOTT, T.L. and SHAMOO, A.E. (1982) Disruption of energy transduction in sarcoplasmic reticulum by trypsin cleavage of Ca²⁺-ATPase. J. Membrane Biol. 64:137–144.
SERRANO, R. (1983) *In vivo* glucose activation of the yeast plasma membrane ATPase. FEBS Letters 156:11–14.
SERRANO, R. (1984a) Plasma membrane ATPase of fungi and plants as a novel type of proton pump. Curr. Topics Cell. Regulation 23:87–126.
SERRANO, R. (1984b) Purification of the proton pumping ATPase from plant plasma membranes. Biochem. Biophys. Res. Comm. 121:735–740.
SERRANO, R. (1985) Plasma membrane ATPase of plants and fungi. CRC Press, Inc., Boca Raton, FL. 174 pp.
SERRANO, R., KIELLAND-BRANDT, M.C. and FINK, G.R.(1986a) Yeast plasma membrane ATPase is essential for growth and has homology with (Na⁺+K⁺)-, K⁺- and Ca²⁺-ATPases. Nature 319:689–693.
SERRANO, R., MONTESINOS, C. and CID, A. (1986a) A temperature-sensitive mutant of the yeast plasma membrane ATPase obtained by *in vitro* mutagenesis. FEBS Letters 208:143–146.
SHULL, G.E., GREEB, J. and LINGREL, J.B. (1986) Molecular cloning of three distinct forms of the Na⁺,K⁺-ATPase α-subunit from rat brain. Biochem. 25:8125–8132.
SHULL, G.E. and LINGREL, J.B. (1986) Molecular cloning of the rat stomach (H⁺,K⁺)-ATPase. J. Biol. Chem. 261:16788–16791.
SHULL, M.M. and LINGREL, J.B. (1987) Multiple genes encode the human Na⁺,K⁺-ATPase catalytic subunit. Proc. Nat. Acad. Sci. USA 84:4039–4043.
SHULL, G.E., SCHWARTZ, A. and LINGREL, J.B. (1985) Amino-acid sequence of the catalytic subunit of the (Na⁺,K⁺)-ATPase deduced from a complementary DNA. Nature 316:691–695.
SLAYMAN, C.L. (1985) Plasma membrane proton pumps in plants and fungi. BioScience 35:34–37.
SLAYMAN, C.L. (1987) The plasma membrane ATPase of Neurospora: a proton-pumping electroenzyme. J. Bioenerg. & Biomemb. 19:1–20.
SUROWY, T.K. and SUSSMAN, M.R. (1986) Immunological cross-reactivity and inhibitor sensitivities of the plasma membrane H⁺-ATPase from plants and fungi. Biochim. Biophys. Acta 848:24–34.
SUROWY, T.K. and SUSSMAN, M.R. (1987) Molecular cloning of the plant plasma membrane H⁺-ATPase. Plant and Soil 99:185–196.
SUSSMAN, M.R. (1984) Purification of the *Neurospora crassa* plasma membrane H⁺-ATPase by HPLC in the presence of SDS. Anal. Biochem. 142:210–214.
SUSSMAN, M.R. and SLAYMAN, C.W. (1983) Modification of the *Neurospora crassa* plasma membrane H⁺-ATPase with DCCD. J. Biol. Chem. 258:1839–1843.

SUSSMAN, M.R., STRICKLER, J.R., HAGER, K.M. and SLAYMAN, C.W. (1987) Location of a DCCD-reactive glutamate residue in the *Neurospora crassa* plasma membrane H$^+$-ATPase. J. Biol. Chem. 262:4569-4573.

SZE, H. (1980) Nigericin-stimulated ATPase activity in microsomal vesicles of tobacco callus. Proc. Natl. Acad. Sci. USA 77:5904-5908.

SZE, H. (1985) H$^+$-translocating ATPases: advances using membrane vesicles. Ann. Rev. Plant Physiol. 36:175-208.

TAKEDA, T., KURKDJIAN, A.C. and KADO, R.T. (1985) Ionic channels, ion transport and plant cell membranes: potential applications of the patch-clamp technique. Protoplasma 127:147-162.

TUDURI, P., NSO, E., DUFOUR, J.P. and GOFFEAU, A. (1985) Decrease of the plasma membrane H$^+$-ATPase activity during late exponential growth of *S. cerevisiae*. Biochem. Biophys. Res. Comm. 133:917-922.

ULASZEWSKI, S., CODDINGTON, A. and GOFFEAU, A. (1986) A new mutation for multiple drug resistance and modified plasma membrane ATPase activity in *Schizosaccharomyces pombe*. Curr. Genet. 10:359-364.

ULASZEWSKI, S., VANHERCK, J.C., DUFOUR, J.P., KULPA, J., NIEUWENHUIS, B. and GOFFEAU, A. (1987) A single mutation confers vanadate resistance to the plasma membrane H$^+$-ATPase from the yeast *Schizosaccharomyces pombe*. J. Biol. Chem. 262:223-228.

VAI, M., POPALO, L. and ALBERGHINA, L. (1986) Immunological cross-reactivity of fungal and yeast plasma membrane H$^+$-ATPase. FEBS Letters 206:135-141.

VEMURA, M. and YOSHIDA, S. (1986) Studies on freezing injury in plant cells. Plant Physiol. 80:187-195.

VILLALOBO, A. (1984) Energy-dependent H$^+$ and K$^+$ translocation by the reconstituted yeast plasma membrane ATPase. Can. J. Biochem. Cell Biol. 62:865-877.

WALDERHAUG, M.O., POST, R.L., SACCOMANI, G., LEONARD, R.T. and BRISKIN, D.P. (1985) Structural relatedness of three ion-transport ATPases around their active sites of phosphorylation. J. Biol. Chem. 260:3852-3859.

WANG, Y. and SZE, H. (1985) Similarities and differences between the tonoplast-type and the mitochondrial H$^+$-ATPases of oat roots. J. Biol. Chem. 260:10434-10443.

WYSE, R.E., ZAMSKI, E. and TOMOS, A.D. (1986) Turgor regulation of sucrose transport in sugar beet tap root tissue. Plant Physiol. 81:478-481.

YANAGITA, Y., ABDEL-GHANY, M., RADEN, D., NELSON, N. and RACKER, E. (1987) Polypeptide-dependent protein kinase from bakers' yeast. Proc. Nat. Acad. Sci. USA 84:925-929.

YAZYU, H., SHIOTA, S., FUTAI, M. and TSUCHIYA, T. (1985) Alteration in cation specificity of the melibiose transport carrier of *E. coli* due to replacement of proline 122 with serine. J. Bacteriol. 162:933-937.

YOSHIDA, S., KAWATA, T., VEMURA, M. and NIKI, T. (1986) Properties of plasma membrane isolated from chilling-sensitive etioluted seedlings of *Vigna radiata*. Plant Physiol. 80:152-160.

ZOCCHI, G. (1985) Phosphorylation-dephosphorylation of membrane proteins controls the microsomal H$^+$-ATPase activity of corn roots. Plant Sci. 40:153-159.

Oxford Surveys of Plant Molecular & Cell Biology Vol.4 (1987) 71-93

ANAEROBICALLY REGULATED GENE EXPRESSION;
MOLECULAR ADAPTATIONS OF PLANTS
FOR SURVIVAL UNDER FLOODED CONDITIONS.

John C.Walker, Danny J.Llewellyn, Luke E.Mitchell
and Elizabeth S.Dennis.

*CSIRO, Division of Plant Industry, G.P.O. Box 1600,
Canberra, ACT, Australia*

INTRODUCTION

The sessile nature of plants has resulted in the evolution of some unique adaptive responses which ensure that these organisms can cope with an ever-changing environment. Shifts in environmental conditions, such as drought, flooding, light quality or periodicity, and temperature, are just a few examples of the external fluctuations that are encountered during the life cycle of a plant. Although plants would normally encounter these conditions in their natural environment many of these responses are referred to as "stress responses" because they involve environmental conditions that fall outside of the range required for optimal growth. The physiological and biochemical responses of plants to these environmental extremes have been studied for many years, but only recently has there been a focus on the molecular events triggered by a change in the environment. The availability of molecular biology techniques, such as gene cloning, *in vitro* mutagenesis and transformation (either through the stable integration of genes into the genome or the transient expression of extra-chromosomal genes) allows the dissection of several of these stress responses and the mechanisms regulating them to be elucidated. This article is concerned with the recent advances in our understanding of the regulatory mechanisms responsible for the pattern of gene expression during anaerobic conditions, which can be encountered during flooding. Because the alcohol dehydrogenase 1 (*Adh*1) gene of maize is particularly well-characterized, emphasis is placed on the analysis of the regulation of *Adh*1, one of the key genes induced by anaerobic stress.

TABLE 1

Metabolic and molecular changes associated with the anaerobic response in maize (compiled from Roberts *et al.*, 1984a,b; Sachs *et al.*, 1980; Gerlach *et al.*, 1982 and Rowland and Strommer, 1986).

Time	Observation
0 min.	Low levels of *Adh1* mRNA, normal cellular metabolism, cytoplasmic pH ≈7.4, no detectable *Adh1* transcription.
2 min.	Rise in NADH levels, cytoplasmic acidification begins.
10 min.	Ethanol first detected, cytoplasmic pH ≈7.0.
20 min.	Cytoplasmic pH stabilizes, pH ≈6.8.
30–60 min.	Polysomes dissociate, normal protein synthesis repressed, transition polypeptides (TPs) major class of proteins synthesized, *Adh1* transcription detected, rRNA and actin transcription rates decline. *Adh1* mRNA levels begin to increase.
1.5–3.0 h	Anaerobic polypeptides (ANPs) first observed, TPs still major class of proteins synthesized.
3.0–5.0 h	ANPs major proteins synthesized (70% of total protein synthesis), aerobic mRNAs can still be translated *in vitro*, maximum rate of *Adh1* transcription.
5.0–10.0 h	*Adh1* transcription rate begins to fall, maximum level of *Adh1* mRNA observed.

THE ANAEROBIC RESPONSE IN MAIZE

Reduced-oxygen stress (commonly referred to as anaerobic, but also described as hypoxic, anoxic and microaerobic) in maize causes a dramatic shift in metabolism from oxidative phosphorylation to predominantly ethanolic fermentation, the repression of normal protein synthesis and the expression of a specific set of approximately twenty gene products, the anaerobic polypeptides (ANPs). Thus, anaerobic stress overrides normal carbohydrate metabolism and interrupts the normal developmental programme of gene expression.

Roberts and co-workers (1984a,b) have examined the metabolic changes associated with the anaerobic response in maize (See Table 1 and Figure 1 for an outline of the metabolic and molecular changes occurring during anaerobiosis). In the absence of oxygen as the terminal electron acceptor there is a rapid inhibition of oxidative phosphorylation, as indicated by an increase in NADH levels. The NADH can, however, be utilized as a reductant by lactate or alcohol dehydrogenase and regenerate NAD^+ , which is essential for continued glycolysis and ATP production. In maize the cytoplasmic pH initially drops, probably due to the production of lactic acid, but soon stabilizes at approximately 0.5 pH units below that found under aerobic conditions (Roberts et al., 1984a). Following the pH drop, ethanol production is detected, indicating a shift from lactic acid to ethanolic fermentation. This metabolic shift is mediated through the activities of pyruvate decarboxylase and alcohol dehydrogenase (ADH). In maize roots that are homozygous for a non-functional Adh1 gene, and hence incapable of ethanolic fermentation, cellular pH is not stabilized, no ethanol is produced, ATP levels drop to undetectable levels and the plants die. Thus, the shift in metabolism is an adaptive response to maintain energy production by ethanolic fermentation rather than by short-term lactic acid fermentation, and so avoids the problems associated with continued cytoplasmic acidification due to the synthesis and accumulation of lactic acid.

Following the switch in metabolism, the pattern of gene expression is dramatically altered. There is an almost complete dissociation of polysomes (Lin and Key, 1967; Dennis and Pryor, unpublished results) and protein synthesis is repressed (Sachs et al., 1980). This repression, like that for the heat shock response, is at the translational level as in vitro translation of mRNAs isolated from anaerobically treated tissue results in the synthesis of both the normal set of aerobic polypeptides as well as the anaerobic polypeptides (Sachs et al., 1980). Although normal protein synthesis is repressed there is not a complete inhibition of gene expression. In vivo labelling experiments show that two specific sets of polypeptides are expressed de novo during anaerobic stress. During the first hour approximately four polypeptides in the 30,000 M_r range are observed. These polypeptides have been designated the transition polypeptides (Sachs et al., 1980) because their synthesis is specific to the early stages of anaerobiosis, although pulse-chase experiments indicate that they are very stable. After approximately 90 minutes of

73

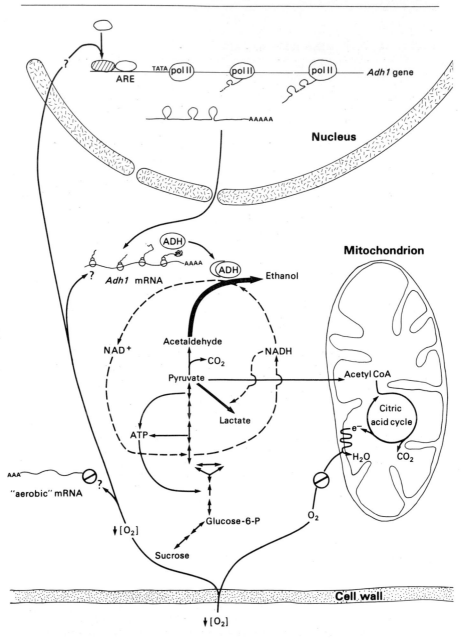

anaerobiosis another set of twenty polypeptides, the anaerobic polypeptides (ANPs), are detected. Greater than 70% of the total label incorporated after 5 hours of anaerobic treatment is found in the ANPs. *In vitro* translation studies show a corresponding increase in the level of translatable mRNA encoding the ANPs. The studies of Sachs et al. (1980) demonstrated that the anaerobic response is a

Figure 1. (opposite). Schematic outline of the biochemical and molecular changes leading to the increased alcohol dehydrogenase (ADH) expression in maize seedlings subjected to reduced oxygen tension. The stop symbol (represented by the crossed circle) indicates processes that are inhibited during anaerobiosis. The arrows indicate the flow of carbohydrates during anaerobic fermentation, with the medium arrow showing the initial, transient step for the regeneration of NAD^+ by lactate dehydrogenase and the heavy arrow indicating the major pathway for the regeneration of NAD^+ via alcohol dehydrogenase. The dashed arrows indicate the pathway of recycling NAD^+ during anaerobic fermentation. ARE, anaerobic regulatory element; TATA, TATA box sequence; pol II, RNA polymerase II; AAAAA, polyadenylated mRNA;, NAD^+ and NADH, oxidised and reduced forms of nicotinamide-adenine dinucleotide, respectively; ATP, adenosine-5'-triphosphate.

significant response in higher plants, with similarities to the heat shock response found in all organisms (Schlessinger et al., 1982; Klemenz et al., 1985; Linquist, 1986). However, it involves the induction of a set of polypeptides different from the heat-shock proteins, presumably by a different mechanism.

Five of the twenty ANPs have so far been identified as ADH1, ADH2, aldolase, sucrose synthase and glucose phosphate isomerase (reviewed by Freeling and Bennett, 1985). Pyruvate decarboxylase enzymatic activity is also induced by anaerobiosis (Lazlo and St. Lawrence, 1983) and may be an ANP. In barley, lactate dehydrogenase is induced about 20-fold by anaerobic conditions (Hoffman et al., 1986). Enhanced fermentation can meet the plant's need for energy under anaerobic conditions (Grable, 1966) and the known ANPs are presumably produced at high levels to facilitate the metabolism of carbohydrates, such as sucrose, by the glycolytic cycle. The physiological function of the transition polypeptides and the other ANPs is unknown but these, too, are likely to contribute to the survival of the plant cell during low oxygen stress.

ALCOHOL DEHYDROGENASE 1 OF MAIZE

The maize alcohol dehydrogenase 1 is the most thoroughly investigated of all the ANPs. Hageman and Flesher (1960) were the first to show ADH activity increased in maize seedlings exposed to an anaerobic environment and Schwartz (1969) demonstrated, by using genetic mutants of *Adh*1, that ADH1 activity is required for maize seedlings to survive anaerobic treatment. This observation is consistent with the hypothesis that the physiological function of ADH1 during anaerobic stress is to regenerate the NAD^+ necessary for continued glycolysis.

The Structure of Alcohol Dehydrogenase 1

The ADH enzymes of many organisms have been well characterized, and of these the horse liver enzyme has been the most extensively studied in terms of structure and catalytic mechanism. The entire amino acid sequence of the horse liver ADH enzyme has been established (Jornvall, 1970), and Eklund et al.(1976) have determined the three dimensional crystal structure of the enzyme to a resolution of 2.4 angstroms. The horse liver enzyme is highly homologous with ADH enzymes from other animals. In particular the mouse ADH (Edenberg et al.,1985) and human B and τ chain enzymes (Hempel et al.,1984) all show more than 85% homology at the amino acid level with the horse liver enzyme. This extensive conservation of amino acid sequence is also found to extend to ADH enzymes from plants with the derived amino acid sequence of maize ADH1 (Dennis et al.,1984) and ADH2 (Dennis et al.,1985) being about 50% homologous to the horse liver ADH.

The maize ADH enzymes, like the mammalian enzymes, are NADH dependent enzymes which reversibly catalyse the oxidation of various alcohols to their corresponding aldehydes and ketones (Sund and Theorell, 1963). The active enzyme has a molecular weight of 80,000 M_r and is a dimer of two identical subunits, each of which has one active site and binds two zinc ions. The extensive amino acid sequence homology between the maize ADH enzymes and horse liver ADH has enabled three dimensional models for the maize polypeptides to be constructed by computer modelling (Eklund, unpublished results). Only minor differences exist between the maize and horse liver ADHs such that all the maize ADH side chain substitutions are compatible with the same folding of the main carbon chain found in the tertiary structure of the horse liver ADH enzyme (Eklund et al., 1976). The homologies with the horse liver ADH enzyme have allowed the identification of residues in the maize ADH proteins with structural and functional properties equivalent to those of horse liver ADH. Within each subunit of the enzyme there is a coenzyme binding domain and a catalytic domain. The coenzyme binding domain consists of a hydrophobic pocket involved in binding the adenine moiety of NADH, as well as a number of residues involved in hydrogen bonding and steric interactions with the coenzyme. The catalytic domain is derived from the amino and carboxyl ends of the polypeptide. One histidine and two cysteine residues within this domain provide ligands to the catalytic zinc atom. This domain also binds the substrate and the nicotinamide moiety of the coenzyme. The second zinc atom is located on the outer surface of the catalytic domain and may be involved in stabilization of the protein structure (Drum et al., 1969).

Evolution of Maize ADH.

Since these structural comparisons were made between the horse liver and maize ADHs, numerous other plant Adh genes have been cloned including barley Adh1, Adh2, Adh3 (M. Trick, personal communication), wheat *Adh*1A (Mitchell et al.,in preparation) , pea *Adh*1 (Llewellyn et al.,1987) and *Arabidopsis Adh* (Chang and Meyerowitz, 1986). The coding region of these genes has been determined by sequence comparison with the maize *Adh*1 gene except in the case of the pea *Adh* where the coding region was verified by cDNA cloning. Each of these genes, with the exception of pea *Adh*1, is predicted to encode a polypeptide of 379 amino acids with a subunit molecular weight of approximately 40,000 M_r. The pea *Adh*1 gene has an additional amino acid in the first exon and is therefore expected to encode a polypeptide of 380 amino acids. Alignment of the predicted amino acid sequence of these genes with the sequence of the maize *Adh*1 enzyme is obvious throughout and shows extensive conservation of those residues predicted by Eklund et al. (1976) to be of functional importance for the horse liver ADH (Fig. 2).

The high degree of amino acid conservation between *Adh* genes from higher eucaryotes is a result of conservation of the nucleotide coding sequence. This conservation is particularly evident in the first and second positions of each codon; due to the degeneracy of the genetic code, essentially random fluctuations are expected in the third position. This can provide a measure of the rate of mutation of the coding region without the constraints of maintaining enzyme function. The difference in third base composition between the monocot and dicot ADH polypeptides is about 10% for purine and 20% for pyrimidine nucleotides. Assuming that the monocot and dicot lineages diverged about 120 million years ago, this suggests a rate of mutation of the order of 0.08–0.16% every million years, in general agreement with calculations by Kimura (1983), who has estimated a mutation rate of 0.1–0.3% per million years based on an examination of mammalian nuclear genes.

Nucleotide sequence analysis of genomic and cDNA clones from the maize *Adh*1, maize *Adh*2 and pea *Adh*1 genes shows that these genes are all interrupted by nine introns. The positions of these nine introns are identical in each gene although their lengths and nucleotide sequences differ. The conservation of amino acid sequence as well as intron number and position between *Adh* genes of monocot and dicot species argues that the *Adh* genes found in these lineages are derived from a common ancestor. Chang and Meyerowitz (1986) have shown that the *Arabidopsis Adh* gene has only 6 introns, whose positions correspond to six of the nine introns in maize and pea *Adh*. Similarly the wheat *Adh*1A, barley *Adh*2 and barley *Adh*3 genes have only eight introns, whose positions are the same as eight of the nine introns in maize and pea *Adh*. This indicates that during evolution there is a process resulting in the loss of introns. The alternative hypothesis that introns are gained during evolution would require the ancestral *Adh* gene to have six introns, similar to *Arabidopsis*, and for the both maize and pea *Adh* to have acquired three introns in identical positions and by independent events.

Gilbert (1978) has suggested that introns may divide the gene coding region into units of protein function and that enzymes are built up by random association of these functional units. This theory may explain why alcohol dehydrogenase (Eklund et al., 1976), glyceraldehyde-3-phosphate (Buehner et al., 1973) and lactate

```
          10        20        30        40        50        60        70        80        90       100
   MATAGKVIKCKAAVAWEAGKPLSIEEVEVAPPQAMEVRVKILFTSLCHTDVYFWEAKGQTPVPRIRFGHEAGGIIESVGEGVTDVAPGDHVLPVFTGECK
1. +++++++++++++++++++++++++++++++++++++++++++++++++++++++++++++++++++++++++++++++++++++++++++++++++++++
2. +++++++++R+++T+++++++++++++I++Y+A+++++++++++++++++++++++++++++++L+++++++++++++++++++++++++++++A+++++
3. ------------------------------------------------------TV++++++++++++++++++++++++++++++++++++++++++++
4. +++++++++++++++++M+++D++++++D++Y+A++++++++N++++++++++++++L++++++++++++++V++++++ELV+++++++++++++++++++
5. +++++++++++++++++++++++++++++++Y+A++++++++++++++++++++++++L++++++++++++++V++++++ELV+++++++++++++++++++
6. ++++++++E++++++++++++++++++H+++++++++++++++++++++++++++++++++++V++++++++++V++++++++++++++++++++++++++G
7. SN+V+QI+++R+++++++++V+++++++G++L+++++++++++++++++++++++++L++++++++++++++++++++++HLK++A++++++I++++++++G
8. +S+T+QI+R++++++++++V+++++KH++I+++++++++++++++++++++++++++L+++++++++++++++++++++++LQ++++++++I++++++++G
                             --      --          --             --                    --           --  N
                              Z       OX          X              Z*                    X            X
                              *O
                              OX

          110       120       130       140       150       160       170       180       190       200
   ECAHCKSAESNMCDLLRINTDRGVMIADGKSRFSINGKPIYHFVGTSTFSEYTVMHVGCVAKINPQAPLDKVCVLSCGISTGLGASINVAKPPKGSTVAV
   ++++P+++++E+++++++++++V++++++G+++++T+S+Q++F++++++++++++++++I++++++L++++E+++++++I+++++++TL+++++A++++++I
   ++++M+E++++L+++++++++++++++++NR+++++F+G+++++++++++++++++++++D+E+++++L++++++++++++++++++++TL++T+K++M+++I
   D+++++E+++L++++++++++++V+++++G+++Q+++T++++F++++++++++++++++I+++++++L+++E+++++++L+++++++++TL+++++K++++++I
   D+++++E+++L++++++++++++V+++++G+++Q+++T++++F++++++++++++++++I+++++++L+++E+++++++L+++++++++TL+++++K++++++I
   ++P++++E++++++++++++++LN+N++++++K+Q+VH++++++++++++++++++++++V+A++++++D+++++I+++C+++++++T++++C++++++KP+S++=
   D+R++Q+E++++++++++++++E++G++H++E++++++++++++++++++V+S+Q++++++D++++++IV++L++++++D++++++++TL++++++K++QS++I
                               --                                                 --                    --
                                X                                                  Z*                    O
                                Z                                                  Z

          210       220       230       240       250       260       270       280       290       300
   FGLGAVGLAAAEGARIAGASRIIGVDLNPSRFEEARKFGCTEFVNPKDHNKPVQEVLAEMTNGGVDRSVECTGNINAMIQAFECVHDGWGVAVLVGVPHK
   +++++++++M+++L++++++I++AKY+Q+K++++++++D+++L++++++++Q++D+++++++T+++IV++D++N+A++++++AD++S++++++++++++++
   ++++S++++T+++++++++I+++AF++++++++++++++A++++++Q++++D++++++++V++++++++++++H+D++AT++++++++++++++++++????
   +++++++M+++MS++++++++++AKH+Q+K+++++++++T++++++Q++++T+++++++V++++++++N+A+++++++++++H+D++A++++++++++?????
   +++++++M++M+++++++++++++AKY+Q+K+++++D+++++++++T+++++++V++++++++A++++H+D+++AT++++++++++++++++++++++++++
   ++++++M+++M+++++S++++++++AKY+Q+K+++++D+++++++V++++E+D+++++++Q+I++++++A++++S+++++++++++++++++++++++S+
   +++++++G+++++++++S+++++F+SK++DQ+KE+V++C++++++D++I+Q+I+++++D++I++++SVQ+++++++++++++++++++++++++++++S+
                 --                --                         --                             --      --
                  Z                 X                          Z*                             O
                  Z

          310       320       330       340       350       360       370       380
   DAEFKTHPMNFLNERTLKGTFFGNYKPRTDLPNVVELYMKKELEVEKFITHSVPFAEINKAFDLMAKGEGIRCIIRMEN*
   +DQ+++++++++S+K++++++++++++++M++++++E+++M+R++DL++++++SQ++T++++++L+++SL+++M+++D+
   +++------------------------------+++++++++++++++++S+++T+++++L++++M++++++++++
   E+V+++++++R+++++++++++G++G++DM++R+++LD+++++++L++SQ+T+++++LR++L++V++S+E+
   E+V+++++++K++++++++++++++E+++M++R++DL+++++++SQ++T++++L+++T+++TDQ+
   E+V++Y++++++++++++++++++++E+++M+++R+++L++++++++SQ+++++++L+++M++DQ+
   +DA+++++++++++++++Y+++++++G++L+++++++T+++S+++++Y+L+++++K+++E+
   +DA+++++++++++++++++++++K+I+G+++K++N++++L+++++++T+++S+++++Y+L+++S+++++T+GA+
                                                                         O
```

78

Figure 2. (opposite). Derived amino acid sequence of the maize Adh1 gene (top line, 1) aligned with maize Adh2 (2), barley Adh1 (3), Adh2 (4) and Adh3 (5), wheat Adh-1A (6), pea Adh1 (7) and Arabidopsis Adh (8). Conserved residues are indicated by a (+), sequence data not available by (-), possible amino acid deletions by (?) and alternative residues as shown. The first residue (methionine) of the pea ADH has been omitted to allow maximum sequence alignment. Residues presumed to be involved in the catalytic cleft (X), cofactor binding (open 0), catalytic (Z) and non-catalytic (Z) zinc binding ligands and subunit interactions (closed 0) are indicated.*

---

dehydrogenase (Adams et al., 1970) all have similar coenzyme binding domains yet their catalytic domains are quite different. Branden et al. (1984) have attempted to correlate the exons of maize *Adh1* with functional domains determined from the 3D crystal structure of the horse liver ADH enzyme. Introns 1,2,3 and 9 are all found within the catalytic domain while intron 7 divides the coenzyme domain into two supersecondary structural units. One of these units is further subdivided by introns 5 and 6 into three structural sub-units. Intron 4 separates the two domains of the enzyme.

In some cases the removal of intron sequences from intron-containing genes has been observed to cause severe destabilization of the messenger activity of transcripts (Hamer and Leder, 1979; Wickens et al., 1980; Gruss and Khoury, 1980). Recent work shows that intron sequences from the maize *Adh1-1S* gene are required for efficient expression of the gene (Callis et al., 1987). Expression of the intact maize *Adh1* gene in maize protoplasts was 50 to 100-fold higher than when introns and exons were replaced by their cDNA equivalents. Addition of the first intron into the *Adh*-cDNA construction was sufficient to restore activity to a level approximately equivalent to that of the genomic *Adh1* gene (Callis et al., 1987). Expression of chimeric gene constructs between the *Adh1* promoter, chloramphenicol acetyl transferase coding region and 3' terminal regions of the nopaline synthase gene was also increased by the addition of sequences containing the *Adh1* intron (Fromm et al., 1987). The *Adh* 1 intron sequences did not stimulate CAT expression when located outside of the transcribed region, indicating that they do not contain a transcriptional enhancer. Since the *Adh1* intron stimulates the expression of other promoters and coding regions it seems likely that introns in general increase gene expression in maize. The effect is probably not transcriptional, but is perhaps due to the increased stability of mRNAs in the prescence of an intron, similar to that seen in animal cells.

Regulation of the Expression of the Maize *Adh*1 Gene.

*Adh*1 expression is controlled by both developmental and environmental signals. ADH1 activity is constitutively expressed in pollen, aleurone and scutellum, but expression is at either low or undetectable levels in roots, mesocotyl and epicotyl (Freeling and Bennett, 1985). Anaerobiosis specifically induces an increased level of expression of *Adh1* and the other ANPs in a variety of organs, such as roots, endosperm and anther wall, but not in mature leaves (Okimoto et al., 1980). However, organs such as the root contain a variety of different cell types and little information is available about the expression of any of the ANPs in different cell types during either aerobic growth or anaerobic stress.

In maize seedlings, ADH1 activity and mRNA can be detected at low levels (Gerlach et al., 1982; Hake et al., 1985) but transcription is undetectable in run-on transcription experiments (Rowland and Strommer, 1986). Transcriptional induction of *Adh1* can be detected within one hour of anaerobic treatment (the shortest time examined) and reaches a peak at about 5 hours (Rowland and Strommer, 1986). Within 5-10 hours of anaerobic stress there is a 20-50 fold induction of *Adh1* mRNA (Gerlach et al., 1982; Rowland and Strommer, 1986). Rowland and Strommer (1986) have also demonstrated that when anaerobically-stressed seedlings are returned to an aerobic environment, transcription of *Adh1* falls to nearly undetectable levels within one hour, but *Adh1* mRNA can be detected for at least 26 hours.

The chromatin structure in the 5' upstream region of the *Adh1* gene has been examined by sensitivity to nucleases.(Ferl, 1985; Vayda and Freeling, 1986; Paul et al., 1987). These studies indicate the *Adh1* gene undergoes a change in chromatin structure upon anaerobic induction and are in agreement with other observations correlating changes in nuclease sensitivity with the transcriptional state of genes (Weintraub and Groudine, 1976; Wu, 1980; Igo-Kemenses et al., 1982).

Initial approaches to identifying and analyzing the *cis*-acting DNA sequences responsible for selective transcription of ANP genes during anaerobiosis have focussed on the *Adh1* gene of maize. Howard et al. (1987) developed a transient expression assay for examining the increase of *Adh1* expression in maize protoplasts (plant cells divested of their cell walls) under conditions of low oxygen tension. A chimeric gene (*Adh/Cat*) containing the *Adh1* promoter linked to the coding sequences of the chloramphenicol acetyl transferase (*Cat*) gene was introduced into maize protoplasts by electroporation (Fromm et al., 1985). ADH or CAT activity was 4-5 fold higher in protoplasts treated with reduced levels of oxygen than in controls. RNA protection experiments demonstrated a corresponding increase in *Adh1* or *Adh/Cat* mRNA level and that both the endogenous and introduced genes initiated transcription from the same position. The anaerobic response in maize protoplasts qualitatively reflects the anaerobic response in maize seedlings and expression of the *Adh/Cat* chimeric gene faithfully reflects the regulation of the endogenous *Adh1* gene.

Using the transient expression assay, Walker et al. (1987) examined the expression of a series of *in vitro* mutagenized *Adh/Cat* constructions. Deletion analysis of the *Adh1* promoter delineated a 40 base pair region (designated the anaerobic regulatory element, ARE;

−140 to −99) essential for anaerobic expression of the *Adh/Cat* construction. The ARE was mapped with linker–scanning (LS) mutants which identified two essential sub–regions (region I, −140 to −124 and region II, −113 to −99) separated by a third segment (−123 to −114) that could be mutated without affecting the low oxygen stimulation of *Adh/Cat* expression. The ARE conferred anaerobic regulation on a heterologous promoter from the 35S transcript of the Cauliflower Mosaic Virus (CaMV); consequently, the ARE is not only necessary but sufficient for anaerobic regulation of gene expression.

Peacock et al. (1987) and Ferl and Nick (1987) have proposed models for the anaerobic regulation of *Adh*1 expression by *trans*-acting regulatory factors that bind specifically to the *Adh*1 promoter. The model put forward by Peacock and co–workers is based on a functional analysis of the *Adh*1 promoter (Howard et al., 1987; Walker et al., 1987). These studies indicated that a positive regulatory factor activates the expression of the *Adh*1 gene during anaerobiosis. The model can be further refined to include a regulatory factor, bound at all oxygen concentrations, that must undergo an "activation" during anaerobiosis (Peacock et al., 1987). Ferl and Nick (1987), using *in vivo* dimethyl sulphate (DMS) protection analyses to locate binding sites of potential regulatory molecules, identified a region of the *Adh*1 promoter that corresponds closely to the position of the ARE determined by Walker et al. (1987). This region has regulatory factors bound under aerobic conditions and the pattern of DMS methylation is altered during anaerobic stress. However, they also identified two additional sites that have an altered sensitivity to DMS during anaerobiosis. Based on these observations Ferl and Nick proposed that there are three binding sites for factors that regulate expression of *Adh*1 during anaerobic stress: site B which is occupied in aerobic cells and undergoes an alteration during anaerobiosis (this site corresponds to the ARE), and sites A and C which are occupied only under anaerobic conditions. Both models are similar and are consistent with what is generally known about transcriptional regulation in prokaryotes and eukaryotes (Ptashne, 1986; Echols, 1986; McKnight and Tjian, 1986; Struhl, 1987; Maniatis et al., 1987).

EXPRESSION AND REGULATION OF OTHER ANAEROBICALLY REGULATED GENES OF MAIZE

Although much of what we know about the anaerobic regulation of gene expression is the result of studies on the expression and regulation of *Adh*1 in maize, there are approximately 20 other anaerobically induced polypeptides (ANPs) (Sachs et al., 1980). Cloned cDNAs corresponding to a number of these genes have been isolated (Gerlach et al., 1982; Hake et al., 1985; Gerlach and Dennis, personal communication) and used in studies of the expression of mRNAs during anaerobic conditions. Hake et al. (1985) have examined the expression of 5 cDNA clones complementary to anaerobically induced mRNAS: *Adh*1 (pZML84); sucrose synthase (pMx96); aldolase (pMx71); and two unidentified ANPs, ANP31 (pMx59) and ANP40 (pMx97). mRNAs hybridizing to these cDNA clones, except pMx97 (ANP40), begin to accumulate within 30 minutes of anaerobiosis, with the accumulation of ANP40 mRNA not detectable until after 3 hours of anaerobic stress. The expression of these five anaerobic mRNAs could be further distinguished on the basis of the kinetics of their accumulation: aldolase and ANP31 reached their plateau levels in 2–3 hours; *Adh*1 and sucrose synthase at 6–12 hours; and ANP40 at greater than 24 hours. Once a plateau level was attained the mRNA levels were maintained for at least 24 hours under anaerobic conditions. However, if the tissues were returned to aerobic conditions the level of all mRNAs declined at approximately the same rate. These results demonstrate that the anaerobic mRNAs do not accumulate simultaneously, but show different kinetics of accumulation after the imposition of an anaerobic stress. It seems likely that these differences are due to both transcriptional and post–transcriptional processes and are related to the different physiological functions of the individual ANPs.

Springer et al. (1986) have examined the expression of the *Shrunken* (*Sh*) gene product, sucrose synthase A (ANP87). Sucrose synthase mRNA is expressed in maize kernels, and the shoots and roots of etiolated maize seedlings, but not in mature leaves. However, after etiolated seedlings are placed in the light for 48 hours there is a 50 fold reduction in the level of *Sh* mRNA. Under anaerobic conditions there is a 10–20 fold induction of sucrose synthase mRNA levels, and anaerobiosis can override the repression of *Sh* mRNA levels in photosynthetically active seedlings.

While six of the anaerobic polypeptides have been identified as enzymes involved in fermentative carbohydrate metabolism, the core glycolytic enzymes such as aldolase show a pattern of inducibility different to those terminating the pathway (pyruvate decarboxylase (PDC), ADH1 and ADH2). ADH and PDC are scarcely present in aerobic roots and are induced about 10 fold after 72 hours of anaerobiosis (Kelly and Freeling, 1984). Aldolase, on the other hand, is synthesized at appreciable levels aerobically and the enzyme level increases less than two fold after 72 hours of anaerobiosis. The mRNA level for aldolase, however, increases approximately 20 fold in response to anaerobic conditions (Hake et al., 1985; Dennis et al., 1987). The discrepancy between mRNA and enzyme induction implies a high stability of the aldolase protein under aerobic conditions and/or a high turnover of protein under anaerobic conditions. Since aldolase is required for aerobic glycolysis and is probably not rate–limiting under these conditions, the two–fold increase in enzyme levels may be sufficient to cope with the required increase in the flux of carbohydrate through the fermentative pathway during anaerobiosis.

The genes of a number of these other anaerobic proteins have now been isolated and sequenced: *Adh*2 (Dennis et al., 1985), sucrose synthase (Werr et al., 1985) and aldolase (Dennis et al., 1987). A preliminary functional analysis of the aldolase gene has provided some insights into the possible mechanisms of co–ordinate regulation of the anaerobically induced genes. Chimeric constructs consisting of aldolase promoter segments, CAT coding region and NOS 3' end were assembled and introduced into maize protoplasts by electroporation. Expression of CAT activity after 24 hours was determined in a manner identical to that described for the *Adh1/CAT* chimeric genes. Deletion analysis of the aldolase promoter showed that the region downstream of -110 was sufficient for activity but removal of the sequences to -60 abolished expression (E.S. Dennis, unpublished observation).

If there is a common regulatory mechanism for all anaerobically induced genes then the sequences shown to be functionally important for the *Adh1* gene should also be present in the other anaerobically induced genes. The aldolase promoter contains a portion of the ARE sequence in a similar position relative to the start of transcription and located in the region shown by deletion analysis to be functionally important for expression of the gene. The *Adh2* gene also has sequences homologous to the maize *Adh1* ARE which are located in a similar position (Fig. 3). In sucrose synthase a sequence in the first intron was previously noted which showed 13 bases out of 13 homology to region II of the ARE of *Adh1* (Springer et al., 1986). We have detected another region with homology located at -162 to -170 (Fig. 3), a position much more analogous to the position of the homologous sequence in *Adh1* and *Adh2* and aldolase. When these homologies are compared a short consensus sequence can be discerned, TGGTTT, located in the region between -170 and -70 relative to the transcription start of all four of these genes.

The anaerobically induced genes of maize differ in the fine tuning of their expression, e.g., the time of induction and the relative level of response between genes (Hake et al., 1985). As is seen in some heavy metal or hormone inducible animal genes (reviewed by Schaffner, 1985) the difference between gene induction may depend on the relationship of the regulatory element to its flanking sequences, the number of core elements in the promoter, the orientation of the different elements and their position in the promoter. Some of these controls may be more precisely defined for the anaerobically regulated genes by mutagenesis of individual bases in and around the ARE.

THE ANAEROBIC RESPONSE IN OTHER PLANT SPECIES.

The anaerobic response has only been well characterized at the molecular level in maize, but probably exists in some form in all plants. In *Arabidopsis thaliana*, *in vivo* labelling during anaerobiosis does not show the drastic switch in the pattern of protein synthesis seen in maize, but there is clearly a quantitative shift towards increased synthesis of several of the proteins seen under aerobic conditions (Dolferus et al., 1985). One of these proteins has been identified as ADH. Induction of ADH enzyme

Organism	Gene	Sequence	Comments
Maize	Adh1-1S (Dennis et al., 1984)	-130 -120 -110 -100 CTGCAGCCCCGGTTCCAAGCCGCGCCCTGGTTTGCTTGC	Identified as functionally important by linker scanner analysis (Walker et al, 1987) and by DMS protection (Ferl & Nick, 1987)
	Ferl & Nick (1987) #	-185 -195 GGACCTGGTTTTCGCTCGT	Identified by DMS protection (Ferl & Nick, 1987)
	Adh2-N (Dennis et al., 1985)	-140 -130 CTCCCTGGTTTCTAACCGCG	Sequence homology to Adh1-1S
	Sucrose synthase (Werr et al., 1985)	+431 +441 CGTGGTTTGCTTG -170 -160 AATTCTGGTTTTGAGTAA	Sequence homology to ARE of Adh1-1S
	Aldolase (Dennis et al., 1987)	-70 -60 TTTCGCTGGTTTCTTTCCCCTT	Lies in functionally important region (-60 to -110)
		-180 -190 CCTTTCTGGTTTTTTGTTTT	Sequence homology, inverted orientation

Wheat   Adh-1A
        (Mitchell et al., 1987)

-525   -490

CAGTTGCTTGCTCGGTCCGCCCCGGTTCGGCCAC   Sequence homology to ARE of Adh1-1S. Numbering is from translation start not transcription start.

Pea   Adh
        (Llewellyn et al., 1987) #

                -100   -110
                *   *
        CCGCTTTTGCTTGTTT   Lies in functionally important region (-51 to -129)

Arabidopsis
        Adh
        (Chang & Meyerowitz, 1986) #

        -150   -160
        *   *
        GTATTTGCTTTTGCCTT   Sequence homology

CONSENSUS SEQUENCE

        TC   TC
        CGTGCTTGT
        GT   CG

#Occurs in inverse orientation

activity during flooding has been demonstrated for a variety of other monocot and dicot plants (Wignarajah et al., 1976, McManmon and Crawford, 1971; Suseelan et al., 1982). Using both homologous and heterologous cDNA probes, *Adh* mRNA induction by anaerobiosis has been shown in wheat (Mitchell et al., in preparation), pea (Llewellyn et al., 1987), carrot, tomato (Llewellyn & Lyfschitz, unpublished), and *Arabidopsis* (Chang and Meyerowitz, 1986).

Pea, which is particularly susceptible to injury during flooding (Jackson, 1979) shows the same drop in cytoplasmic pH seen in maize (Roberts et al., 1984b), as well as the high levels of induced ADH activity, but the pH is only stabilized briefly and the seedlings die within 24 hours due to excessive cytoplasmic acidosis. ADH induction *per se* is not sufficient for survival over prolonged periods of hypoxia, without other compensatory metabolic changes.

Expression and Regulation of Pea *Adh* During Anaerobiosis.

A number of *Adh* genes have been cloned from *Pisum sativum* and the promoter of one of these has been characterized functionally using plant transformation techniques (Llewellyn et al., 1987). This gene, designated as the pea *Adh*1 gene, like the maize *Adh*1 gene, is expressed constitutively in pea seeds and is induced in seedlings by anaerobic conditions. In the first three hours of flooding *Adh*-specific mRNAs are induced about 50-fold (Llewellyn et al., 1987).

A chimeric gene was constructed incorporating the first 350 bp of the pea *Adh*1 gene immediately upstream from the initiator ATG, linked in a transcriptional fusion to the bacterial *Cat* reporter gene. Transcription termination and polyadenylation signals were supplied by the 3' end of the nopaline synthase gene. The chimeric gene was introduced into protoplasts of *Nicotiana plumbaginifolia* by electroporation and transient CAT enzyme activity measured after 20 hours culture under various oxygen concentrations. An increase in pea *Adh* promoter activity was seen with decreasing oxygen tension (Llewellyn et al., 1987). Induction of CAT from 20% to 0% oxygen was approximately 10-fold. Under the same conditions, the constitutive 35S/*Cat* chimeric gene showed a slight drop in activity (Llewellyn et al., 1987). The pea promoter fragment (−290 to +57) appeared to contain all of the DNA sequences necessary for both expression and anaerobic regulation in tobacco protoplasts.

Preliminary deletion studies from the 5' end of the pea promoter fragment have narrowed the important sequences to those downstream of −129 (Llewellyn, unpublished) and hybrid promoter fusions with a deleted *Nos* promoter, to above −51 (Llewellyn & Tokuhisa, unpublished). As indicated later, this 80bp region contains some homology to the "core" sequence of the maize ARE, although on the complementary strand. Further refinements of the hybrid promoter constructions are being tested to define more precisely the pea ARE and to confirm functionally, in tobacco, this observation of DNA sequence homology.

86

The pea *Adh* promoter is recognized in tobacco, another dicot plant, and regulated in the correct manner. We have tested the same chimeric gene constructs in maize protoplasts using electroporation, but have observed no CAT enzyme activity under either aerobic or anaerobic conditions (Llewellyn & Walker, unpublished). The sequence homologies that we have observed between the pea and maize *Adh* promoters (Llewellyn et al., 1987) are clearly not sufficient to allow the pea promoter to function efficiently, if at all, in maize cells. Addition of a transcriptional enhancer from the octopine synthase gene (Ellis et al., 1987) does not improve the expression of the pea promoter in maize, although it considerably enhances expression in tobacco both transiently and in stable transformants (Llewellyn, unpublished).

We have begun an analysis of the inability of this dicot gene to function in maize by constructing a reciprocal set of hybrid promoters between the pea and maize *Adh* promoters. One preliminary result indicates that the upstream region of the maize promoter (−1094 to −78), which contains the ARE, when fused to the downstream part (−51 to +57) of the pea promoter (containing the TATA, transcription start and leader sequences) cannot drive the expression of the CAT gene in maize protoplasts. The same construct is expressed in tobacco protoplasts. Sequences below −51 of the pea promoter are clearly not efficiently recognized in maize. Further hybrids should indicate whether it is the TATA, leader or some other sequence which is responsible. Gene expression is clearly dependent on complex interactions between several factors which recognise and bind at specific DNA sequence elements within the promoter and we are only now beginning to understand some of these interactions through promoter manipulations. Such studies will have important implications for the genetic engineering of plants where gene transfers between different species or genera are contemplated.

A COMMON ANAEROBIC RESPONSE WITHIN THE PLANT KINGDOM?

The ability to respond to anaerobic conditions by a switch in metabolism and gene expression appears to be widespread in plants but little is known about the mechanisms controlling these processes or their conservation within the Plant Kingdom. Two ways of addressing the conservation of the anaerobic control mechanisms are first, to introduce anaerobically induced genes into another species and determine whether they can still be anaerobically regulated, and second, to look for homologous sequences in the upstream regions of the promoters of anaerobically induced genes in many different species.

When the maize *Adh1Cat* chimeric construct is introduced into tobacco using the stable transformation system of the Ti–plasmid of *Agrobacterium tumefaciens* it is expressed at a very low level under aerobic or anaerobic conditions (Ellis et al.,1987a). To increase expression to a detectable level a 176 bp fragment (from −292 to −116) which we know to contain a transcriptional enhancer (Ellis et al., 1987b) was placed upstream of position −140 of the maize gene. This hybrid construct was expressed at high levels in tobacco and was anaerobically inducible. The region of the maize *Adh*1 gene between −140 and +106 containing the ARE was sufficient for anaerobic induction in tobacco, but other important interactions essential for expression were not efficiently recognised in the absence of the added enhancer sequences. The ARE region from −140 to −100 of the maize *Adh*1 gene was transferred to a constitutively expressed but truncated nopaline synthase promoter−*Cat* gene, which, in the presence of the octopine synthase enhancer became anaerobically regulated in transgenic tobacco plants (Ellis, unpublished). This confirms that the maize ARE is recognized in tobacco and that the regulatory binding factors presumed to be responsible for anaerobic induction of genes are sufficiently similar in maize and tobacco to function across wide species barriers. The same appears to be true between pea and tobacco, since the pea promoter functions efficiently and correctly in tobacco (Llewellyn et al., 1987).

We have examined the sequence of a number of promoters from different anaerobically induced genes from a variety of species to search for homology to the maize ARE. The wheat *Adh*1A gene shows extensive homology to the ARE of maize *Adh*1, although the relative orientation of the two sub−regions are reversed (Fig. 3). There are insufficient sequence data from the 5' regions of the barley *Adh*1, *Adh*2 or *Adh*3 genes to determine whether they have any homology to the ARE.

Functional analysis of the pea gene promoter in tobacco has identified the region between −129 and −51 as critical for anaerobic expression. This sequence was searched for homology to the ARE of the maize *Adh*1 gene and in particular for homology to the "core" sequence TGGTTT which we suspect to be critical for anaerobic induction. No homologous sequence was found but the complement of the sequence, AAACCA, was found in the region shown to be functionally important. Another anaerobically induced *Adh* gene which has been analyzed at the nucleotide sequence level is the *Adh* gene of *Arabidopsis thaliana* (Cheng and Myerowitz, 1986). Examination of the upstream region of this gene also shows an inverse complement to the core ARE sequence. The sequence AAACCA is present in the region −166 to −158, similar to that of the pea ADH gene. These data suggest that the core sequence is conserved between plant species, or at least between anaerobically induced *Adh* genes of different plant species, and that it may work in either orientation similar to many enhancer sequences.

Figure 3 summarizes the data available from all anaerobically induced plant genes that have so far been sequenced. All of the genes contain the core sequence in the promoter, but only in the case of the *Adh*1−1S gene has this region been precisely defined by LS mutations (Walker et al., 1987) and DMS footprinting studies (Ferl and Nick, 1987). However, functional analyses of the pea *Adh*1 and maize aldolase genes have located critical sequences to relatively small regions which contain the core ARE seqeuence. We are currently testing the hypothesis that the consensus core sequence is primarily responsible for the anaerobically regulated expression of all of these genes.

SUMMARY OF THE ANAEROBIC RESPONSE.

The anaerobic response, as we currently understand it in maize, is summarized in Figure 1. At the metabolic level the reduced concentration of oxygen results in the inhibition of oxidative phosphorylation and an increase in the ratio of NADH/NAD$^+$. The NADH is utilized first by lactate dehydrogenase and subsequently by ADH as the terminal dehydrogenase. These reactions regenerate NAD$^+$ which is required for continued glycolysis and substrate-level phosphorylation. At the molecular level the reduced oxygen concentration is perceived by the translational machinery of the cell, by some as yet unknown mechanism, and normal protein synthesis is repressed. Concomitant with the repression of normal protein synthesis, the expression of the ANP genes is transcriptionally induced. Newly synthesized anaerobic mRNAs are specifically translated by a modified translational apparatus that is yet to be characterised.

Although the details of the transcriptional induction of ANP genes are not known, it appears that *trans*-acting transcription factors activate the co-ordinate expression of this set of genes. A common factor, or factors, may be involved in the activation of all the ANP genes, or at least some subset thereof, through their binding to one or more core ARE sequences within the promoters of each of the anaerobically regulated genes. The exact nature of the interactions between these regulatory proteins, the ARE core, and other transcription factors may determine the differences in the rate of accumulation and the maximum levels observed for several of the ANP mRNAs. These differences in the accumulation of the ANP mRNAs may also be a function of differential rates of processing and mRNA stability.

Attention is now turning to the molecular analysis of the *trans*-acting DNA binding factors that regulate anaerobic gene expression. The nature and extent of the binding of such proteins to key regulatory DNA sequences can be quantified and used in an assay for the purification of the factors (eg. Johnson et al., 1987). Such analyses should bring us closer to a complete understanding of the molecular processes that regulate gene expression by environmental signals.

REFERENCES

ADAMS, M.J., FORD, G.C., KOEFOK, R., LENTZ, P.J. Jr., McPHERSON, A. Jr., ROSSMANN, M.G., SMILEY, I.E., SCHEVITZ, R.W. and WONCOTT, A.J. (1970). Structure of lactate dehydrogenase at 2.8Å resolution. Nature. 227, 1098–1103.
BRANDEN, C-I., EKLUND, H., CAMBILLAU, C. and PRYOR, A.J. (1984). Correlation of exons with structural domains in alcohol dehydrogenase. EMBO J. 3, 1307–1310.
BUEHNER, M., FORD, G.C., MORAS, D., OLSEN, K.W. and ROSSMANN, M.G. (1973). D–glyceraldehyde-3-phosphate dehydrogenase: three dimensional structure and evolutionary significance. Proc. Natl. Acad. Sci. U.S.A. 70, 3052–3054.
CALLIS, J., FROMM, M. and WALBOT, V. (1987). Efficient expression of the maize alcohol dehydrogenase-1 gene: Requirement for intervening sequences. Genes & Dev., Submitted.
CHANG, C. and MEYEROWITZ, E.M. (1986). Molecular cloning and DNA sequence of the Arabidopsis thaliana alcohol dehydrogenase gene. Proc. Natl. Acad. Sci. U.S.A., 83, 1408–1412.
DENNIS, E.S., GERLACH, W.L., PRYOR, A.J., BENNETZEN, J.L., INGLIS, A., LLEWELLYN, D., SACHS, M.M., FERL, R.J. and PEACOCK, W.J. (1984). Molecular analysis of the alcohol dehydrogenase (Adh1) gene of maize. Nucl. Acids Res. 12, 3983–4000.
DENNIS, E.S., SACHS, M.M., GERLACH, W.L., FINNEGAN, E.J. and PEACOCK, W.J. (1985). Molecular analysis of the alcohol dehydrogenase 2 (Adh2) gene of maize. Nucl. Acids Res. 13, 727–743.
DENNIS, E.S., GERLACH, W.L., WALKER, J.C., LAVIN, M. and PEACOCK, W.J. (1987) An anaerobically regulated aldolase gene of maize. J. Mol. Biol., Submitted.
DRUM, D.E., LI, T.-K. and VALLEE, B.L. (1969). Zinc isotope exchange in horse liver alcohol dehydrogenase. Biochemistry 8, 3792–3797.
DOLFERUS, R., MARBAIX, G. and JACOBS, M. (1985). Alcohol dehydrogenase in Arabidopsis: analysis of the induction phenomenon in plantlets and tissue cultures. Mol. Gen. Genet. 199, 256–264.
ECHOLS, H. (1986). Multiple DNA–Protein interactions governing high–precision DNA transactions. Science 233,1050–1056.
EDENBERG, H.M., ZHANG, K., FONG, K., BOSRON, W.F. and LI, T.-K. (1985). Cloning and sequencing of a cDNA encoding the complete mouse liver alcohol dehydrogenase. Proc. Natl. Acad. Sci. U.S.A. 82, 2262–2266.
EKLUND, H., NORDSTRÖM, B., ZEPPEZAUER, E., SÖDERLUND, G., OHLSSON, I., BOIWE, T., SÖDERBERG, B-O., TAPIA, O., BRÄNDEN, C.-I., AKESON, A. (1976). Three dimensional structure of horse liver alcohol dehydrogenase at 2.4Å resolution. J. Mol. Biol. 102, 2 7–59.
ELLIS, J.G., LLEWELLYN, D.L., DENNIS, E.S. and PEACOCK, W.J. (1987a). Maize Adh1 promoter sequences control anaerobic regulation: addition of upstream promoter elements from constitutive genes is necessary for expression in tobacco. EMBO J. 6, 11–16.
ELLIS, J.G., LLEWELLYN, D.L., WALKER, J.C., DENNIS, E.S. and PEACOCK, W.J. (1987b). The ocs element: a 16 base pair palindrome essential for activity of the octopine synthase enhancer. EMBO J. In press.
FERL, R.J. (1985). Modulation of chromatin structure in the regulation of the maize Adh1 gene. Mol. Gen. Genet. 200, 207–210.

FERL, R.J. and NICK, H.S. (1987). *In vivo* detection of regulatory factor binding sites in the 5' flanking region of maize *Adh*1. J. Biol. Chem. 262, 7947-7950.

FREELING, M. and BENNETT, D.C. (1985). Maize *Adh*1. Ann. Rev. Genet. 19, 297-323.

FROMM, M., TAYLOR, L.P. and WALBOT, V. (1985). Expression of genes transferred into monocot and dicot plant cells by electroporation. Proc. Natl. Acad. Sci. U.S.A. 82, 5824-5828.

FROMM, M., CALLIS, J. and WALBOT, V. (1987). Introns increase chimeric gene expression in maize. Genes & Dev., Submitted.

GERLACH, W.L., PRYOR, A.J., DENNIS, E.S., FERL, R.J., SACHS, M.M. and PEACOCK, W.J. (1982). cDNA cloning and induction of the alcohol dehydrogenase gene (*Adh*1) of maize. Proc. Natl. Acad. Sci. USA 79, 2981-2985.

GILBERT, W. (1978). Why genes in pieces? Nature 271, 501.

GRABLE, A.R. (1966) Soil aeration and plant growth. Adv. Agron. 18, 57-106.

GRUSS, P. and KHOURY, G. (1980). Rescue of a splicing defective mutant by insertion of an heterologous intron. Nature 286, 634-637.

HAGEMAN, R.H. and FLESHER, D. (1960). The effect of an anaerobic environment on the activity of alcohol dehydrogenase and other enzymes of corn seedlings. Arch. Biochem. Biophys. 87, 203-209.

HAKE, S., KELLEY, P., TAYLOR, W.C. and FREELING, M. (1985). Coordinate induction of alcohol dehydrogenase 1, aldolase, and other anaerobic RNAs in maize. J. Biol. Chem. 260, 5050-5054.

HAMER, D.H. and LEDER, P. (1979). Splicing and the formation of stable mRNAs. Cell 18, 1299-1302.

HEMPEL, J., BUHLER, R., KAISER, R., HOLMQUIST, B., De ZALENSKI, C., von WARTBURG, J.-P., VALLEE, B. and JÖRNVALL, H. (1984). Human liver alcohol dehydrogenase 1. The primary structure of the β.β. isoenzyme. Eur. J. Biochem. 145, 437-445.

HOFFMAN, N.E., BENT, A.F. and HANSON, A.D. (1986). Induction of lactate dehydrogenase isozymes by oxygen deficit in barley root tissues. Plant Physiol. 82, 658-663.

HOWARD, E.A., WALKER, J.C., DENNIS, E.S. and PEACOCK, W.J. (1987). Regulated expression of an alcohol dehydrogenase 1 chimeric gene introduced into maize protoplasts. Planta 170, 535-540.

IGO-KEMENES, T., HORZ, C. and ZACHAU, H.G. (1982). Chromatin. Ann. Rev. Biochem. 51, 89-121.

JACKSON, M. (1979). Rapid injury to peas by soil waterlogging. J. Sci. Food Agric. 30, 143-152.

JOHNSON, P.F., LANDSCHULTZ W.H. GRAVES, B.J. and McKNIGHT, S.L. (1987). Identification of a rat liver protein that binds to the enhancer core elements of three animal viruses. Genes & Dev. 1, 133-146.

JÖRNVALL, H. (1970). Horse liver alcohol dehydrogenase. The primary structure of the protein chain of the ethanol active isozyme. Eur. J. Biochem. 16, 25-40.

KELLEY, P.M. and FREELING, M. (1984). Anaerobic expression of maize fructose-1,6-diphosphate aldolase. J. Biol Chem. 259, 673-678.

KIMURA, M. (1983). The neutral theory of molecular evolution (M. Nei and R.K. Koehn, eds.), Sinauer Associates, Sunderland, Massachusetts.

KLEMENZ, R., HULTMARK, D. and GEHRING, W.J. (1985). Selective translation of heat shock mRNA in *Drosophila melanogaster* depends on sequence information in the leader. EMBO J. 4, 2053-2060.

LAZLO, A. and St. LAWRENCE, P. (1983). Parallel induction and synthesis of PDC and ADH in anoxic maize roots. Molec. Gen. Genet. 192, 110–117.

LIN, C.Y. and KEY, J.L. (1967). Dissociation and reassembly of polyribosomes in relation to protein synthesis in the soybean root. J. Mol. Biol. 26, 37–247.

LINQUIST, S. (1986). The heat shock response. Ann. Rev. Biochem. 55, 1151–1191.

LLEWELLYN, D.J., FINNEGAN, E.J., ELLIS, J.G., DENNIS, E.S. and PEACOCK, W.J. (1987). Structure and expression of an alcohol dehydrogenase 1 gene from *Pisum sativum* (cv. Greenfeast). J. Mol. Biol. 195, 115–123.

MANIATIS, T., GOODBOURN, S. and FISHER, J.A, (1987). Regulation of inducible and tissue-specific gene expression. Science 236, 1237–1245.

McKNIGHT, S. and TJIAN, R. (1986). Transcriptional selectivity of viral genes in mammalian cells. Cell 46, 795–805.

McMANMON, M. and CRAWFORD R.M.M. (1971). Metabolic theory of flooding tolerance: The significance of enzyme distribution and behaviour. New Phytol. 70, 299–306.

MITCHELL, L.E., DENNIS, E.S. and PEACOCK, W.J. Molecular analyses of an alcohol dehydrogenase gene from chromosome 1 of wheat. (In preparation.) OKIMOTO, R., SACHS, M.M., PORTER, E.K. and FREELING, M. (1980). Patterns of polypeptide synthesis in various maize organs under anaerobiosis. Planta 150, 89–94.

PAUL, A.-L., VASIL, V., VASIL, I.K. and FERL, R.J. (1987). Constitutive and anaerobically induced DNase-I-hypersensitive sites in the 5' region of the maize *Adh*1 gene. Proc. Natl. Acad. Sci. USA 84, 799–803.

PEACOCK, W.J., WOLSTENHOLME, D., WALKER, J., SINGH, K., LLEWELLYN, D., ELLIS, J. and DENNIS, E. (1987). Developmental and environmental regulation of the maize alcohol dehydrogenase 1 (*Adh*1) gene: promoter–enhancer interactions. CETUS-UCLA Symposium (In press).

PTASHNE, M. (1986). A Genetic Switch: gene control by phage lambda . Cell Press and Blackwell Scientific Publication, Palo Alto, CA.

ROBERTS, J.K., CALLIS, J., WEMMER, D., WALBOT, V. and JARDETZKY, O. (1984a). Mechanisms of cytoplasmic pH regulation in hypoxic maize root tips and its role in survival under hypoxia. Proc. Natl. Acad. Sci. USA. 81, 3379–3383.

ROBERTS, J.K., CALLIS, J., JARDETSKY, O., WALBOT, V. and FREELING, M. (1984b). Cytoplasmic acidosis as a determinant of flooding intolerance in plants. Proc. Natl. Acad. Sci. USA. 81, 6029–6033.

ROWLAND, L.J. and STROMMER, J.N. (1986) Anaerobic treatment of maize roots affects transcription and transcript stability. Mol. Cell. Biol. 6, 3368–3372.

SACHS, M.M., FREELING, M. and OKIMOTO, R. (1980) The anaerobic proteins of maize. Cell 20, 761–767.

SCHAFFNER, W. (1985) Introduction. In Eukaryotic transcription: the role of *cis-* and *trans*-acting elements in initiation. (Y. Gluzman, ed.) Cold Spring Harbor Laboratory, Cold Spring Harbor, pp 1–18.

SCHLESSINGER, M.J., ASHBURNER, M. and TISSIERES, A. (1982) Heat shock: from bacteria to man. Cold Spring Harbor Laboratory, Cold Spring Harbor, N.Y.

SPRINGER, B., WERR, W., STARLINGER, P., BENNETT, D.C., ZOKOLICA, M. and FREELING, M. (1986) The *Shrunken* gene on chromosome 9 of *Zea mays*

L is expressed in various plant tissues and encodes an anaerobic protein. Mol. Gen. Genet. 205, 461–468.

STRUHL, K. (1987) Promoters, activator proteins, and the mechanisms of transcription initiation in yeast. Cell 49, 295–297.

SCHWARTZ, D. (1969). An example of gene fixation resulting from selective advantage in sub–optimal conditions. Amer. Nat. 103, 479–481.

SUND, H. and THEORELL, H. (1963). The enzymes (Boyer, P.D., Lardy, H. and Myrback, K., eds). 2nd edn, vol. 7, pp. 25–83. Academic Press, New York and London.

SUSEELAN, K.N., BHATIA, C.R. and EAPEN, S. (1982). Two NAD–dependent alcohol dehydrogenases (E.C. 1.1.1.1) in callus cultures of wheat, rye and Triticale. Theor. Appl. Genet. 62, 45–48.

VAYDA, M.E. and FREELING, M. (1986) Insertion of the Mu1 transposable element into the first intron of maize Adh1 interferes with transcript elongation but does not disrupt chromatin structure. Plant Mol. Biol. 6, 441–454.

WALKER J.C., HOWARD E.A., DENNIS E.S., PEACOCK W.J. (1987). DNA sequences required for anaerobic expression of the maize alcohol dehydrogenase–1 gene. Proc. Natl. Acad. Sci. USA. (in press).

WEINTRAUB, H. and GROUNDINE, M. (1976) Chromatin subunits in active genes have an altered conformation. Science 193, 848–856.

WERR, W., FROMMER, W.-B., MAAS, C. and STARLINGER, P. (1985) Structure of the sucrose synthase gene on chromosome 9 of Zea mays L. EMBO J. 4, 1373–1380.

WICKENS, M.P., WOO, S., O'MALLEY, B.W., GURDON, J.B. (1980). Expression of a chicken chromosomal gene injected into frog oocyte nuclei. Nature 285, 628–634.

WIGNARAJAH, K., GREENWAY, H. and JOHN, C.D. (1976). Effect of waterlogging on growth and activity of alcohol dehydrogenase in barley and rice. New Phytol. 77, 585–592.

Oxford Surveys of Plant Molecular & Cell Biology Vol.4 (1987) 95-135

THE MOLECULAR ANALYSIS OF THE ASSEMBLY,
STRUCTURE AND FUNCTION OF RUBISCO

Steven Gutteridge and Anthony A. Gatenby

*Central Research and Development Department, Experimental Station,
E.I. DuPont de Nemours and Company,
Wilmington, Delaware 19898 U.S.A.*

THE CENTRAL ROLE OF RUBISCO

From the perspective of the developing plant, the reliance on ribulose bisphosphate carboxylase/oxygenase (Rubisco) must be a daunting prospect. For C3 species, the quantities of enzyme that must be synthesised to sustain growth are prodigious requiring the synchronisation of two separate genomic systems. The nucleus has to supply enough of the small subunit to serve some 100 chloroplasts per cell, each of which synthesise enough large subunit to constitute 90% of the stroma. Somehow the two subunits associate to form a complex hexadecameric heterologous (L_8S_8) protein aggregate.

For those plants, like the C4 species, that have been able to avoid the massive investment in protein bulk, the prospect has been no less herculean: a whole different cell architecture with attendant CO_2 concentrating mechanisms has evolved to surmount the unavoidable limitations intrinsic to the nature of the Rubisco.

The major limitation is the ever present, almost parasitic, oxygenase activity that ensures a constant drain on the pool of ribulose-P_2 (see Figure 1) that would normally be consumed during CO_2 fixation breeding more bisphosphate substrate and triose phosphate. The photorespiratory cycle attempts to limit the damage by recapturing the lost carbon but can only do so by evolving CO_2 at further energy cost to the plant. CO_2 fixation may be such difficult chemistry that it may always be accompanied by some oxygenation but it is clear that Rubisco has not yet attained the highest specificity possible.

Whether there is a means of increasing the efficiency of the process in a plant depends on our understanding of at least the reaction chemistry of the enzyme, and then on how this relates to the enzyme in the context of the overall control and productivity of photosynthetic carbon fixation. From the point of view of the catalytic mechanism of the carboxylase, there is always a limitation on the amount of CO_2 fixation, irrespective of the natural environmental conditions of growth, but how the biogenesis of the enzyme, and those regulatory features in the chloroplast that have just been discovered, contribute to these limitations of development and productivity are extremely complex and not well understood. In this review we relate what is known about most of these aspects at the molecular level and highlight areas of interest that require much more insight.

ASSEMBLY OF THE HOLOENZYME

The multi-subunit structure of Rubisco, the dual coding sites for the two subunit types, large(L) and small(S), and the trans-membrane import of pre-S into chloroplasts indicates that a complex assembly mechanism exists in plants that is, at present, incompletely understood. In attempts to understand how all the components involved in assembly function together, several types of experimental approaches have been used. These range from assembly of the enzyme in isolated chloroplast, import studies in chloroplasts, reconstitution of assembly in *E. coli*, and the *in vitro* dissociation and re-association of subunits. In this section we shall outline the various experimental systems, what has been learnt from each, and to what extent an overall model for assembly can be deduced.

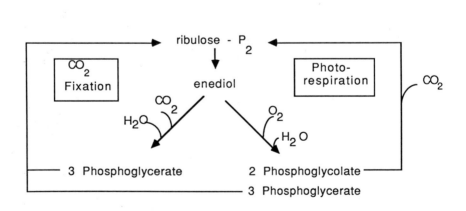

Figure 1. The partitioning of ribulose-P_2 by Rubisco through carboxylation and oxygenation reactions and the initiation of CO_2 fixation and photorespiration, respectively.

In Organello Synthesis and Assembly.

If isolated chloroplasts are incubated in an appropriate medium containing radioactive amino acids, the synthesis of chloroplast proteins can be observed. The first *in vitro* system to produce discreet polypeptides used light energy to drive protein synthesis in pea (*Pisum sativum*) (Blair and Ellis, 1973). These authors also showed that one of the major products synthesized was the L-subunit of Rubisco. Several groups then examined the fate of the newly synthesized L-subunit to see if it would assemble into holoenzyme. Bottomley et al. (1974) and Mendiola-Morganthaler et al. (1976) reported that *in vitro* synthesized L-subunit does assemble into holoenzyme, but this work was criticized because the analysis of assembly was carried out at only a single gel concentration on non-denaturing gels (see discussion by Ellis et al., 1979), and there was also conflicting data showing that the newly-synthesized L-subunit did not enter pre-existing Rubisco (Blair and Ellis, 1973). The difference of opinion, however, was resolved when it was realized that the osmoticum used during the incubation was of critical importance in obtaining assembly (Barraclough and Ellis, 1980). The use of KCl prevents the incorporation of newly-synthesized L-subunit into holoenzyme, whereas a sorbitol medium enables transfer of L-subunit into the holoenzyme.

The assembly of holoenzyme in isolated chloroplasts would indicate that either a pool of free S-subunit exists in the organelle, or that S-subunit is being released from the holoenzyme during dissociation and is then able to re-assemble with the newly-synthesized L-subunit. There are data that indicate the presence of free S-subunit in both pea leaves (Roy et al., 1978; 1979) and in isolated leaf cells from soybean (*Glycine max.*) (Barraclough and Ellis, 1979). The pool sizes of unassembled S-subunit, however, are presumably small because the half life of free S-subunit in, e.g. *Chlamydomonas* is less than 7.5 min. (Schmidt and Mishkind, 1983). This assumes that S-subunit stability in *Chlamydomonas* and higher plant chloroplasts are similar. The wheat mature S-subunit synthesized in *E. coli* is also unstable with a half-life of <10 min., suggesting that unassembled S-subunit is a very labile protein (Gatenby et al., 1987).

When Barraclough and Ellis (1980) were examining the synthesis and assembly of L-subunit into Rubisco holoenzyme, they observed that newly-synthesized L-subunits which do not assemble with holoenzyme are associated with a protein of subunit molecular weight 60,000 (polypeptide 60). These authors noted that the protein band containing the non-assembled L-subunits, together with the unlabelled 60,000 subunits, had a molecular weight of greater than 600,000. The results of time course experiments indicated that as radioactive L-subunit became assembled into the holoenzyme, the radioactivity in the aggregate band declined. These observations raised the possibility that newly-synthesized L-subunits are specifically aggregated with polypeptide 60 subunits prior to assembly, and that this complex is an obligatory intermediate in the assembly of Rubisco. Polypeptide 60 was subsequently referred to as the large subunit binding protein (LSBP) and it was suggested that it was likely to be a nuclear gene product since, although it could be labelled *in vivo* with radioactive amino acids, it was not labelled in isolated chloroplast (Ellis, 1981). A model for the role of LSBP in holoenzyme formation was also presented.

Roy et al. (1982) reported that L-subunit synthesized *in vivo* or *in organello* could be recovered from intact chloroplasts in the form of two different sedimentation complexes of 7S and 29S on sucrose gradients. The 29S complex contained unassembled L-subunit and LSBP and represents the complex of 600,000 M_r observed by Barraclough and Ellis (1980). These latter authors did not detect the 7S complex, which is suggested to be a dimer or heterodimer of L-subunits. According to Roy et al. (1982), this failure to observe the 7S complex was because different methods of analysis were used (sucrose gradients vs non-denaturing gels). With illumination of chloroplasts it was observed that the newly-synthesized L-subunit present in both the 7S and 29S complexes disappeared and was subsequently found in the assembled 18S Rubisco holoenzyme (Roy et al., 1982). It should be noted that both Barraclough and Ellis (1980) and Roy et al. (1982) commented on the solubility of newly-synthesized L-subunit when associated with the LSBP, in contrast to the insolubility of L-subunit isolated from the holoenzyme. In a subsequent paper Bloom et al. (1983) showed that light can promote the post-translational assembly of L-subunit into Rubisco. Rubisco assembly was also accelerated when extracts of chloroplasts were treated with ATP or GTP, but the 29S complex was maintained. However, in the presence of Mg^{2+}, ATP caused an almost complete dissociation of the 29S complex whereas GTP and a non-hydrolyzable analogue of ATP had no effect. Bloom et al. (1983) postulated a complex set of reactions involved in assembly requiring nucleotides, Mg^{2+} and putative intermediates; they also suggested that the 29S complex may not serve as the final donor of L-subunit in assembly.

The large subunit binding protein.

The LSBP has been purified from pea, it has a molecular weight of 720,000 and is composed of two types of subunits of 60,000 and 61,000 (Hemmingsen and Ellis, 1986). Highly specific antisera were raised against the LSBP and used to show that a high molecular weight form is immunoprecipitated from products of pea polysomes translated in a wheat-germ system, thus confirming that LSBP is synthesized by cytoplasmic ribosomes. The protein was found in a wide range of plant species. These authors also confirmed the results of Bloom et al., (1983) that addition of ATP *in vitro* caused dissociation of the 29S complex. They also proposed that *in vivo* the LSBP is in an equilibrium between a dodecamer and monomers, and that the equilibrium is influenced by the ATP concentration. It was also pointed out that under *in vivo* conditions the bulk of the LSBP may not be in the high molecular weight form, since reported concentrations of ATP in chloroplast stroma would cause dissociation of the complex (Hemmingsen and Ellis, 1986). Antisera raised against the LSBP has also been shown to inhibit Rubisco assembly in an *in vitro* assembly reaction, further supporting the suggested role of this protein in assembly (Cannon et al., 1986). Musgrove et al. (1987) have demonstrated that the LSBP of pea is an oligomer with a composition of $\alpha_6\beta_6$, in which the α and β subunits are immunologically distinct, show different partial protease digestion patterns and have different amino-terminal sequences.

In addition to the synthesis and assembly of L-subunit into holoenzyme in isolated chloroplasts, it is also possible to import the S-subunit into the isolated organelles and to obtain assembly. The rbcS gene of photosynthetic eukaryotes encodes a pre-S protein containing an N-terminal transit peptide essential for transport of S-subunit into the chloroplast, at which time the transit peptide is removed (reviewed by Schmidt and Mishkind, 1986). Chua and Schmidt (1978) demonstrated that imported S-subunit from pea or spinach (Spinacea oleracea) could be assembled into holoenzyme in isolated chloroplasts. Assembly of imported S-subunit was also reported by Smith and Ellis (1979) who, in addition, found that assembly of holoenzyme occurred in the stroma. It is apparently not possible to assemble a heterologous enzyme if the imported S-subunit originates from an alga and higher plant chloroplasts are used (Mishkind et al., 1985). In unpublished experiments it has been found that if the rbcL gene is fused to the rbcS transit peptide coding sequence, the in vitro transcribed and translated fusion protein can be imported into chloroplasts and the L-subunit assembled into the holoenzyme (T.H.Lubben, A.A.Gatenby, P. Ahlquist and K.Keegstra, submitted).

Synthesis of Plant Rubisco Subunits in Escherichia coli.

An alternative approach to examining Rubisco assembly in isolated chloroplasts is to express the appropriate genes in E. coli. If successful assembly could be demonstrated, then it would be very convenient to mutate and modify the system further to gain an understanding of the mechanism involved. Because the translation initiation signals of chloroplast and prokaryotic genes are highly conserved, it was relatively straightforward to obtain the initial expression of the chloroplast encoded rbcL genes in E. coli (Gatenby et al., 1981). Unmodified levels of expression were low, but could be significantly enhanced during transient expression in a bacteriophage lambda vector. It was subsequently observed that high levels of expression of the maize (Zea mays) rbcL gene could be obtained by making a transcriptional fusion with part of a bacteriophage operon on a multicopy plasmid (Gatenby and Castleton, 1982), provided that a transcription anti-termination protein was present (Gatenby and Cuellar, 1985). With such a system, maize L-subunit accumulated to levels of 2% of total bacterial cell protein. An alternative system to synthesize maize L-subunit using a lac promoter and ribosome binding site has been described by Somerville et al. (1986).

With the levels of expression obtained it was possible to examine some of the properties of L-subunit that had been synthesized in the absence of S-subunit. This was an important point, since it had not been possible previously to examine the properties of L-subunit without denaturing the holoenzyme to separate L- and S- subunit. It was found that the maize L-subunit synthesized in *E. coli* was insoluble, lacked ribulose-P_2 carboxylase activity, and failed to form a stable enzyme.metal.CO_2.^{14}C 2'-carboxyarabinitol-P_2 quaternary complex (Gatenby, 1984). Attempts to solubilize L-subunit with ribulose-P_2 or the positive effector fructose-P_2 were unsuccessful, and L-subunit was solubilized only at a slow rate at pH11-12, suggesting the formation of aggregates. This could account for the remarkable stability of L-subunit during pulse chase experiments which revealed that there is no degradation over a period of 4 h. Since L-subunit synthesized by itself showed no indication of self-assembly or activity, it was decided to express both the L- and S- subunits together in *E. coli*. The maize *rbcL* gene was expressed from a temperature inducible P_L promoter, and a wheat (*Triticum aestivum*) *rbcS* cDNA clone was modified to enable synthesis of a mature S-subunit under the control of a *lac* promoter (Gatenby et al., 1987). These combinations of genes and promoters were present on a single replicon such that L- and S- subunit synthesis could be initiated independently (Figure 2). When wheat S-subunit was synthesized by itself it was soluble but unstable, with a half-life of about 10 min. Maize L-subunit by itself was insoluble but very stable. When L-subunit and S were synthesized together, the S-subunit was now converted to an insoluble, but very stable form. Assembly of the two subunits did not occur, and it appeared that an aggregate had formed which had sequestered the S-subunit into a protected form, suggesting some interaction between the L-subunit and S-subunits. In related experiments it was found that the wheat L-subunit expressed in *E. coli* was more soluble than maize L-subunit (Bradley et al., 1986). However, even when soluble wheat L- and S- subunits were expressed together, neither assembly nor activity could be detected.

Because of the problems associated with higher-plant Rubisco assembly, several groups have developed experimental systems in which a plant-like L_8S_8 Rubisco can be expressed and assembled in *E. coli* to give an active enzyme. Genes have been used from cyanobacteria (Christeller et al., 1985; Gatenby et al., 1985; Gurevitz et al., 1985; Tabita and Small, 1985, Gutteridge et al., 1986a), purple sulphur bacteria (Viale et al., 1985), and purple non-sulphur bacteria (Gibson and Tabita, 1986). These could be valuable experimental systems for examining Rubisco function, but they still do not address the problem of the mechanism of assembly of the higher plant enzyme. van der Vies et al. (1986) have extended the cyanobacterial Rubisco assembly experiments by demonstrating the formation of a hybrid Rubisco in *E. coli* comprising both cyanobacterial L-subunit and wheat S-subunits. The hybrid was active and this experiment also demon- strated that the wheat S-subunit was competent for assembly. One implication from this result is that the barrier to successful assembly resides in the structure of the plant L-subunit, which could support the idea that the LSBP is required at that stage.

pLSS308

30°c 41°c

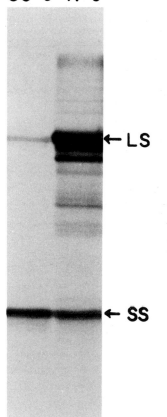

←LS

← SS

Figure 2. Synthesis of both the S- and L- subunits of higher plant Rubisco in E. coli . The plasmid pLSS308, which encodes the wheat S-subunit transcribed from P_{lac} and the maize L-subunit transcribed from P_L, was transformed into E. coli and grown in a medium containing ^{35}S-methionine to label the proteins, and IPTG to induce wheat S-subunit synthesis. Cultures were maintained either at 30° or 41°C to obtain repression or induction respectively of maize L-subunit synthesis. The radiolabelled proteins were incubated with anti-wheat Rubisco serum and protein-A Sepharose, and the immuno-reactive Rubisco subunits resolved on a 10% gel containing SDS. The position of the L- and S- subunits are marked LS and SS respectively. Reprinted with permission from Eur. J. Biochem. (Gatenby et al., 1987).

The Complexity of Assembly in Photosynthetic Prokaryotes and Eukaryotes.

In the prokaryotic cyanobacteria the *rbc*L and *rbc*S genes are closely linked together on a contiguous piece of DNA and are co-transcribed (Shinozaki and Sugiura, 1985). The gene products are present together from their time of synthesis in the same cell compartment and do not need to be transported through membranes. This arrangement has presumably favoured a facile self-assembly mechanism which enables the holoenzyme to form. This is borne out by the ease with which cyanobacterial Rubisco can self-assemble into an active enzyme when the genes are expressed in *E. coli* (van der Vies et al., 1986). It is also known that a cyanobacterial Rubisco can be

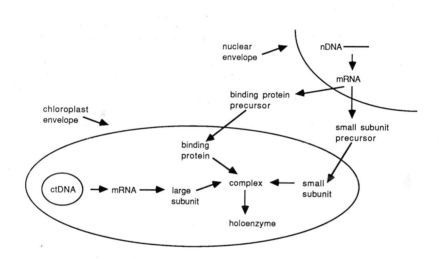

Figure 3. A suggested assembly pathway for higher plant Rubisco. The chloroplast encoded L-subunit is synthesised in the chloroplast on 70S ribosomes prior to assembly into the stromal holoenzyme. The nuclear encoded S-subunit is synthesised on 80S cytoplasmic ribosomes as a precursor protein, which is then imported into the chloroplast post-translationally. The transit peptide is then removed in a two step process before assembly into the holoenzyme. A third protein, LSBP, may also be involved in the formation of the complex suspected of being an intermediate in the assembly. The binding protein asssociated in the complex has two different subunits which are thought to be nuclear encoded and then imported into the chloroplast. It is not yet clear if the assembled Rubisco resides in the stroma as a stable molecule or whether there is a dynamic process which allows the subunits to dissociate, exchange and reassemble.

dissociated *in vitro* to form an L_8 core devoid of S, and that S-subunit can be added back to give a functional enzyme, using either homologous (Andrews and Ballment, 1983) or heterologous S-subunits (Andrews and Lorimer, 1985). From these data and the work of van der Vies et al. (1986) in which self-assembly of an L_8 core occurs in *E. coli*, it would appear that formation of an L_8 core may be the critical first step in assembly. It is to this L_8 core that S-subunit could be attached to obtain high levels of activity. An alternative model for assembly has been suggested by Gurevitz et al. (1985) in which assembly proceeds via L- and S- subunit heterodimers.

In plants the assembly of Rubisco appears more complex. The *rbc*L and *rbc*S genes have different coding sites and the *rbc*S gene product has to be transported through the chloroplast membrane before Rubisco assembly can occur. Some of the steps thought to be involved in holoenzyme assembly are outlined in Figure 3. In addition, *in vitro* dissociation studies of plant Rubisco reveal that it is not possible to obtain an L_8 core that remains soluble in aqueous solutions (Voordouw et al., 1984), and maize L-subunit synthesized in *E. coli* is insoluble (Gatenby, 1984). It has also not been possible to obtain assembly of the plant L- and S- subunits expressed in *E. coli* (Bradley et al., 1986; Gatenby et al., 1987). For these reasons, and others cited earlier, the hypothesis has been invoked that the chloroplast synthesized L-subunit requires the LSBP to assist in holoenzyme formation. Formal proof of the role of the LSBP in assembly has not yet been presented, but its suggested function becomes increasingly attractive as a mechanism for mediating Rubisco assembly.

REGULATION OF GENE EXPRESSION

The dual coding sites for the *rbc*S and *rbc*L genes, in the nucleus and chloroplasts respectively, indicates that transcription and translation of the genes will use different types of RNA polymerases and ribosomes. Expression of *rbc*S involves the use of eukaryotic RNA polymerase and ribosomes, whereas expression of *rbc*L in the chloroplast will be more prokaryotic-like in nature. The regulation of these two genes could occur at many different steps, including the obvious ones of transcription, translation and subunit stability. In this section we shall describe what is known about the structure and regulation of these very different genes.

*rbc*S Genes.

The S-subunit polypeptide is encoded in higher plants by a small multigene family (Berry-Lowe et al., 1982; Coruzzi et al., 1983; Broglie et al., 1983; Dean et al., 1985; McKnight et al., 1986; Dean

et al., 1987). All *rbc*S genes examined to date contain introns but
the number present is different for monocots (one) and dicots (two or
three) (Nagy et al., 1986). The *rbc*S genes encode transit peptides
for chloroplast targeting, in addition to the mature protein sequence
information. The transit peptide varies in size from 46 to 57 amino
acids, has a more variable sequence than the mature protein, and is
basic with a conserved Cys-Met-Gln processing site.

The transcription of *rbc*S can be regulated by light in flowering
plants, although there are exceptions (reviewed by Tobin and
Silverthorne, 1985). Smith and Ellis (1981) observed that light
stimulated the accumulation of *rbc*S transcripts, when comparing the
amount of mRNA in light- or dark- grown pea leaves. This transcription was subsequently shown to have a typical phytochrome
mediated response (Tobin, 1981; Thompson et al., 1983; Coruzzi et
al., 1984; Gallagher et al., 1985). It was also observed that the
transcriptional control of individual *rbc*S genes vary (Coruzzi et
al., 1984). Using isolated nuclei from both light-grown and dark-
grown leaves of pea, Gallagher and Ellis (1982) demonstrated that
light stimulated the rate of transcription of *rbc*S, rather than the
accumulation of mRNA being due to a decrease in its degradation.
Translational regulation of light-induced Rubisco expression has also
been detected (Berry et al., 1986). When dark-grown cotyledons of
amaranth *(Amaranthus hypochondriacus)* were transferred into light,
synthesis of S-subunit was initiated very rapidly and before any
increase in the levels of mRNA. On transfer to darkness, synthesis
was rapidly depressed in the absence of mRNA reduction. These results
suggested that regulation of *rbc*S at the translational level can
occur.

The ability to transform plants and transfer DNA into chromosomes
using *Agrobacterium* plasmids has enabled the DNA sequences
responsible for the photoregulation of *rbc*S to be analyzed.   A pea
*rbc*S gene, containing about 1 kb of 5' flanking sequence before the
transcription start site, was transformed into petunia (*Petunia
hybrida*) cells grown in culture (Broglie et al., 1984a). The pea *rbc*S
gene was expressed in the petunia cells under the transcriptional
control of its own promoter in a light-dependent fashion similar to
that observed in pea leaves. The mRNA was spliced accurately and was
polyadenylated at the correct site. Herrera-Estrella et al. (1984)
found that this light-dependent regulation of pea *rbc*S was located in
the 970bp 5' region upstream of the transcription start site. This
was achieved by using the 970bp sequence to obtain the light-
inducible regulation of a bacterial chloramphenicol acetyl
transferase gene in transformed tobacco (*Nicotiana tabacum*) tissue.
The soybean *rbc*S gene is also light regulated in tobacco (Facciotti
et al., 1985). Using deletion mutants, Morelli et al. (1985) showed
that a 33bp sequence from −35 to −2, that also includes the TATA
box, contained the necessary information for responsiveness to light.
It was also shown that the region from −1052 to −352bp was needed
for maximum transcription, but not for photoresponsiveness. In a
subsequent paper Timko et al. (1985) demonstrated that the region
from −973 to −90 not only contained the information required for
maximum expression, but also conferred photoregulation. The upstream
regulatory sequences functioned independently of orientation and

behaved as an enhancer–like element.

The expression of rbcS is tissue specific. Coruzzi et al. (1984) used Northern blot analysis of pea RNA to show that compared with green leaves, the level of rbcS mRNA is reduced to 50% in pericarps, 8% in petals and seeds, and 1-3% in etiolated leaves, stems and roots. A similar pattern of expression in leaves, stems and roots was observed when the pea rbcS gene was transformed in to petunia plants (Nagy et al., 1985). In maize the S-subunit is present in bundle sheath cells but not in mesophyll cells. This differential expression of rbcS has been shown to be controlled at the level of mRNA (Broglie et al., 1984b).

rbcL Genes.

Since the isolation and sequencing of the rbcL gene from maize (McIntosh et al., 1980), the corresponding genes from a variety of other plant species have been isolated and characterized (reviewed by Rochaix, 1985). The rbcL genes are present as a single copy on the circular chloroplast chromosome and they do not contain introns (see Koller et al., 1984 for an exception). Although the nuclear genome contains 6-12 copies of the rbcS gene, depending on the species, the number of copies of rbcL genes in the plant cell can approach 10,000 (Lamppa and Bendich, 1979), because of the polyploid nature of the chloroplast DNA and the number of chloroplast in the cell.

The chloroplast rbcL gene is transcribed and translated using prokaryotic-like sequences, RNA polymerase and ribosomes. From a comparison of the nucleotide sequences near chloroplast transcription initiation sites, and the use of deletion and base substitution mutants, it has been shown that the promoter structures and their interaction with the chloroplast RNA polymerase is similar to that observed in prokaryotes (Bradley and Gatenby, 1985; Gruissem and Zurawski, 1985). Many chloroplast genes also have prokaryotic 'Shine-Dalgarno' sequences which enable bacterial ribosomes to initiate translation at chloroplast ribosome binding sites (Gatenby et al., 1981).

When etiolated pea seedlings are illuminated there is a light-stimulated accumulation of rbcL mRNA (Smith and Ellis, 1981). It was also shown that the amount of Rubisco holoenzyme present was correlated with the amount of rbcL mRNA in the tissue. In contrast, using mustard seedlings, Link (1982) found that the rbcL mRNA is not increased drastically in light-grown as compared to dark-grown tissue. Similarly, Berry et al. (1986) found that with amaranth cotyledons a transfer from dark to light did not increase the level of rbcL mRNA, but caused an increased synthesis of L-subunit by a rapid adjustment at the level of translation. In barley, the synthesis of L-subunit is most active in young dark-grown seedlings, but synthesis of L-subunit and the rbcL transcript levels decrease in illuminated seedlings (Klein and Mullet, 1987). It is clear, from the examples cited here, that the mechanisms of regulation of rbcL can vary with different plant species. Differential expression of the *rbcL* gene in maize leaf cell types has also been reported (Link et al., 1978). The mRNA for L–subunit can be detected in bundle sheath cells, but is absent, or at low levels, in mesophyll cells. Light stimulates the accumulation of *rbcL* mRNA in bundle sheath cells but not in mesophyll cells, and the mRNA detected in mesophyll cells of dark–grown seedlings declines and disappears with illumination (Sheen and Bogorad, 1986).

PROPERTIES OF THE PURIFIED ENZYME

Carbamylation and Co–ordination.

Rubisco isolated from all photosynthetic organisms in simple buffer solutions at neutral pHs is an inactive protein. Supplement the buffer with bicarbonate as a source of CO_2, and a metal ion (Mg^{2+} is the physiological choice) and the enzyme responds by turning over bisphosphate substrate. The reason is that CO_2 acts not only as a substrate in these reactions (without CO_2 the enzyme is still incapable of oxygenase activity) but performs a dual function. In the presence of the metal ion it activates Rubisco. The reaction is the carbamylation of one specific lysine ϵ–amino group at the active site of the enzyme (Lorimer et al., 1976a). The location of the lysine has been determined in the L–subunit of spinach Rubisco at position 201 (Lorimer and Miziorko, 1980) and 191 in the *Rhodospirullum rubrum* enzyme (Donnelly et al., 1983). The kinetics of activation indicate that the slower process of carbamylation precedes the rapid reversible binding of the metal ion (Lorimer, 1981).

The appearance of enzyme activity may not reveal all the details involved in the formation of a competent active site. For example, the region that surrounds the important lysine residue is highly acidic which would stabilise the protonated form of the amino group. This, in turn, should inhibit or prevent carbamylation unless the metal also serves to neutralise these charges allowing carbamylation to proceed on the deprotonated amino group.

The replacement of a positive charge of a lysine with a negative charge and the proximity of adjacent acidic amino acids should generate an acceptable binding site for a Mg^{2+} ion. Unfortunately, the evidence that the metal does interact directly with either the oxygen or the nitrogen atoms of the carbamate group has been tantalisingly elusive. Both NMR(nuclear magnetic resonance) and EPR(electron paramagnetic resonance) techniques have been used to probe the metal binding site on the enzyme. In tight binding quaternary complexes of enzyme–CO_2–Me^{2+} and the intermediate analogue 2'–carboxyarabinitol bisphosphate(2CABP), specific distance information between isotopically enriched groups of the intermediate and the metal have been obtained. Certainly the form of the EPR spectra of Mn^{2+} or Cu^{2+} substituted complexes indicate that many of the ligands have oxygen character. However, no direct interaction of the carbamate with the metal has been detected. The data from NMR experiments (Pierce and Reddy, 1985) using the *R.rubrum* enzyme suggest that the metal is within 5-6Å of the carbamate.

The investigations of the interaction of substrates and intermediate analogues with the metal using paramagnetic probes has been more fruitful. Substitution of the Mg^{2+} with Mn^{2+} has shown that ^{13}C enriched CO_2 substrate has enhanced relaxation (Miziorko and Mildvan, 1974). With the same paramagnetic probe, ^{17}O enriched water and the oxygens, specifically the 2-hydroxyl and 2'carboxy groups of 2CABP, were shown to interact with the metal (Miziorko and Sealy, 1980 and 1984). Unfortunately, Mn^{2+} as a probe produces a complicated spectrum that requires Q band frequencies to collapse the 2nd order features. In so doing the ^{17}O hyperfine become intermixed with the anisotropic g values characteristic of these tightly bound complexes, complicating serious quantification. Substitution of the Mg^{2+} with Cu^{2+} has allowed quantification because the less complex nuclear interaction and with well spaced g_{II} region produces resolved ^{17}O superhyperfine. Styring and Branden (1985) and Branden et al.,(1984) were able to confirm the earlier Mn^{2+} work that in the 2CABP complexes one H_2O and the oxygens at the 2 and 2' positions of the inhibitor bind to the metal.

On the basis of these results, it is clear the metal must play a vital role in catalysis by interacting with specific groups on substrates and intermediates, but the function of the carbamate is still unknown. It might be that the carbamate contributes to the catalytic process by either modulating the acidic nature of the metal or being directly involved in the chemistry. It is unlikely that it is simply a means of ensuring the enzyme is responsive to environmental concentrations of the gas and thus a regulatory feature, because the activating CO_2 on the enzyme *in vivo* seems particularly reluctant to turnover (Gutteridge and Parry, unpublished results).

The effect of inhibitors

In the absence of CO_2 and Mg^{2+} nearly all species of Rubisco show some affinity for the bisphosphate substrate or its analogues. Obviously groups at the active site that interact with the phosphates of the substrate or contact the other atoms of the pentulose backbone still retain the ability to do so on the non-active enzyme. An interesting exception is the enzyme purified from wheat. At normal temperatures this Rubisco does not activate rapidly with the addition of CO_2 and Mg^{2+}, it takes some hours for only about 50% to become active (Machler et al., 1979). With moderate heat however, the enzyme can be activated almost to optimum levels within 30 minutes. Compared to other purified carboxylases, e.g. spinach, this is an order of magnitude slower (Gutteridge et al., 1982). The reason is that the wheat enzyme is unable to bind bisphosphates with any affinity in the non-active state (Parry et al., 1985). Presumably the time and temperature required to generate an active carboxylase is associated with the formation of a substrate binding site. It is not clear whether these preparations still retain the ability to form a metal carbamate ternary complex with CO_2 and Mg^{2+}. Applying the same purification procedures to other sources of higher plant enzyme also showed that a significant proportion could be isolated as slow activating species. An apparent explanation for the differences between the standard method of purifying the wheat enzyme compared with other methods that generate mainly fast activating carboxylases, might be the presence of low molecular weight effectors, e.g. phosphate or sulphate in final preparations. Addition of these oxyanions to the normally slow form of the wheat Rubisco enhances considerably the rate and extent of activation (Schmidt et al., 1984a,b).

Changes in conformation associated with activation have been reported for other species of Rubisco. The reactivity of specific amino acids is different in the activated enzyme compared with the non active form, as are the binding constants for ribulose-P_2 and some analogues. Although these may be a consequence of the formation of the metal site, physical techniques such as circular dichroism indicate changes in protein conformation (Grebanier et al., 1978, Schmidt et al., 1984a).

The ability of bisphosphates and other ribulose-P_2 analogues to bind to the active site, with or without activating cofactors, is of relevance to the conditions in vivo of higher plants.   Many of these favour the active form of the enzyme, i.e. the ternary complex of enzyme-CO_2-Mg^{2+}, and thus stabilise this species. The concentrations of the cofactors required to achieve full activity is therefore significantly reduced (Lorimer and Badger, 1981, McCurry et al., 1981, and Gutteridge et al., 1982). However, the concentration range over which these analogues are effective ensure that they also inhibit catalysis severely. One significant exception is 6-phospho gluconate where complete activation with suboptimal cofactor concentrations using the spinach enzyme can be mediated with only sub-stoichiometric concentrations of the inhibitor (Jordan et al., 1985). This would clearly be the desired mode of action of an effector in vivo, i.e. the ability to generate the activated form of the enzyme in physiological concentrations of CO_2 and Mg^{2+}, but at the same time never in amounts to inhibit enzyme activity. The mechanism of this 6-phosphogluconate mediated activation is not known, except it is assumed that all eight sites of the enzyme are carbamylated and bind Mg^{2+} . It might be possible that only one of the active sites requires to bind the effector but by co-operative means allows the other seven to activate with CO_2 and Mg^{2+}.

Whether 6-phosphogluconate acts this way in vivo is not known. An analogue that might function in a similar way that resembles 6-phosphogluconate is the natural inhibitor 2'-carboxy arabinitol-1-phosphate, normally only detected in plants during or emerging from the dark. However, if only sub-stoichiometric concentrations of the effector are required to generate full activity, then it is likely these would not have been detected in the light.

One further aspect of co-operativity associated with activation has emerged with the wheat enzyme. Effector mediated activation was also accompanied by a doubling of the specific activity with suboptimal, i.e. physiological concentrations of the cofactors. An attempt was made to determine the numbers of sites that had formed the active ternary complex. Unexpectedly, there was no difference between the numbers of sites carbamylated with or without effector (Parry et al., 1985). It would be assumed that with effector all eight sites would be carbamylated and thus catalytically competent especially since full activity was not only achieved but exceeded. Without effector only some 25% of sites should bind CO_2 and Mg^{2+}. Increasing the CO_2 and Mg^{2+} to concentrations normally required by the purified carboxylase to carbamylate all eight sites, induced a negative response such that the stimulation did not increase as more sites were activated. The contradiction with previous quantification (see McCurry et al., 1981) can be partially rationalised when it is realised that the concentrations of the effector, in this case phoshate, were lower than that required to inhibit the wheat enzyme significantly. That is, the conditions are those demonstrated for 6-phosphogluconate and spinach enzyme, namely sub-stoichiometric in terms of the numbers of sites occupied by the effector.

The Role of the Metal.

The details of the specific role of the essential metal are only now becoming clear. Broadly, these might be classed as twofold. The first is in activation, assuming a metal is required at all for carbamylation or at the very least to stabilise the carbamate. For this particular function almost any metal seems acceptable . The second is to mediate in catalysis where there is a specific requirement for Mg^{2+}. There are at least four functions that might be tentatively ascribed to the metal, although there is some spectroscopic and other evidence to support some of these roles. The particular steps in catalysis at which the Mg^{2+} ion, or metal substitutes that support activity, might be described as:

a) directing the binding of the bisphosphate substrate and presumably analogues, the 'right way round' at the active site. In the case of ribulose-P_2 and intermediates, this means ensuring the correct stereochemical orientation.
b) Initiating the catalytic cycle in concert with at least one other group at the active site, by polarising the C2 carbonyl of ribulose-P_2 and rendering the C3 proton more acidic, thereby generating the all important enediol.
c) Potentially activating the gaseous substrates (in the case of transition metal ions) as well as directing their approach to the enediol, and then stabilising the intermediates thus formed.
d) Activating H_2O for the hydrolysis or hydration of these intermediates to generate products.

Substitution of the Mg^{2+} by other metals has proved useful for investigating the active site, but few allow the enzyme to retain functionality. Apart from Ca^{2+} (Parry et al., 1983), those that support some function interfere in the rates of oxyge nation relative

109

to carboxylation and substrate binding. With Mn^{2+}, the enzyme is still capable of carboxylation and oxygenation of ribulose-P_2 (Wildner, and Henkel 1978). However, the chemistry is disrupted because the partitioning now favours oxygenation. Co^{2+} substituted in the *R.rubrum* carboxylase (Robison et al., 1979) only catalyses the oxygenase reaction as does Cu^{2+} substituted spinach Rubisco (Branden et al., 1984).

THE REACTIONS CATALYSED BY RUBISCO

Partitioning of Ribulose-P_2

Understanding the details of the reactions between Rubisco and substrates has been aided by the ability to obtain large, often gram, quantities of the enzyme in a highly purified form relatively quickly. This has meant that the reaction steps leading from substrate to product can be investigated in some detail by direct spectroscopic observation or, alternatively, trapping and identifying the individual reaction intermediates. The reactions catalysed by Rubisco are shown in Figure 4. The source and fate of the various atoms involved in the formation of product from substrate have been traced for carboxylation and oxygenation. Only some of the intermediates shown in the Figure have been identified directly.

Intermediates in the reactions

Although it was suspected that an enediol intermediate of ribulose-P_2 was the reactive form of the substrate (Calvin, 1956), only recently has evidence for this intermediate been obtained. The carboxylase reaction of the *R.rubrum* enzyme was carried out in ^3H_2O or with C3 tritiated ribulose-P_2. In both cases the C3 group exchanged with the proton or tritium of the medium, clear evidence for the participation of an enediol during turnover (Sue and Knowles,1983; Saver and Knowles, 1983).

A different approach used NMR techniques, to follow the fate of the C3 proton of the bisphosphate, in experiments designed to determine whether Rubisco catalyses the formation of the enediol without the other substrates CO_2 and O_2 (Gutteridge et al., 1984a). A major problem is that the CO_2 molecule required for activation often interferes by acting as a substrate. This was avoided by incubating the nonactive enzyme with sub-optimal CO_2 in the presence of Pi. In these conditions, the enzyme requires much less CO_2 to be fully competent. Thus the effect of activated enzyme on the ribulose-P_2 can be investigated in the presence of CO_2 at concentrations below its Km. These experiments showed that the C3 proton of ribulose-P_2 was exchanged out of the substrate and replaced with deuterium from the medium. The constancy of the C1 proton resonances and the absence of product features indicated there was little conversion of the substrate to product. The rate of exchange of the C3 proton was calculated to be just within the expected rate for overall substrate turnover during carboxylation. Increasing the CO_2 to substrate concentrations showed that it effectively competed against the deuterium for the enediol and the loss of the C3 proton was this time concomitant with the formation of products. Thus, it was concluded that the enzyme could bind the bisphosphate substrate and catalyse the exchange of the C3 proton and hence generate the enediol intermediate without other substrates. The particular hypothesis, subsequently proposed and based on these results, was that the enzyme may simply generate the enediol form of the bisphosphate and then present the appropriate 'face' of the intermediate for the two gaseous substrates to compete over. There was potentially no requirement for formal binding sites for the second substrates.

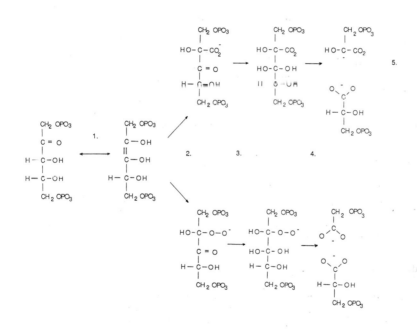

Figure 4. The reactions catalysed by Rubisco. 1) Enediol formation by abstraction of the proton at C3 of ribulose-P_2. 2) Carboxylation or oxygenation by electrophilic addition of gaseous CO_2 or oxygen. 3) Hydration of the C3 carbonyl by H_2O. 4) Cleavage of the intermediate generating product 3P-glycerate from the bottom of the bisphosphate and 2P-glycolate from the top after oxygenation. Or 5) an aci-acid derivative that is protonated on the enzyme to produce the second molecule of 3 P-glycerate of carboxylation.

These experiments were applied to the bacterial enzyme from *R.rubrum* (Pierce et al., 1986) which had the advantage that CO_2 concentrations well below the substrate Km could be used. This meant that the partitioning of the enediol between substrate and subsequent product formation could be studied in various concentrations of CO_2. The rate of enediol formation could then be determined by extrapolation to zero CO_2 concentrations. Again the rates of formation were within those expected for the overall catalytic reaction. A careful study was also undertaken to detect any binding of the second substrates CO_2 or O_2 to the protein. These were unsuccessful leading the authors to conclude that no Michaelis complex is formed with these substrates. The result of these two NMR studies indicated that the mutual inhibition of the gaseous substrates must arise from their competition for the enediol intermediate. Furthermore, both the wheat enzyme and the *R. rubrum* enzymes must have similar mechanism, at least up to this point in the reactions. What then is the origin of the factor of ten difference in specificity?

Species variation in specificity

Assuming that the inhibition of carboxylation of ribulose-P_2 by oxygen is due simply to the oxygenation of the substrate and,likewise, the inhibition of oxygenation by CO_2 is due to the carboxylation reaction, then the ratio of this mutual competition is a measure of the partitioning of the substrate between the two reactions. The steady state equations for this ratio of competitive inhibition simplify readily to a linear relationship that describes the initial rates of carboxylation and oxygenation in terms of the concentrations of the respective gases (Laing et al., 1974), i.e.

$$vc/vo = VcKo/VoKc \cdot CO_2/O_2$$

where the ratio of the kinetic constants Vmax for carboxylation (Vc) times Km oxygenation (Ko) over the Vmax oxygenation (Vo) times Km for carboxylation (Kc) defines τ the specificity factor. A plot of vc/vo against CO_2/O_2 produces a straight line , the slope of which is τ. The specificity factor for a number of carboxylases from a diverse range of photosynthetic organisms shows that this parameter can differ by almost an order of magnitude (Jordan and Ogren, 1981 and 1983). Even between closely related higher plants with presumably the most highly evolved enzyme, show small but significant variations (Gutteridge et al., 1986a). An essential consideration if it is hoped that the enzyme as it presently exists has not achieved the maximum possible carboxylation efficiency.

These data immediately raise a further question. How then do different species function with different specificity factors? The difference between form I and II enzymes might be rationalised on the basis of structure. However, even though the form II species are devoid of S-subunits and function without requiring a complicated quaternary structure, they must have evolved from the same primaeval ancestor as the form I species. The acquisition of an L_8S_8 functional unit apparently endowed the enzyme with a significant enhancement of carboxylation specificity, but did not exhaust the potential for further gain. The specificity of, for example, a cyanobacterial L_8S_8 is only half of the value achieved by many higher plants. It should be emphasised that the cyanobacterial enzyme turns over substrate five times the rate of spinach Rubisco, suggesting that there may be some price to pay for the higher efficiency; but this is pure conjecture, just possibly overdoing the role of devil's advocate.

No binding site for CO_2 or O_2, presumably means no discrimination between the gases and that either are equally likely to reach the enediol. It is difficult to envisage an approach path for carbon dioxide that could prevent the passage of the smaller molecular oxygen. Two other possibilities might be operating. Firstly, that different species of the enzyme generate different amounts of the enediol, thus allowing the less reactive substrate to see more of the intermediate during each cycle. Alternatively, the intermediate generated immediately after reaction with CO_2 or O_2 is more or less stabilised. Thus in the case of the superior carboxylases the 2'-carboxy 3-keto arabinitol bisphosphate (2C3KABP) is better stabilised than in the inferior species. Tentative evidence supporting these proposals might be obtained from the relative binding constants for the substrates and transition state analogues. It is generally the case that those enzymes exhibiting the highest affinity for substrate CO_2 (as reflected in the Km) and 2CABP are the better carboxylases. One other parameter that varies between species

---

Table 1. A comparison of the structural forms of Rubisco, the efficiency of carboxylation and the specificity factors.

Source	Form	Vc/Kc	τ
		$(M^{-1}.s^{-1})$	
Higher Plants			
Wheat	I	20×10^4	100
Tobacco	I	18×10^4	90
Bacteria			
Synechococcus	I	6×10^4	52
R. rubrum	II	7×10^4	15

and modulates specificity is the kcat for carboxylation. Table 1 shows that the cyanobacterial carboxylases have been able to increase their efficiency of carboxylation relative to oxygenation and still retain a low affinity for CO_2. The reasons for the differences between the enzyme species is of course difficult to explain without details of the contribution of reaction intermediate structure and groups at the active site. It is possible that going from a planar type conformation, in the case of the enediol intermediate, to a tetrahedral geometry, as the enediol becomes substituted at the 2' position (see fig. 5), is favoured in some species of the enzyme over others. Alternatively, there may be groups of the protein structure that are well placed, in say the better carboxylases, to offset the charge density that must occur on the newly acquired 2'-carboxyl group. Just preceding this stage of the reaction there is the proton abstraction from C3 of the bisphosphate to some group at the active site and it might be this protonation/exchange that dictates, a) the number of active sites that carry the enediol form of the bisphos-phate and b) the stability of the six carbon intermediate, 2'-carboxy 3-keto arabinitol bisphosphate (2C3KABP).

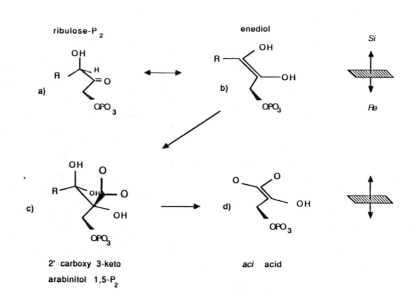

Figure 5. The stereochemical course of the carboxylation of ribulose-P_2 at the active site of Rubisco. The Re face of the bisphosphate substrate and intermediates is unavailable for reaction within the site and the approach of the second substrate CO_2 to form (c) and protons to generate product from (d) is toward the Si face.

One limitation is that the rate-limiting step of the reaction is not well defined. There is a point of no return somewhere in the reaction process. This appears to be the steps after the formation of the six carbon intermediate. Addition of this intermediate to the activated enzyme goes only forward to product, there is no evidence that the reverse reaction to substrates occurs (Pierce et al., 1986). This implies that once the intermediate has fully formed then the reaction is all downhill. Therefore the intervention of the protein structure that gives rise to discrimination between the gaseous substrates, and thus partitioning, must happen at an earlier step of the reactions. One flaw in this reasoning is the assumption that the form of the enzyme accepting 2C3KABP when it is supplied in solution, is the same as the species that exists after generation of the intermediate at the active site. During the initial exposure of the activated Rubisco to the intermediate this cannot be the case, because the enzyme (or the intermediate) has to undergo some conformational change, as observed with 2CABP (Pierce et al.,1980), to achieve tight binding at the active site. At present the influence of these changes on the rate of product formation, or whether some adjustment to our vague concepts of the mechanism of partitioning is required, has yet to be determined.

Stereochemical considerations.

D-ribulose 1,5-bisphosphate is carboxylated to form two molecules of 3-phospho D glycerate. Figure 5 depicts this process in stereochemical terms showing firstly, that addition, hydration and protonation can all be directed on one face of the bisphosphate molecule. The movement of protons from substrate to enzyme would generate active site conformations appropriate to the intermediate to be stabilised. Thus when deprotonated, tetrahedral conformations around C3 might be favoured, whereas protonation of enzyme groups may stabilise the planar intermediates as in Figure 5b and d. The formation of only D-isomeric product requires at some stage of the reaction inversion round C2 of the molecule, i.e. the 3PGA that originates from the lower three carbons of the bisphosphate retains the original geometry, but that arising from the top two must invert. The stereochemistry around these two carbons is known up to the hydrolysis step that splits the 2C3KABP. The use of borohydride to reduce the carbonyl groups that sequentially occupy the C2 and C3 positions of the molecule have not only enlightened aspects of the stereochemistry of the reactions but also highlighted the protein contribution to these reactions.

The bisphosphate substrate, when bound at the active site, is susceptible to reduction by borohydride. Analysis of the products of this reaction, as shown in Figure 6, after enzyme bound substrate is quenched into solutions containing the reducing agent, indicates that, as the ratio of enzyme to substrate increases, the ratio of the arabinitol to ribitol stereoisomers also increases (Lorimer and Gutteridge, unpublished results and Lorimer et al., 1986). Extrapolation to infinite concentrations of the enzyme, i.e. when all the bisphosphate would be bound, indicates that only the arabinitol isomer would be formed. Thus at the enzyme active site only one side

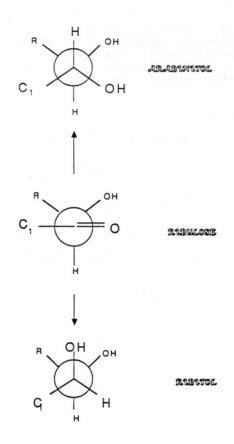

*Figure 6. The stereoisomeric products of borohydride reduction of
enzyme bound ribulose-P_2. In solution the ratio of ribitol to
arabinitol is slightly more than unity. With increasing enzyme and
thus binding of the substrate, only the arabinitol isomer is formed.*

or 'face' of the bisphosphate is available for reaction, not just
toward borohydride but also the second substrates CO_2 or O_2. The
Figure shows that it is the *Si* face that is oriented for reaction.
When the same reaction is done without the presence of the activating
cofactors, then the discrimination toward arabinitol is not as great.
This suggests that the metal ion at the active site acts to orient
the bisphosphate correctly for reaction.

It could be argued that, since the enzyme has to generate an
enediol before reaction with CO_2 or O_2, the stereochemical
orientation might change. However, reduction of the newly formed
2C3KABP (see Figure 7.) with borohydride again generates only the
arabinitol stereoisomer (Schloss and Lorimer, 1982). Thus the

stereochemistry is maintained to this point in the reactions. One difference between the reducibility of the C2 carbonyl of ribulose-P_2 and C3 carbonyl of the 6 carbon intermediate with borohydride, is that the latter has to be released from the active site before the reduction, presumably because the C3 carbonyl is not accessible for reaction.

The scheme in Figure 5 shows at which point in the chemistry the inversion might occur. Just after the hydrolysis and release of the lower 3PGA, the top part of the molecule becomes an *aci*-acid derivative that must remain at the active site to become protonated. If the pattern of approach toward enzyme bound intermediates is maintained at this stage, i.e. only *Si* faces are available for reaction, then the correct orientation of the product is maintained and only D-3-PGA is generated.

There is as yet, no information whether the enediol generated during turnover is *cis* or *trans*. The position and hence identity of the group at the active site that abstracts the proton from the C3 position of the bisphosphate will depend on which isomer is formed.

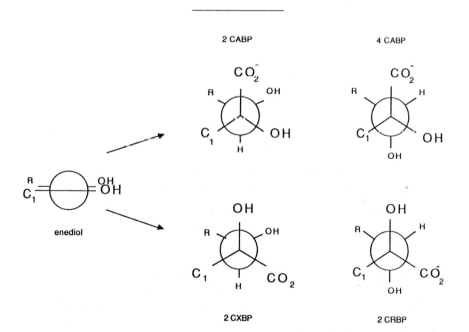

Figure 7. The four isomers expected without stereochemical direction of the reactions of Rubisco. The 6 carbon intermediate released from the enzyme and reduced with borohydride could potentially be 2'-carboxy arabinitol (2CABP), ribitol (2CRBP), xylitol (2CXBP) or 4 carboxy arabinitol (4CABP) bisphosphates, but only 2CABP and 4CABP are detected as products of the reduction.

Oxygenation

The finer details of the immediate fate of the enediol intermediate that sees an oxygen molecule at the active site of Rubisco
rather than CO_2 is not known. In fact the reaction itself is rather
unique when compared with other oxygenase enzymes that catalyse the
insertion of an atom of molecular O_2 into a substrate and reduce the
other atom to H_2O. In nearly every case a transition metal ion is
required to activate the unreactive oxygen molecule. Careful attempts
have been made to quantify the amount of transition metal ions that
accompany purification of the carboxylase, at least those suspected
of mediating oxygenase type activities (Schmidt et al., 1984). All
were significantly below protomer concentrations. Thus activated
Rubisco requires only the enediol and Mg^{2+} to catalyse the reaction.
Substitution of the Mg^{2+} by some transition metals that catalyse
ribulose-P_2 enolisation also favour the oxygenase reaction.
Presumably the availability of unfilled d orbitals on the metal aids
the triplet – singlet transitions required for activating O_2.

Chemically the reaction is not unusual because enediol forms of
ketose sugars generated at high pHs are susceptible to autoxidation.
Very often the intermediates generated in solution have arisen from
peroxy and hydroperoxy derivatives that degrade to various products
via epoxy mechanisms. In the case of the reaction catalysed by
Rubisco this does not occur and products arise as a result of the
insertion (a step common with carboxylation) and then elimination of
H_2O, leaving an atom of oxygen in the 2-phosphoglycolate product.
These products are most easily explained if it is assumed that a
2'-hydroperoxy derivative of the enediol is first generated with
molecular O_2 (Lorimer,1981). At the enzyme active site, the presence
of a metal and the inaccessibility of the C3 centre would prevent the
same breakdown observed in solution systems ensuring only 2-phospho-
glycolate was released. There is some evidence for a putative
hydroperoxy intermediate using EPR techniques and Rubisco activated
with Cu^{2+} (Branden et al., 1984). The enzyme was allowed to bind
ribulose-P_2 in the presence of ^{17}O enriched O_2 generating a small
amount of 2-phosphoglycolate. The resulting Cu^{2+} spectrum showed
broadening due to the interaction of an atom of ^{17}O with the metal
that was depleted with the formation of product.

THE STRUCTURE OF THE PROTEIN

Forms of Rubisco

There are two distinct forms of Rubisco that can be identified
from their quaternary structure. Form I is the most commonly
occurring found in photosynthetic bacteria and higher plants. It has
a complicated hexadecameric structure (L_8S_8) and an M_r of about

550,000 daltons. Form II Rubisco is the simplest Rubisco, composed of only L-subunits, usually as a dimer, and present only amongst some photosynthetic bacteria. Apart from the subunit composition, differences also extend to their specificity factors. The form II enzymes are the poorest carboxylases, but also have some of the highest turnover rates. In the case of the L_2 from *R. rubrum*, the specificity favours the oxygenase reaction so much over carboxylation, that the organism is unable to fix enough CO_2 if grown in air concentrations of the gases. The organism is therefore limited to anaerobic growth or environments with elevated CO_2 concentrations.

There is very low homology between the L-subunits of the form I and II enzymes (Nargang et al., 1984), but there are clearly regions and individual amino acids that are highly conserved. A number of these amino acids have been identified as part of the active site using site directed affinity labels and mutagenesis. Both forms have produced crystals of enough quality to generate X-ray diffraction patterns, the L_2 recombinant form of *R.rubrum* (Pierce and Gutteridge, 1985) producing the highest resolution and a structural model solved to 2.9Å (Schneider et al., 1986). The data for the form I enzyme (Branden et al.,1986, Chapman et al., 1986, Bowien et al., 1980) from a number of sources and laboratories is rapidly approaching a similar level of resolution, particularly for the tobacco and spinach Rubisco. From a preliminary model of the tobacco enzyme compared with *R.rubrum* enzyme, there may only be some 25% homology at the primary level but a very close relationship at the tertiary and quaternary levels.

The form II Rubisco

The simplest Rubisco, the L_2 dimer from *R.rubrum* has been the first of the carboxylases whose structure is now solved to a resolution of 2.9Å (Schneider et al., 1986). This gives enough structural detail to be able to chain trace and recognise the main secondary structural motifs. Each L-subunit is composed of two domains. The N-terminal domain accounts for about the first one third of the amino acids in the primary sequence, the remaining two thirds at the C-terminal end of the subunit forming a familiar barrel tertiary conformation. The particular organisation of the secondary elements of the barrel is similar to that found in glycolate oxidase (Lindquist et al., 1985) and triose phosphate isomerase (Banner et al., 1975) with a central ring of 8 parallel sheets surrounded by, and alternating with, 8 helices. The connection between each sheet and helix is by a loop of amino acids that extend above and over the 'top' surface of the barrel. The active site is located on this particular surface and the essential amino acids involved in catalysis are located in these loops.

The interactions between the N-terminal domain and the barrel of one subunit are small, the connection being through two small helices, one of which closes off the bottom 'N-terminal' end of the barrel. However the interactions between the N-terminal domain of one subunit with the barrel of the second subunit are extensive. From both the crystallographic studies and now mutagenesis (Larimer and

Hartman, 1987) it is clear that one particular region of the N–domain contributes amino acids essential for catalysis in combination with residues of the barrel of the opposite subunit. Thus the active site residues are contributed by both subunits and Figure 8 is an attempt to display this interaction.

Form I Rubisco

Just recently, with the information available from the structure of the *R.rubrum* enzyme, much of the chain for the more complicated form I Rubisco has been followed (Chapman et al., 1986 and 1987). The barrel structure of the large subunits was best fitted using the barrel of pyruvate kinase, and then the *R. rubrum* N-terminal domain was used to follow much of the chain of the remaining sequence. The remaining density due to the small subunits positioned them at the top and bottom of the L_8S_8 structure with extensive contact over and between the L-subunits penetrating into the centre of the aggregate. The centre of the molecule down the four–fold axis is dominated by a 20 – 25Å channel accessible to solvent molecules, a feature very prominent in electron micrographs.

The refinement of structural models of both forms of Rubisco is essential so that a comparison can be made of Rubisco species with very different specificities. Unfortunately, the structures that are being refined presently are of the non–activated state of the enzyme

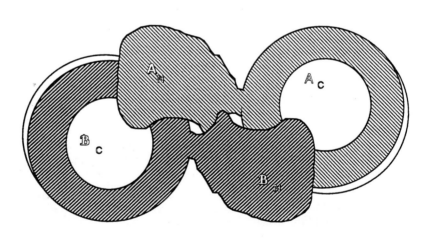

Figure 8. An illustration of the dimeric structure of form II Rubisco from Rhodospirillum rubrum *based on the crystallographic data. The active site is shared between the surface of the barrels (Ac and Bc) and an N–terminal conserved octapeptide of the opposite subunit (Bn and An, respectively).*

that, around the active site, may bear only a superficial resemblance to the functional protein. Ideally, it is the structure of the stable quaternary complex of the activated enzyme with 2CABP that will be most revealing. Nevertheless, the combination of information from affinity labelling, sequence homology and the three dimensional structure has stimulated the application of mutagenic techniques to identify the function of active site amino acids.

Regions of the Structure that form the Active Site.

Those regions of high homology that are part of the L–subunits of form I and II enzymes are shown in Figure 9. The amino acids that have been labelled covalently with active site affinity reagents (see Hartman et al., 1986 and references cited therein) are marked within these regions of homology. One sequence from form I enzyme included in the Figure has no counterpart in the form II version. This is from the N–terminus of the L–subunit of the L_8S_8 enzyme that can be readily removed by limited proteolysis, but which is accompanied by a significant loss of activity (Gutteridge et al., 1985). The specific function of any of the residues in this N–terminus region is unknown, particularly as the sequence from different species is of relatively low homology. Therefore, rather than contribute any single amino acid

```
1              MDQSSRYVN
1        MSPQTETKASVEFKAGVKDYK
1         MPKTQSAAGYKAGVKDYK


               *                          *
40     ATAAHFAAESSTGT          162    GTIIKPKLGL
53     AGAA VAAESSTGT          171    GCTIKPKLGL

         *  *  **                          *
186    GG DFIKNDE              284    LHYHRAG
195    GGLDFTKDDE              291    LHIHRAM

           *
328    GKMEGE
333    GKLEGE
```

Figure 9. Regions of the primary sequence of the higher plant L–subunit and R. rubrum with the highest homology. The amino acids that are implicated in catalysis by in vitro mutagenesis are so marked. The low homology around the N–terminus of the L–subunit is illustrated by comparing the R. rubrum sequence with that of spinach and Synechoccus.

critical for functionality, it maybe a region of the protein essential for maintaining active site integrity, albeit some distance from the site. The absence of an equivalent region in the form II enzyme tentatively suggests that some of the amino acids might be associated with the interaction of the S–subunit and the L–subunit in the form I enzyme. The low homology would then not be critical because each sequence would be matched to the particular S–subunit for that species of enzyme, i.e. tailored to the lower homology that exists amongst the different S–subunits (Nagy et al., 1986). This would also explain the similar characteristics of the proteolysed L_8S_8 and the S–subunit-less L_8 core of, for example $Synechococcus$ Rubisco (Andrews and Ballment., 1984). Both still retain the ability to carbamylate and bind 2CABP. Thus removal of the S–subunits from $Synechococcus$ Rubisco might disrupt the structure at the N–terminus of the L–subunit with subsequent loss of activity.

The function of specific amino acids

Although active site directed affinity labels have located a number of amino acids that must reside close to the active site, their specific function is unknown. Only one amino acid has known function and that is Lys191 (Lys201 spn)[*] which is carbamylated during activation. Nevertheless, the details of the reactions catalysed by Rubisco suggest the involvement of two or more acid base groups in proton movements and some of the affinity labels have pinpointed prime candidates for these roles. Lys166 buried within a highly conserved septapeptide (Lys175 spn) has been identified after reaction with pyridoxal phosphate and bromobisphosphate analogues (Hartman et al., 1986). It is a residue that also exhibits a lower pKa (7.9) than might be expected for a lysine residue. A study of the pH dependency of the catalytic reactions requires the involvement of an amino acid with a pKa of this order (Van Dyk and Schloss, 1986). One other lysine also identified was Lys334 in the spinach enzyme. These are clearly prime candidates for site specific mutagenesis as are any acid base groups located in the conserved sequences shown in Figure 8.

Mutagenesis

There are three main directions that in $vitro$ mutagenesis studies are developing. The first is to change individual amino acids that are essential for catalysis and by analysis of the mutant enzymes determine the function of that group. The second strategy is to attempt to identify those regions of the primary sequence that determine the specificity of Rubisco by replacing them with the same regions from superior carboxylases generating chimaeric genes and thus hybrid enzyme. A third is to chemically mutagenise the genes or

---

[*] *Residues are numbered according to their position in the R. rubrum enzyme sequence the residue number in spinach is denoted by spn after the number*

selected regions of the genes of Rubisco and then select for a superior enzyme, using an organism that is devoid of its own carboxylase and sensitive to the conditions of growth.

Site-specific mutagenesis

Most of the *in vitro* mutagenesis studies have involved the form II Rubisco from *R.rubrum*. The first expression vectors containing the gene for the enzyme were low copy number plasmids (Somerville and Somerville., 1984 and Nargang et al., 1984) that were unstable and thus produced inconsistent amounts of the protein. Once this had been recognised, then the growth selection strategy could be arranged to produce the best levels of the recombinant enzyme (Pierce and Gutteridge., 1985). In fact under the best conditions up to 25% of the *E. coli* host cell contents were the *R. rubrum* carboxylase. The enzyme was extended by some 25 amino acids from the N terminus of β-galactosidase to which it was fused. Nevertheless, the characistics of the enzyme were identical to the authentic Rubisco. This was the expression vector first chosen for site specific mutagenesis. When the first changes were made with this clone there was no advantage of a crystal structure to limit the potentially large number of mutations based on the DNA comparisons. However, knowing that the essential Lys191, once carbamylated, might be involved in metal binding, and that this group was surrounded by a number of highly conserved acidic amino acids that may also contribute to metal binding, it was proposed that one should be changed. Asp188 is one of these amino acids which could be altered to Glu188 with a single base change in the DNA that also introduced a new restriction site into the *rbcL* gene and readily allowed easy identification of the mutant (Gutteridge et al., 1984b). Spectroscopy indicated that the modified sequence perturbed the metal binding enough to reduce the activity of the enzyme by approximately 30%. The specificity of the protein was unchanged .

The crystal structure now indicates the reason for this small change to the enzyme. Asp188 is positioned at the wrong end of the barrel to be directly involved in catalysis or metal binding. However, it is conserved in all species of the enzyme sequenced to date and also lies in close proximity to another small highly conserved region of the primary sequence of the L-subunit. Thus, this is a second region of the protein divorced from the active site that has some, as yet undefined function. Furthermore, it also illustrates that a small subtle mutation outside the active site can disturb the geometry of those groups involved in catalysis enough to modify the activity.

The other mutations that have been introduced, have been designed to render the enzyme inactive and thus at least define their essential character. Analysis of the partial reactions, carbamylation and 2CABP binding are then used to characterise the mutants more fully and to assign function to these groups. The status of the metal site can also be investigated by spectroscopy.

The form of the reaction catalysed by the enzyme predicts that at least two groups should be present at the active site that are involved in proton movements to and from reaction intermediates. One of prime importance is the group that acts in concert with the metal ion to labilise the C3 proton and initiate the formation of the enediol. Crystallography is unable to help in identifying these groups directly because the structure is of the non-activated enzyme. One clear candidate is Lys166 that exhibits an unusually low pKa of 7.9. The amino acid is part of a loop region around the barrel and in close proximity to the substrate bound at the active site and site specific changes at this position prevent catalysis. The mutations so far constructed are consistent with Lys166 having a function as a proton abstraction group (Larimer et al., 1985).

Mutagenesis studies of the form I enzyme have been delayed because of the problems associated with assembling recombinant L- and S- subunits into a functional enzyme (see this review). Fortunately, the genes for the two subunits of cyanobacterium *Synechococcus* PCC6301 are located close together under the control of the same promoter (Shinozaki and Sigura., 1985). The S-subunit gene was located some 90 bases downstream of the L-subunit gene, both transcribed as a discistronic message.

Insertion of a 2.2kb *Pst*I fragment of PCC6301 in a pUC9 vector produced acceptable amounts of active Rubisco under the control of the *lac* promoter, however some 50% of the L-subunit was obtained as a fusion protein with 34 extra amino acids extending from the N-terminus because the insert was 'in-frame' with β–galactosidase. Transfer of the *Pst*I fragment to pUC18 and thus out of frame produced the correct construct with high expression and the desired characteristics of this species of protein (Phillips et al., 1987). The specific activity of the recombinant product was somewhat lower than authentic enzyme although the quantification of the L- and S-subunits (Andrews and Ballment, 1984) suggested that most of the enzyme must be present as L_8S_8. The specificity factor at 45 was that expected for cyanobacterial Rubisco. Thus these constructs are suitable for manipulation using *in vitro* methods.

The major approach has been to replace parts of the sequences of the L–subunit with the same region of a higher plant enzyme that has a superior specificity factor. The methods so far have simply exploited common restriction sites in the genes. The homology between the L-subunit of cyanobacteria and higher plants is quite low particularly at the level of DNA. Transfer of stretches of the subunit therefore necessarily involve large replacements with the introduction of many mutations in one chimaeric construction. This has ensured that many of the disadvantages of higher plant constructs have also been transferred. This, in many cases has resulted in low expression and insoluble protein material with little or no assembly. Thus it is clear that determining regions of the L–subunit involved in dictating the specificity factor will require a more systematic approach.

One particular series of changes investigated the importance of the N-terminal region of the L-subunit on activity and specificity.

The first 15 amino acids of the N-terminus encompass a region of very low homology amongst form I Rubisco species. Thus, assuming differences in specificity may arise from such non-conserved sequences, the N-terminus of *Synechococcus* L-subunit was replaced with the same region from wheat L-subunit, or with a number of amino acids from β–galactosidase. The wheat – *Synechococcus* chimaera was still active but retained a *Synechococcus* like specificity. The β–galactosidase – *Synechococcus* chimaera was of low activity and similar to the species generated by limited proteolysis.

REALITY

Finally, has the molecular analysis of the purified enzyme told us anything about the enzyme as it exists in photosynthetic organisms and more particularly in the chloroplast? Clearly, the preceding sections have indicated that the analysis of the enzyme is moving with some pace. However, extrapolation to the situation that exists in the stroma highlights some glaring anomalies. A major one is the discrepancy between the higher concentrations of CO_2 required to activate the purified enzyme compared with those likely to exist in the chloroplast. The difference is at least a factor of ten. A second problem is that during turnover of ribulose-P_2, either during oxygenation or carboxylation, there is a relatively rapid loss of enzyme activity to a significantly lower rate than initial turnover. This loss of activity is not associated with a decrease in the activated state of the enzyme (Lorimer and Pierce, unpublished work) or loss of the S subunit (Andrews, personal communication), rather it is a consequence of the chemistry during substrate consumption. Inhibitor studies indicate that the effect is exacerbated by SO_3^{2-} (Parry and Gutteridge, 1983) but somewhat alleviated by H_2O_2 (Lorimer et al., 1973). In both cases the response of the enzyme is superimposed on, and complicated by, the normal reversible inhibition of activity caused by both compounds. Presumably the enzyme in the stroma is equally susceptible to this form of inhibition but has been shown to recover relatively quickly (Perchorowitz and Jensen., 1984). The purified enzyme also recovers but only after protracted periods (Gutteridge and Parry, unpublished results). Finally, it has been documented on a number of occasions that ribulose-P_2 is an extremely effective inhibitor of activation. Preincubation of the inactive carboxylase with the bisphosphate and then addition of the activating cofactors produces very little turnover. Restoration of the activity by carbamylation of the active site requires some hours.

The carboxylase of the stroma must be able to circumvent all of these problems. Two regulatory processes, that have only recently been detected in the chloroplast, may act to prevent these undesirable traits. The best characterised of these is the isolation and structural determination of the natural inhibitor of the enzyme 2'carboxy arabinitol 1-phosphate (2CA1P) that binds at the active site during periods of reduced light. The inhibitor favours the activated form of the carboxylase preferentially and thus maintains the enzyme in this state, presumably when the CO_2 in the stroma is low. Although it is difficult to assign a function for 2CA1P, its preference for activated Rubisco suggests that the enzyme will be active on emergence of the plant from darkness. Ribulose-P_2 will therefore be synthesised in the presence of a functional enzyme that is primed to fix CO_2 as soon as the substrate appears.

The occurrence of the natural inhibitor was discovered by at least two laboratories almost simultaneously and deduced by comparing the activity of the carboxylase isolated from light and dark treated leaves (Seeman et al., 1985; Servaites, 1985). The synthesis of the inhibitor in leaves of *Phaseolus vulgaris* was relatively rapid with onset of darkness (t1/2 ≈ 60min) and it was released equally quickly upon exposure to light (Gutteridge et al., 1986b). The tight binding to the enzyme (Ki ≈ 60nM) was exploited to isolate and purify the compound away from other low molecular weight components of the chloroplast. The application of high resolution NMR solved the structure of the inhibitor, initially from potato (Gutteridge et al., 1986c) and then *Phaseolus* (Berry et al., 1987), and indicated that it was a 2'-carboxy pentitol 1-phosphate. The NMR of the stereo pentitol monophosphate isomers identified the natural compound as 2'-carboxy arabinitol-1-phosphate . The structure shown in Figure 10 is almost identical to the intermediate of the carboxylase reaction and thus ideally designed to stabilise the activated form of the enzyme. The absence of the phosphate at the 5 position ensures that it is not bound irreversibly at the active site and unable to dissociate, e.g. as with 2CABP. As yet the biosynthetic processes that produce the inhibitor (in considerable concentrations in some plants, see Servaites et al., 1986) have not been identified. Similarly, the reason for the variable distribution amongst plants is not understood.

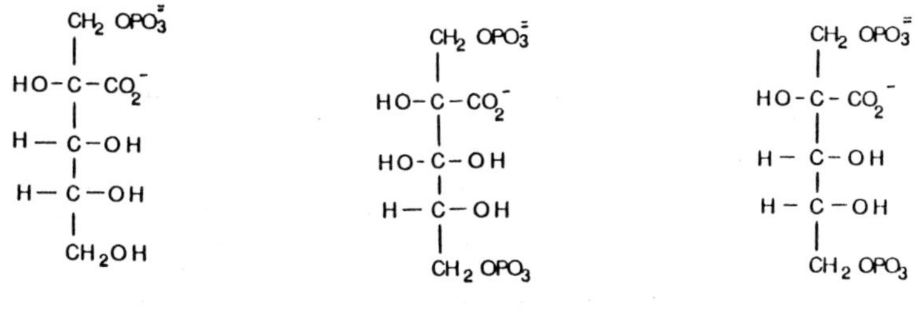

Figure 10. The structure of the natural inhibitor/intermediate analogue 2CA1P showing the close resemblance in structure with the 6 carbon intermediate of carboxylation and the irreversible inhibitor 2CABP.

The second regulatory process involves the interaction of Rubisco with a protein that mediates activation (Portis et al., 1986). This protein was discovered to be absent in a Rubisco activation mutant of *Arabidopsis* (Somerville et al., 1985). The subsequent isolation and purification of the protein has indicated that it has kinase function, requiring apart from Rubisco and ATP as substrate, thylakoid membranes and light to be most effective. The activity of this composite is to produce active Rubisco in physiological concentrations of CO_2. The ultimate fate of the phosphate or what becomes phosphorylated is not known. However, this is a system which explains the discrepancy in the amounts of CO_2 required for activation in chloroplasts compared to those required for purified enzyme. Furthermore, it is a second process that, like 2CA1P inhibition, potentially closely links the regulation of Rubisco activity to changes in light irradiance in the chloroplast.

In describing the biogenesis of Rubisco we often assume that once assembled the enzyme constitutes a relatively stable, imperturbable pool of protein. The investigations of the import of L-subunit into chloroplast with attached transit sequences has indicated that it is capable of associating with pre-existing indigenous subunits. This suggests that there may always be a fraction of the holoenzyme in some dynamic equilibrium with smaller aggregates of the two subunits. Clearly, this is an important discovery if L-subunit manipulation is the only means of obtaining more efficient forms of the enzyme. In the absence of chloroplast mutagenesis, any transferred L-subunit gene would have to reside in the nucleus and the protein transported into the chloroplast, presumably to compete with the resident species. A dynamic process would ensure that some of the nuclear encoded material was incorporated into holoenzyme.

REFERENCES

ANDREWS, T.J. and BALLMENT, B. (1983) The function of the small subunits of ribulose bisphosphate carboxylase-oxygenase. J. Biol. Chem. 258: 7514-7518

ANDREWS, T.J. and BALLMENT, B. (1984) Active site carbamate formation and reaction intermediate analogue binding by ribulose bisphosphate carboxylase-oxygenase. J. Biol. Chem. 260: 4632-4636

ANDREWS, T.J. and BALLMENT, B. (1984) A rapid, sensitive method for quantitating subunits in purified ribulose bisphosphate carboxylase preparations. Plant Physiol. 75: 503-510

ANDREWS, T.J. and LORIMER, G.H. (1985) Catalytic properties of a hybrid between cyanobacterial large subunits and higher plant small subunits of ribulose bisphosphate carboxylase-oxygenase. J.Biol.Chem. 260: 4632-4636

BADGER, M.R. and LORIMER, G.H. (1981) Interaction of sugar phosphate with the catalytic site of ribulose-1,5-bisphosphate carboxylase. Biochemistry 20: 2219-2225

BANNER, D.W., BLOOMER, H.C., PETSKO, G.A., PHILLIPS, D.C., POGSON,

C.I. and WILSON, A.I. (1975) Structure of chicken muscle triose phosphate isomerase. Nature 255: 609–614.
BARRACLOUGH, R. and ELLIS, R.J. (1979) The biosynthesis of ribulose bisphosphate carboxylase. Uncoupling of the synthesis of the large and small subunits in isolated soybean leaf cells. Eur. J. Biochem. 94: 165–177.
BARRACLOUGH, R. and ELLIS, R.J. (1980) Protein synthesis in chloroplasts. IX. Assembly of newly synthesised large subunits into ribulose bisphosphate carboxylase in isolated intact pea chloroplasts. Biochim. Biophys, Acta. 608: 19–31.
BERRY, J.D., NIKOLAU, B.J., CARR, J.P. and KLESSIG, D.F. (1986) Translational regulation of light induced ribulose-1,5-bisphosphate carboxylase gene expression in Amaranth. Mol. Cell. Biol. 6: 2347–2353.
BERRY, J.A., LORIMER, G.H., PIERCE, J., SEEMAN, J.R., MEEK, J. and FREAS, S. (1987) Isolation, identification and synthesis of 2'-carboxy arabinitol-1-phosphate, a diurnal regulator of ribulose bisphosphate carboxylase activity. Proc.Natl.Acad.Sci. U.S.A. 84: 734–738.
BERRY-LOWE, S.L., McKNIGHT, T.D., SHAH, D.M. and MEAGHER, R.B. (1982) The nucleotide sequence, expression and evolution of one member of a multigene family encoding the small subunit of ribulose-1,5-bisphosphate carboxylase in soybean. J. Mol. App 1. Genet. 1: 483–498
BLAIR, G.E. and ELLIS, J.R. (1973) Protein synthesis in chloroplasts. I. Light–driven synthesis of the large subunit of Fraction I protein by isolated pea chloroplasts. Biochim. Biophys. Acta. 319: 223–234.
BLOOM, M.V., MILOS, P. and ROY, H. (1983) Light dependent assembly of ribulose-1,5-bisphosphate carboxylase. Proc. Natl. Acad. Sci.U.S.A. 80: 1013–1017.
BOTTOMLEY, W., SPENCER, D. and WHITFIELD, P.R. (1974) Protein synthesis in isolated spinach chloroplasts: comparison of light driven and ATP driven synthesis. Arch. Biochem. Biophys. 164: 106–117.
BRADLEY, D. and GATENBY, A.A. (1985) Mutational analysis of the maize chloroplast ATPase–β subunit gene promoter: the isolation of promoter mutants in E. coli and their characterisation in a chloroplast in vitro transcription system. EMBO. J. 4: 3641–3648
BRADLEY, D., VAN DER VIES, S.M. and GATENBY, A.A. (1986) Expression of cyanobacterial and higher plant ribulose 1,5-bisphosphate carboxylase genes in E. coli. Phil. Trans. R. Soc. Lond. B313: 447–458.
BOWEIN, B., MEYER, F., SPEISS,E., PAHLER, A., ENGLISCH, U. and SAENGER, W. (1980) On the structure of crystalline ribulose bisphosphate carboxylase from Alcaligenes eutrophus. Eur. J. Biochem. 106: 405–410.
BRANDEN, C.I., SCHNEIDER, G., LINDQVIST, Y., ANDERSSON, I., KNIGHT, S. and LORIMER, G.H. (1986) X-ray structural studies of Rubisco from Rhodospirillum rubrum and spinach. Phil. Trans. R. Soc. B313: 359–365.
BRANDEN, R., NILSSON, T. and STYRING, S. (1984) Ribulose-1,5-bisphosphate carboxylase/oxygenase incubated with Cu^{2+} and studied by electron paramagnetic spectroscopy. Biochemistry 23: 4373–4382.
BROGLIE, R., CORUZZI,G., FRALEY, R.T., ROGERS, S.G., HORSCH, R.B., NIEDERMEYER,J.G., FINK, C.L., FLICK, J.S. and CHUA, N. –H (1984a) Light-regulated expression of a pea ribulose-1,5-bisphosphate

carboxylase small subunit gene in transformed plant cells. Science 224: 838–843.

BROGLIE, R., CORUZZI,G., KEITH, B. and CHUA, N. -H. (1984b) Molecular biology of C4 photosynthesis in Zea mays: differential localisation of proteins and mRNAs in the two leaf cell types. Plant Mol. Biol. 3:431–444.

BROGLIE, R., CORUZZI,G., LAMPPA, G., KEITH, B. and CHUA, N. -H. (1983) Structural analysis of nuclear genes coding for the precursor to the small subunit of wheat ribulose-1,5-bisphosphate carboxylase. Biotechnology 1: 55–61

CALVIN., M. (1956) The photosynthetic carbon cycle. J. Chem. Soc. pt 2: 1895–1915

CANNON, S., WANG, P. and ROY, H. (1986) Inhibition of ribulose bisphosphate carboxylase assembly by antibody to a binding protein. J. Cell Biol. 103: 1327–1335.

CHAPMAN,M.S., SMITH,W.W., SUH, S.W., CASCIO, D., HOWARD, A., HAMLIN,R., XUONG,N.-H. and EISENBERG, D. (1986) Structural studies of Rubisco from tobacco. Phil. Trans. R. Soc. B313: 367–378.

CHRISTELLER, J.T., TERZAGHI, B.E., HILL, D.F. and LAING, W.A. (1985) Activity expressed from cloned Anacystis nidulans large and small subunit ribulose bisphosphate carboxylase genes. Plant Mol. Biol. 5: 257–263.

CHUA, N.-H. and SCHMIDT, G.W. (1978) Post-translational transport into intact chloroplasts of a precursor to the small subunit of ribulose-1,5-bisphosphate carboxylase. Proc. Natl. Acad. Sci. U.S.A. 75: 6110–6114.

CORUZZI, G., BROGLIE, R., CASHMORE, A. and CHUA, N.-H. (1983) Nucleotide sequences of the pea cDNA clones encoding the small subunit of ribulose-1,5-bisphosphate carboxylase and the major chlorophyll a/b binding thylakoid polypeptide. J. Biol. Chem . 258: 1399–1402.

CORUZZI, G., BROGLIE, R., EDWARDS, C. and CHUA, N.-H. (1984) Tissue specific and light regulated expression of a pea nuclear gene encoding the small subunit of ribulose-1,5-bisphosphate carboxylase. EMBO. J. 3: 1671–1679.

DEAN, C., VAN DEN ELZEN, P., TAMAKI, S., BLACK, M., DUNSMUIR, P, and BEDBROOK, J. (1987) Molecular characterisation of the rbcS multi-gene family of Petunia (Mitchell). Mol. gen. Genet. 206: 465–474

DEAN, C., VAN DEN ELZEN, P., TAMAKI, S., BLACK, M., DUNSMUIR, P, and BEDBROOK, J. (1985) Linkage and homology analysis divides the eight genes for the small subunit of petunia ribulose-1,5-bisphosphate carboxylase into three gene families. Proc. Natl. Acad. Sci. U.S.A. 82: 4964–4968.

DONNELLY, M.I. STRINGER, C.D. and HARTMAN, F.C. (1983) Characterisation of the activator site of Rhodospirillum rubrum ribulose bisphosphate carboxylase/oxygenase. Biochemistry 22: 4346–4352.

ELLIS, R.J. (1981) Chloroplast proteins: synthesis, transport and assembly. Ann. Rev. Plant Physiol. 32:111–137.

ELLIS, R.J., SMITH, S.M. and BARRACLOUGH, R. (1979). Nuclear-plastid interactions in plastid protein synthesis. Proc. 4th John Innes Symp., eds. D.R. Davies, D.A. Hopwood. Norwich: John Innes Charity. pp 147–160.

FACCIOTTI, D., O'NEAL, J.K., LEE, S. and SHEWMAKER, C.K. (1985) Light inducible expression of a chimaeric gene in soybean tissue transformed with *Agrobacterium*. Biotechnology 3: 241–246.

GALLAGHER, T.F. and ELLIS, R.J. (1982) Light stimulated transcription of genes for two chloroplast polypeptides in isolated pea leaf nuclei. EMBO J. 1 : 1493–1498.

GALLAGHER, T.F., JENKINS, G.I. and ELLIS, R.J. (1985) Rapid modulation of transcription of nuclear genes encoding chloroplast proteins by light. FEBS Lett. 186: 241–245.

GATENBY, A.A. (1984) The properties of the large subunit of maize ribulose bisphosphate carboxylase/oxygenase synthesised in *Escherichia coli*. Eur. J. Biochem. 144: 361–366.

GATENBY, A.A. and CASTLETON, J.A. (1982) Amplification of maize ribulose bisphosphate carboxylase large subunit synthesis in *E. coli* by transcriptional fusion with the lambda *N* operon. Mol. gen.   Genet. 185: 424–429.

GATENBY, A.A., CASTLETON, J.A. and SAUL, M.W. (1981) Expression in *E. coli* of maize and wheat chloroplast genes for large subunit of ribulose bisphosphate carboxylase. Nature 291: 117–121.

GATENBY, A.A. and CUELLAR, R.E. (1985) Antitermination is required for readthrough transcription of the maize *rbcL* gene by a bacteriophage promoter in *Escherichia coli*. Eur.J. Biochem. 153: 355–359.

GATENBY, A.A., VAN DER VIES, S.M. and BRADLEY, D. (1985) Assembly in *E. coli* of a functional multi–subunit ribulose bisphosphate from a blue–green alga. Nature 314: 617–620.

GATENBY, A.A., VAN DER VIES, S.M. and ROTHSTEIN, S.J. (1987) Co-expression of both the maize large and wheat small subunit genes of ribulose bisphosphate carboxylase in *E. coli*. Eur. J. Biochem. in press.

GIBSON, J.L. and TABITA, F.R. (1986) Isolation of the *Rhodopseudomonas sphaeroides* form I ribulose-1,5-bisphosphate carboxylase/oxygenase large and small subunit genes and expression of the active hexadecameric enzyme in *E. coli*. Gene 44: 272–278 .

GREBANIER,A.E., CHAMPAGNE, D. and ROY, H. (1978). Effects of magnesium ion and substrate on the conformation of ribulose-1, 5-bisphosphate carboxylase. Biochemistry 17: 5150–5155.

GRUISSEM, W. and ZURAWSKI, G. (1985) Analysis of promoter regions for the spinach chloroplast *rbcL*, *atpB* and *psbA* genes. EMBO J. 4: 3375–3383

GUREVITZ, M., SOMERVILLE, C.R. and McINTOSH, L. (1985) Pathway of assembly of ribulose bisphosphate carboxylase/oxygenase from *Anabaena* 7120 expressed in *E. coli*. Proc. Natl. Acad. Sci.U.S.A. 82: 6546–6550.

GUTTERIDGE, S., MILLARD, B.N. and PARRY, M.A.J. (1985) Inactivation of ribulose-P_2 carboxylase by limited proteolysis: loss of catalytic activity without disruption of bisphosphate binding or carbamylation. FEBS Lett 196: 263–268.

GUTTERIDGE, S., PARRY, M.A. and SCHMIDT, C.N.G. (1982) The reactions between active and inactive forms of wheat ribulose bisphosphate carboxylase and effectors. Eur. J. Biochem. 126: 597–602.

GUTTERIDGE, S., PARRY, M.A.J., SCHMIDT, C.N.G. and FEENEY, J. (1984a) An investigation of ribulose bisphosphate carboxylase activity by high resolution ^{1}H NMR. FEBS Lett 170: 355–359.

GUTTERIDGE, S., SIGAL, I., THOMAS, B., ARENTZEN,R., CORDOVA, A. and LORIMER, G.H. (1984b) A site–specific mutation within the active site of ribulose-1,5-bisphosphate carboxylase of *Rhodospirillum rubrum*.

EMBO. J. 3: 2737-2743.

GUTTERIDGE, S., PHILLIPS, A.L., KETTLEBOROUGH, C.A., PARRY, M.A.J. and KEYS, A.J. (1986a) Expression of bacterial Rubisco genes in *E. coli*. Phil. Trans. R. Soc. Lond. B313: 433-445.

GUTTERIDGE, S., PARRY, M.A.J., KEYS, A.J., SERVAITES, J.C. and FEENEY, J. (1986b) The structure of the naturally occurring inhibitor of Rubisco that accumulates in the chloroplast during the dark is 2'-carboxy arabinitol-1-phosphate. Progress in Photosyn. Res. III. ed Biggins, J., 5:395-398.

GUTTERIDGE, S., PARRY, M.A.J., BURTON, S., KEYS, A.J., MUDD, A., FEENEY, J., SERVAITES, J.C. and PIERCE, J. (1986c) A nocturnal inhibitor of carboxylation in leaves. Nature 324: 274-276.

HARTMAN, F.C., STRINGER, C.D., MILANEZ, S. and LEE, E.H. (1986) The active site of Rubisco. Phil. Trans. R. Soc. B313: 379-395.

HEMMINGSEN, S.M. and ELLIS, R.J. (1986) Purification and properties of ribulose bisphosphate carboxylase large subunit binding protein. Plant Physiol. 80: 269-276

HERRERA-ESTRELLA, L., VAN DEN BROECK, G., MAENHAUT, R., VAN MONTAGU, M., SCHELL, J., TIMKO, M., and CASHMORE, A. (1984) Light inducible and chloroplast associated expressiom of a chimaeric gene introduced into *Nicotiana tabacum* using a Ti plasmid vector. Nature 310: 115-120

JORDAN, D.B. and OGREN, W.J. (1981) Species variation in the specificity of ribulose bisphosphate carboxylase/oxygenase. Nature 291: 513-515.

JORDAN, D.B. and OGREN, W.J. (1983) Species variation in the kinetic properties of ribulose-1,5-bisphosphate carboxylase/oxygenase. Arch Biochem. Biophys. 227: 425-433

JORDAN, D.B., CHOLLET, R. and OGREN, W.J. (1983) Binding of phosphor ylated effectors by active and inactive forms of ribulose-1,5-bisphosphate carboxylase. Biochemistry 22: 3410-3418.

KLEIN, R.R. and MULLET, J.E. (1987) Control of gene expression during higher plant chloroplast development. J.Biol. Chem. 262: 4341-4348

KOLLER, B., GINGRICH, J.C., STEIGLER, G.L., FARLEY, M.A., DELIUS, H. and HALLICK, R.B. (1984) Nine introns with conserved boundary sequences in the *Euglena gracilis* chloroplast ribulose-1,5-bisphosphate carboxylase gene. Cell 36:545-553.

LAING, W.A., OGREN, W.L. and HAGEMAN, R.H. (1976) Regulation of soybean net photosynthetic CO_2 fixation by the interaction of CO_2, O_2 and ribulose 1,5-diphosphate carboxylase. Biochem. J. 159: 563-570.

LAMPPKA, G.K. and BENDICH, A.J. (1979) Changes in chloroplast DNA levels during development of pea (*Pisum sativum*). Plant Physiol. 64: 126-130.

LARIMER, F.W., MACHANOFF, R., FOOTE, R.S., MITRA, S., SOPER,T.S., FUJIMURA, R.K. and HARTMAN, F.C. (1985) Site-directed mutagenesis of the active site of ribulose bisphosphate carboxylase. Biochemistry 24:3374-3379.

LINK, E. (1982) Phytochrome control of plastid mRNA in mustard. (*Sinapis alba* L.). Planta 154: 81-86.

LINDQVIST, Y. and BRANDEN, C.I. (1985) Structure of glycolate oxidase from spinach. Proc. Natl. Acad. Sci. U.S.A. 82: 6855-6859.

LINK, G., COEN, D.M. and BOGORAD, L. (1978). Differential expression of the gene for the large subunit of ribulose bisphosphate carboxylase in maize leaf cell types. Cell 15: 725-731.

LORIMER, G.H. (1981) The carboxylation and oxygenation of ribulose-1,5-bisphosphate. The primary events in photosynthesis and photorespiration. Ann. Rev. Pl. Physiol. 32: 349–383.

LORIMER, G.H. (1981) Ribulose bisphosphate carboxylase – amino acid sequence of a peptide bearing the activator carbon dioxide. Biochemistry 20: 1236–1240.

LORIMER, G.H. and MIZIORKO, H. (1980) Carbamate formation on the-amino group of a lysyl residue as the basis for activation of ribulose bisphosphate carboxylase by CO_2 and Mg^{2+}. Biochemistry 19: 5321–5328.

LORIMER, G.H., BADGER, M.R. and ANDREWS, T.J. (1976) The activation of ribulose-1,5-bisphosphate carboxylase by CO_2 and Mg ions. Equilibria,kinetics, a suggested mechanism and physiological implications. Biochemistry 15: 529–536.

LORIMER, G.H., ANDREWS, T.J., PIERCE, J. and SCLOSS, J.V. (1986) 2'carboxy -3keto-D-arabinitol 1,5-bisphosphate, the six carbon intermediate of the ribulose bisphosphate carboxylase reaction. Phil. Trans. R. Soc. B313: 397–407.

MACHLER, F., CORNELIUS, M. and KEYS, A.J. (1980) Activation of ribulose bisphosphate carboxylase purified from wheat leaves. J.Exp.Bot. 31: 7–14.

McCURRY, S.D., PIERCE, J., TOLBERT, N.E. and ORME-JOHNSON, W.H. (1981) The mechanism of effector mediated activation of ribulose bisphosphate carboxylase/oxygenase. J. Biol.Chem. 256: 6623–6628.

MENDIOLA-MORGENTHALER, L.R. MORGANTHALER, J.J. and PRICE, C.A. (1976) Synthesis of coupling factor CF1 protein by isolated spinach chloroplasts. FEBS Lett 62: 96–100.

McKNIGHT, T.D., ALEXANDER, D.C., BABCOCK, M.S. and SIMPSON, R.B. (1986) Nucleotide sequence and molecular evolution of two tomato genes encoding the small subunit of ribulose-1,5-bisphosphate carboxylase. Gene 48: 23–32.

McINTOSH, L., POULSEN, C. and BOGORAD, L. (1980) Chloroplast gene sequence for the large subunit of ribulose-1,5-bisphosphate carboxylase of maize. Nature 288: 556–560.

MISHKIND, M.L. WESSLER, S.R. and SCHMIDT, G.W. (1985) Functional determinants in transit sequences: Import and partial maturation by vascular plant chloroplasts of the ribulose-1,5-bisphosphate carboxylase small subunit of *Chlamydomonas*. J.Cell Bio l. 100: 226–234.

MIZIORKO, H.M. and MILDVAN, A.S. (1974) Electron paramagnetic resonance, ^1H and ^{13}C magnetic resonance studies of the interaction of manganese and bicarbonate with ribulose-1,5-bisphosphate carboxylase. J. Biol. Chem. 249: 2743–2750.

MIZIORKO, H.M. and SEALY, R.C. (1980) Characterisation of the ribulose bisphosphate carboxylase-carbon dioxide-divalent cation-carboxy pentitol bisphosphate complex. Biochemistry 19: 1167–1171.

MIZIORKO, H.M. and SEALY, R.C. (1984) Electron resonance studies of ribulose bisphosphate carboxylase : identification of activation cation ligands. Biochemistry 23: 479–485.

MORELLI, G., NAGY, F., FRALEY, R.T., ROGERS, S.G. and CHUA, N.-H. (1985) A short conserved sequence is involved in the light inducibility of a gene encoding ribulose-1,5-bisphosphate carboxylase small subunit of pea. Nature 315: 200–204.

MUSGROVE, J.E. JOHNSON, R.A. and ELLIS R.J. (1987) Dissociation of the ribulose bisphosphate carboxylase large subunit binding protein into dissimilar subunits. Eur. J. Biochem. 163: 529–534.

NAGY, F., FLUHR, R., MORELLI, G., KUHLEMEIR,C., POULSEN,C., KEITH,B., BOUTRY, M. and CHUA, N.-H. (1986) The Rubisco small subunit gene as a paradigm for studies on the differential gene expression during plant development. Phil. Trans. R. Soc. B3 13: 409–417.

NAGY, F., MORELLI, G., FRALEY, R.T., ROGERS, S.G. and CHUA, N.-H. (1985) Photoregulated expression of a pea *rbc*S gene in leaves of transgenic plants. EMBO J. 4: 3063–3068.

NARGANG, F., McINTOSH, L. and SOMERVILLE, C.R. (1984) Nucleotide sequence of the ribulose bisphosphate carboxylase gene from *Rhodospirillum rubrum*. Molec. gen. Genet. 193: 220–224.

PARRY, M.A.J. and GUTTERIDGE, S. (1984) The effect of SO_3^{2-} and SO_4^{2-} ions on the reactions of ribulose bisphosphate carboxylase. J.Exp. Bot. 35: 157–168.

PARRY, M.A.J., SCHMIDT, C.N.G., KEYS, A.J. and GUTTERIDGE, S. (1983) Activation of ribulose-1,5-bisphosphate carboxylase by Ca^{2+}. FEBS Lett. 159: 107–111.

PARRY, M.A.J., SCHMIDT, C.N.G., CORNELIUS, M., KEYS, A.J., MILLARD, B.N. and GUTTERIDGE, S. (1985) Stimulation of ribulose bisphosphate carboxylase activity by inorganic orthophosphate without an increase in bound activating CO_2: co-operativity between the subunits of the enzyme. J. Exp. Bot. 36: 1396–1404.

PIERCE, J. and GUTTERIDGE, S. (1985) Large scale preparation of ribulose bisphosphate carboxylase from a recombinant system in *E. coli* charaterised by extreme plasmid instability. Appl. Env. Micrbiol. 49: 1094–1100.

PIERCE, J. and REDDY, G.E. (1986) The sites for catalysis and activation of ribulose bisphosphate carboxylase share a common domain. Arch. Biochem. Biophys. 245: 483–493.

PIERCE, J. LORIMER, G.H. and REDDY, G. S. (1986) The kinetic mechanism of ribulose bisphosphate carboxylase: Evidence for an ordered, sequential reaction. Biochemistry 25: 1636–1644.

PIERCE, J., TOLBERT, N.E. and BARKER, R. (1980) Interaction of ribulose bisphosphate carboxylase/oxygenase with transition state analogues. Biochemistry 19: 934–942.

PORTIS, A.R., SALVUCCI, M.E. and OGREN, W.L. (1986) Activation of ribulose bisphosphate carboxylase/oxygenase at physiological CO_2 and ribulose bisphosphate concentrations by Rubisco activase. Plant Physiol. 82: 967–971.

ROBISON, P.D., MARTIN, M.N. TABITA, F.R. (1979) Differential effects of metal ions on *Rhodospirillum rubrum* ribulose bisphosphate carboxylase/oxygenase and stoichiometric incorporation of HCO_3^- into a cobalt(III)–enzyme complex. Biochemistry 18: 4453–4458.

ROCHAIX, J.D. (1985) Genetic organisation of the chloroplast. Int. Rev. Cytol. 93: 57–91.

ROY, H. ADARI, H. and COSTA, K.A. (1979) Characterisation of free subunits of ribulose bisphosphate carboxylase. Plant Sci. Lett. 16: 305–318.

ROY, H., BLOOM, M., MILOS, P. and MONROE, M. (1982) Studies on the assembly of large subunits of ribulose bisphosphate carboxylase in isolated pea chloroplasts. J. Cell Biol. 94: 20–27.

ROY, H., COSTA, K.A. and ADARI, H. (1978) Free subunits of ribulose bisphosphate carboxylase in pea leaves. Plant Sci. Lett. 11: 159–168.

SAVER, B.C. and KNOWLES, J.R. (1982) Ribulose-1,5-bisphosphate carboxylase: Enzyme catalysed appearance of solvent tritium at carbon-3 of ribulose bisphosphate reisolated after partial reaction.

Biochemistry 21: 5398-5403.
SCLOSS, J.V. and LORIMER, G.H. (1982) The stereochemical course of ribulose bisphosphate carboxylase. J. Biol. Chem. 257: 4691-4694.
SCHMIDT, C.N.G., GUTTERIDGE,S., PARRY, M.A.J. and KEYS, A.J.(1984) Inactive forms of wheat ribulose bisphosphate carboxylase. Conversion from the slowly activating into the rapidly activating form. Biochem. J. 220: 781-785.
SCHMIDT, C.N.G., CORNELIUS, M.J., BURTON, S.,PARRY, M.A.J.,MILLARD, B.N. KEYS, A.J. and GUTTERIDGE,S. (1984) Purified ribulose bisphosphate carboxylase from wheat with high specific activity and fast activation. Photosyn. Res. 5:47-62.
SCHMIDT, G.W. and MISHKIND, M.L. (1983) Rapid degradation of unassembled ribulose-1,5-bisphosphate carboxylase small subunits in chloroplast. Proc. Natl. Acad. Sci. U.S.A. 80: 2632-2636.
SCHMIDT, G.W. and MISHKIND, M.L. (1986) The transport of protein into chloroplasts. Ann. Rev. Biochem. 55: 879-912.
SCHNEIDER, G., LINDQVIST, Y., BRANDEN, C.I. and LORIMER, G.H. (1986) The three dimensional structure of ribulose-1,5-bisphosphate carboxylase/oxygenase from *Rhodospirillum rubrum* at 2.9Å resolution. EMBO J. 5: 3409-3415.
SEEMAN, J.R., BERRY, J.A., FREAS, S. and KRUMP, M.A. (1985) Regulation of ribulose bisphosphate carboxylase activity *in vivo* by a light-modulated inhibitor of catalysis. Proc.Natl.Acad.Sci. U.S.A. 82: 8024-8028
SERVAITES, J.C. (1985) Binding of a physiological inhibitor to ribulose bisphosphate carboxylase/oxygenase during the night. Plant Physiol. 78: 839-843.
SERVAITES, J.C., PARRY, M.A.J., GUTTERIDGE, S. and KEYS, A.J. (1986) Species variation in the predawn inhibition of ribulose-1,5-bisphosphate carboxylase/oxygenase. Plant Physiol. 79:1161-1163
SHEEN, J-Y. and BOGORAD, L. (1986) Expression of the ribulose bisphosphate carboxylase large subunit gene and three small subunit genes in two cell types of maize leaves. EMBO J. 5: 3417-3422.
SHINOZAKI, K. and SUGIURA, M. (1985) Genes for the large and small subunit of ribulose bisphosphate carboxylase/oxygenase constitute a single operon in a cyanobacterium *Anacystis nidulans* 6301. Mol. gen. Genet. 200: 27-32.
SMITH, S.M. and ELLIS, R.J. (1979) Processing of the small subunit precursor of ribulose bisphosphate carboxylase and its assembly into whole enzyme are stromal events. Nature 278: 662-664.
SMITH, S.M. and ELLIS, R.J. (1981) Light stimulated accumulation of transcripts of nuclear and chloroplast genes for ribulose bisphosphate carboxylase. J. Mol. Appl. Genet. 1: 127-137.
SOMERVILLE, C.R., PORTIS, A.R. and OGREN, W.L. (1982) A mutant of *Arabidopsis thaliana* which lacks activation of RuBP carboxylase *in vivo*. Plant. Physiol. 70: 381-387.
SOMERVILLE, C.R., McINTOSH, L., FITCHEN, J. and GUREVITZ, M. (1986) The cloning and expression in *E. coli* of RuBP carboxylase/oxygenase large subunit genes. Meth. Enzymol. 118: 419-433.
SOMERVILLE, C.R. and SOMERVILLE, S.C. (1984) Cloning and expression of the *Rhodospirillum rubrum* ribulose bisphosphate carboxylase gene in *E. coli*. Mol. gen. Genet. 193: 214-219.
STYRING, S. and BRANDEN, R.R. (1985) Identification of ligands to the metal ion in copper(II) activated ribulose bisphosphate carboxylase oxygenase by electron paramagnetic resonance spectroscopy and ^{17}O

labelled ligands. Biochemistry 24: 6011–6 019.

SUE, J.R. and KNOWLES, J.R. (1982) Ribulose-1,5–bisphosphate carboxylase: Fate of the tritium label in the (3-^{3}H)ribulose 1,5–bisphosphate during enzyme catalysed reactions. Biochemistry 21: 5404–5410.

TABITA, F.R. and SMALL, C.L. (1985) Expression and assembly of active cyanobacterial ribulose bisphosphate carboxylase/oxygenase in *E. coli* containing stoichiometric amounts of large and small subunits. Proc. Natl. Acad. Sci. U.S. A. 82: 6100–61 03.

THOMPSON, W.F., EVERETT, M., POLANS, N.O., JORGENSEN, R.A and PALMER, J.D. (1983) Phytochrome control of RNA levels in developing pea and mung-bean leaves. Planta 158: 487–500.

TIMKO, M.P. KAUSCH, A.P., CASTRSANA, C., FASSLER, J., HERRERA-ESTRELLA, L., VAN DEN BROECK, G., VAN MONTAGU, M., SCHELL, J. and CASHMORE, A.R. (1985) Light regulation of plant gene expression by an upstream enhancer like element. Nature 318: 579 –582.

TOBIN, E.M. (1981) Phytochrome mediated regulation of mRNAs for the small subunit of ribulose bisphosphate carboxylase and the light harvesting chlorophyll a/b-protein in *Lemna gibba*. Plant Mol. Biol. 1: 35–51.

TOBIN, E.M. and SILVERTHORNE, J. (1985) Light regulation of gene expression in higher plants. Ann. Rev. Plant Physiol. 36: 569–593.

VAN DER VIES, S.M., BRADLEY, D. and GATENBY, A.A. (1986) Assembly of cyanobacterial and higher plant ribulose bisphosphate carboxylase subunits into functional homologous and heterologous enzyme molecules in *E. coli*. EMBO J. 5: 2439–2444.

VAN DYK, D.E. and SCHLOSS, J.V. (1986) Deuterium isotope effects in the carboxylase reaction of ribulose-1,5–bisphosphate carboxylase/oxygenase. Biochemistry 25: 5145–5156.

VIALE, A.M. KOBAYASHI, H., TAKEBE, T. and AKAZAWA, T. (1985) Expression of genes for subunits of plant-type Rubisco from *Chromatium* and production of the enzymically active molecule in *E. coli*. FEBS Lett. 192: 283–288.

VOORDOUW, G., VAN DER VIES, S.M. and BOUWMEISTER, P.P. (1984) Dissociation of ribulose bisphosphate carboxylase/oxygenase from spinach by urea. Eur. J. Biochem. 141: 313–318.

WILDNER, G.F. and HENKEL, J. (1978) Differential reactivation of ribulose-1,5–bisphosphate oxygenase with low carboxylase activity by Mn^{2+}. FEBS Lett 91: 99–103.

Oxford Surveys of Plant Molecular & Cell Biology Vol.4 (1987) 137–166

GENE–TRANSFER AND HOST–VECTOR SYSTEMS
OF CYANOBACTERIA.

Sergey V. Shestakov[1] and John Reaston[2].

[1]*Department of Genetics, Moscow State University, MOSCOW 119899,
U.S.S.R. and* [2]*Department of Biological Sciences,
University of Warwick, COVENTRY CV4 7AL, UK.*

INTRODUCTION

Cyanobacteria are the only prokaryotes capable of oxygenic
"plant-like" photosynthesis and their photosynthetic apparatus is
similar to that found in higher plants. The resemblance of cyanobacterial and chloroplast genomes helps support the endosymbiotic
hypothesis, which postulates that chloroplasts evolved after the
uptake of cyanobacteria by some ancestral eukaryotic cell (see Grey
and Doolittle, 1982). Thus, cyanobacteria are useful for the study of
the mechanisms and genetic control of plant photosynthesis. They are
also attractive model organisms for the examination of many other
fundamental biological problems (Doolittle, 1979).

Active studies on the molecular biology of cyanobacteria deal with
their mechanisms of cell differentiation, membrane biogenesis and a
number of other complex problems connected with the regulation of
gene action, including light dependent control (chromatic
adaptation), carbon and nitrogen fixation and gene rearrangement.
Investigations with cyanobacteria can also provide important
information resolving the nature of the processes which ensure
thermophilic properties, high resistance to radiation and DNA
restriction and modification. Readers unfamiliar with the
cyanobacteria should consult the excellent review articles by
Doolittle (1979); and Tandeau de Marsac and Houmard (1987), which
give very good insights into their molecular biology and genetics.

Recently, these microorganisms have attracted interest as
potential objects for biotechnology (see Craig and Reichelt, 1986).
Cells of many cyanobacterial strains serve as renewable protein,
sources of feed and nutrition. They are also producers of many other
valuable biologically active compounds. In bioreactors, immobilised
cells of certain cyanobacteria may provide the means for the
production of ammonia on an industrial scale (Kirby et al., 1986).

Cyanobacteria are potentially the most economic producers of useful products because they do not require any organic source of nutrition and accomplish cell biosynthesis using solar energy.

Taking into consideration all the characteristics which cyanobacteria possess, the importance of studies in genetics and genetic engineering are obvious. It is now clear that most of the powerful techniques of bacterial genetics are applicable to cyanobacteria. The elaboration of efficient systems of gene-transfer has opened the door for gene-cloning in cyanobacteria and for the construction of strains with new characteristics. Cyanobacteria are of special interest as recipients for plant genes and the possible introduction of useful genetic information into plants. It appears likely that genetic manipulations aimed at the alteration of plant genes will be easier in the comparatively simply organised cells of cyanobacteria, whose prokaryotic nature is unquestionable (see Stanier and Cohen-Bazire, 1977).

A large number of cyanobacterial genes have been cloned into *E. coli*. There is now sufficient information on many of these clones that we shall not reproduce here as it is dealt with in more detail by Haselkorn (1985b); Tandeau de Marsac and Houmard (1987). The aims of this review are to consider the processes of gene-transfer (advantages and shortcomings) in cyanobacteria, including the characteristics of cyanobacterial host-vector systems which allow cloning of homologous genes and the introduction of foreign DNA into cyanobacteria.

It should be noted that in the literature concerning cyanobacteria (formally known as blue-green algae) different names for the same strain are often used, mainly because of the absence, for a long time, of a formally established and authorised taxonomy. In this article we will normally employ the original designations given to the strains by the authors whose papers are discussed. Alternative names, based on the excellent Pasteur Culture Collection (PCC) classification system (Rippka et al., 1979) will be presented, where appropriate.

GENETIC TRANSFORMATION

Optimisation of the conditions for rapid and quantitative growth of single colonies on solid media (Van Baalan, 1965; Stoletov et al.,1965) and the application of drug concentration gradient plates (Szybalski and Bryson, 1952) allowed the discovery of genetic transformation in unicellular cyanobacteria.

The first report of successful and reproducible transformation with the unicellular cyanobacterium *Anacystis nidulans* 602 (*Synechococcus* PCC7943) was by Shestakov and Khyen (1970). Wild-type and filamentous mutants of the strain were transformed with purified

DNA from erythromycin (Em^r) or streptomycin (Sm^r) resistant donor strains. This process was completely inhibited by DNAase. The transfer frequency was approximately 10^{-4} per cell, more than 10^4 fold higher than the spontaneous mutation rate. The highest yield of transformants was obtained with cells at the end of log phase. Transformation effectively occurs on the surface of agar by the better contact between the cells and donor DNA and possibly as a result of the increase in cell competence.

The mode of addition of the antibiotic to the agar medium was crucial to reveal resistant clones, as recombinants could not be recovered when the transforming mixture was plated directly onto antibiotic containing agar. Transformants were easily found if the selective agent was added underneath the agar after 1-2 days of non-selective incubation, on solid medium, which supported the appearance of a slight green lawn. This technique provided the growth of resistant colonies while sensitive cells were lysed. This "neighbour effect" of surrounding lysed cells may supply the surviving resistant cells with an unidentified growth/competence factor which plays a role in the establishment of recombinants (Shestakov and Mitronova, 1971). Phenotypic expression of the selected marker was necessary, as a minimal incubation period of 16-20 h was required before antibiotic selection.

Another plating method was developed by Gendel et al. (1983a), cells were plated onto nitrocellulose filters on non-selective solid medium and after 1-2 days the filters were transferred onto antibiotic containing agar. This procedure allowed resistant clones to be revealed and counted on the filter efficiently. The non-selective lag period, essential for the expression of resistant phenotypes, is characteristic of all the reliable gene-transfer systems described for cyanobacteria at present.

The system of chromosomal transformation in *A. nidulans* 602 was successfully exploited for the analysis of mutations leading to filamentous and radiosensitive phenotypes (Mitronova et al., 1973), to methionine auxotrophy (Romanova and Shestakov, 1977) and also for the investigation of the DNA repair processes in this strain (Shestakov et al., 1975; Polukhina and Shestakov, 1977). Several mutants of *A. nidulans* 602 were isolated which lacked the ability to be transformed, most were deficient in an early stage of DNA binding or uptake by the recipient cells. A Rec$^-$ mutant, FR-15, highly sensitive to UV-light and X-rays was unable to repair DNA breaks induced by X-rays (Shestakov et al., 1975) and was characterised by a block in the stage of integration of donor DNA into the recipient chromosome (Shestakov et al., 1982).

Herdman and Carr (1971) reported on the transformation of *A. nidulans* UTEX625 (*Synechococcus* PCC6301) by cell-free filtrates from a mutant donor strain. Gene-transfer mediated by the filtrate was partially sensitive to DNase and RNase which indicated the involvement of a DNA/RNA complex in transformation. A similar phenomenon was observed by Devilly and Houghton (1977) with *Gloeocapsa alpicoli* CCAP1430/1 (*Synechocystis* PCC6308). Thus, it is possible that RNA tightly associated with DNA in the folded chromosome may protect DNA fragments released or extracted from the cells. Herdman (1973) described the transformation of several auxotrophic mutants of *A. nidulans* UTEX625 with cell-free filtrate and chemically extracted wild-type DNA and the order of five markers were roughly mapped on the chromosome. In transformation with purified DNA only the *bio*-1 and *str*-2 markers showed a high degree of linkage, transformation mediated by the DNA/RNA complex exhibited more extensive linkage for *nit*-1 and *met*-2, which could not be co-transformed with chemically extracted DNA. In the same strain Orkwiszewski and Kaney (1974) obtained transfer of Sm^r, phe^+ and arg^+ markers with frequences as high as 0.1% using chemically extracted donor DNA. However, such transformation systems do not allow the large scale genetic analysis or accurate mapping of the whole chromosome. Gene mapping in cyanobacteria may be easier when an Hfr-like system has been developed or by the technique of "chromosome-walking" using an ordered DNA cosmid library. Recently, this technique has been used successfully by Herrero and Wolk (1986) to map 40 linkage groups on the chromosome of *Anabaena variabilis* (PCC7937), this method should be applicable to other cyanobacteria.

At present two highly transformable strains of *Synechococcus* PCC7942 (*Anacystis nidulans* R2) and PCC7002 (*Agmenellum quadruplicatum* PR-6) are available for gene-transfer and cloning studies. Transformation systems for chromosomal and plasmid DNA have now been well characterised in both strains and are similar to the heterotrophic bacterial systems in most respects thus far tested (Porter, 1986). Transformation in *A. nidulans* R2 was originally found by Grigorieva and Shestakov (1976). This strain was obtained from K. Floyd (California University, 1973) as a natural fresh water isolate, maintained in the cyanobacterial collection of Moscow State University and later deposited in the Pasteur Culture Collection as *Synechococcus* PCC7942. *A. nidulans* R2 is very closely related to *Synechococcus* PCC6301 in terms of its DNA base composition, DNA/DNA hybridisation homology (Wilmotte and Stam, 1984) and plasmid content (Van den Hondel and Van Arkel, 1980). Also, it was shown that heterospecific transformation of *A. nidulans* R2 by DNA from *Synechococcus* 6301 occurred at the same high frequency as homologous transformation (Grigorieva and Shestakov, 1976; Grigorieva, 1985). *A. nidulans* R2 can aslo be transformed with plasmid DNA, initially found by Van den Hondel et al. (1980). Systematic studies of the parameters influencing plasmid transformation efficiencies in this strain have been performed by two groups (Chauvat et al., 1983; Golden and Sherman, 1984). There are some differences in the findings obtained by these independent workers which are discussed excellently by Porter (1986). A study of ^{32}P-labelled plasmid (pBR322) binding and uptake by permeaplasts of *Synechococcus* PCC6301 has been reported by Daniell and McFadden (1986). These workers found that ATP (2.5-10mM) inhibited both the binding and uptake of DNA by permeaplasts in this strain. Stevens and Porter (1980) observed the transfer of Sm^r and nir^+ markers into *A. quadruplicatum* PR-6 at frequences of $5x10^{-4}$ to $4x10^{-3}$ per cell with purified DNA. A lag period of 40-48 h was allowed for the maximum phenotypic expression of Sm^r before the cells were challenged by spraying with a sterile solution of streptomycin sulphate. Later, Buzby et al. (1983) developed plasmid transformation in this strain.

There have been a number of reports of transformation in other unicellular cyanobacteria. Devilly and Houghton (1977) transferred a Sm^r-marker into *Gloeocapsa alpicola* using cold-shock and $CaCl_2$. This is one of the few reports of transformation in cyanobacteria where a pretreatment with $CaCl_2$ was efficient, Stevens and Porter (1980) showed that this compound actually decreased the transformation frequency in *A. quadruplicatum* PR-6. Transformation in *Aphanocapsa* 6714 (*Synechocystis* PCC6714) for parafluorophenylalanine (pfa) resistance was described by Astier and Espardellier (1976). An efficient system of chromosomal (Grigorieva and Shestakov, 1982) and plasmid (Shestakov et al., 1985a,b) DNA transformation for *Synechocystis* PCC6803 has been described. Purified DNA from erythromycin, chloramphenicol (Cm) or ethionine (Eth) resistant donors transformed the wild-type with a frequency of between $5x10^{-4}$ to $2x10^{-5}$ recombinants per recipient cell. The Em^r-marker became fully expressed phenotypically after 18-20 h, the period required for maximum expression of the Eth^r and Cm^r markers was 12h and 24h, respectively. Cells of 6803 were competent throughout exponential growth and competence was reduced in stationary phase, as appears to be the case for other transformable strains of cyanobacteria. The problems of optimisation of the various chromosomal and plasmid transformation systems (DNA concentration, competence, etc.) of cyanobacteria are analysed perfectly in the recent comprehensive review by Porter (1986).

There is limited information on heterospecific transformation in cyanobacteria. It was shown by Grigorieva, (1985) that DNA from some strains with a similar base composition near 55 mol% GC in the genus *Synechococcus* (6311, 6718, 6908, 7943, R3, R6) transformed *A. nidulans* R2 at levels below the homologous system. *Synechocystis* 6803 is capable of being transformed with a high frequency to Em^r by heterologous DNA from *Synechocystis* 6702 and 6704, confirming the high genetic homology between these strains having a DNA base composition of 47 mol% GC. Heterologous DNA from Em^r mutants of *A. nidulans* R2 (55.6 mol% GC) or *Synechocystis* 6701 (35.7 mol% GC) could not transform *Synechocystis* 6803 cells and Em^r DNA from 6803 failed to transform *A. nidulans* R2 (Grigorieva and Shestakov, 1982). It has been suggested that *A. nidulans* R2 and *Synechocystis* 6803 may be useful as standard test organisms for taxonomic analysis via heterospecific transformation of new isolates belonging to these genera (Grigorieva and Shestakov, 1976; 1982; Grigorieva, 1985). Intergeneric transformation of Sm^r and rif^r markers between *Gloeocapsa alpicola* and *Synechococcus* PCC6301, having considerable different GC ratios, has been reported (Devilly and Houghton, 1977) but this observation awaits confirmation. Porter (1986) mentioned that *A. quadruplicatum* PR-6 can be transformed to Sm^r with donor DNA from some cyanobacteria in the genus *Synechocystis*. This work was described further by Stevens and Porter (1986), who also reported the first transformation of *Synechococcus* PCC73109 (*Agmenellum* quadruplicatum BG-1) and *Synechococcus* PCC6906. They suggested "that the taxonomic boundary between the genera *Synechococcus* and *Synechocystis* is either misplaced or, less likely, does not exist," further taxonomic studies may be required to verify this. The differences found for heterospecific transformation in cyanobacteria are notable and require further investigation.

At present no transformation system of practical use has been developed for filamentous cyanobacteria. Trehan and Sinha (1981) reported gene-transfer by extracellular DNA in mixed cultures of *Nostoc muscorum*. This system was further described by the same authors (Trehan and Sinha, 1982), with purified DNA. Transfer of valine-requiring and *fpa*[r] markers were primarily mediated by DNA but these systems can not be used for gene-cloning. Tripathi and Kumar (1985) described high frequency transformation in two heterocystous *Nostoc* strains with purified DNA from mutants of *Nostoc linckia* resistant to propionate or aminotriazole. Recipient cells were pretreated with CaCl$_2$, incubated with donor DNA and polyethylene glycoll in cold conditions and then heat-shocked. Unfortunately, we know of no further investigations using these systems. Recently, Singh et al. (1987) reported on the transfer of herbicide resistance gene(s) from a *Gloeocapsa* strain into *Nostoc muscorum* in mixed cultures and by DNA-mediated transformation. However, such heterospecific gene-transfer from a unicellular cyanobacterium into a filamentous strain appears unusual and awaits confirmation. Purified vectors previously mobilized into filamentous cyanobacteria, could not be transformed (Wolk, personal communication). There are possibly several reasons for the lack of a reliable and reproducible transformation system for filamentous cyanobacteria: mucilaginous or sheath material, host-restriction (see below), and/or the production of extracellular nucleases (Wolk and Kraus, 1982). The conditions essential for the development of an efficient transformation system in filamentous cyanobacteria require further investigation.

In summary, while transformation can not easily be used for accurate gene mapping of the whole chromosome, it has great potential for the elaboration of gene-cloning systems and for mutant generation by gene insertion. The availability of unicellular cyanobacteria able to be transformed by vectors carrying cyanobacterial or foreign genes will allow molecular genetic studies of their biological processes and help to create strains valuable to biotechnology.

CONJUGATION AND TRANSDUCTION

A high-frequency conjugation system between cyanobacteria has never been demonstrated genetically. There are indications that sexuality exists in cyanobacteria and gene-transfer may occur by such a system. Early studies on genetic recombination were performed in liquid media and therefore it was not possible to distinguish between mutants and recombinants. Kumar (1962) reported on the production of double-resistant recombinants in *A. nidulans*. Similar experiments were performed by Singh and Sinha (1965) with the filamentous cyanobacterium *Cylindrospermum majus* and found double-resistant cells in mixed culture. Pikálek (1967) criticised such approaches because penicillin is unstable under the conditions employed. Using the techniques of quantitative surface plating and clonally derived polymixin B and Sm[r] mutants of *A. nidulans*, Bazin (1968) demonstrated that gene-transfer occurred but at very low frequences (5×10^{-9} – 5×10^{-7}).

The transfer of genes in mixed cultures of different mutants has been reported in *Anabaena doliolum* (Singh, 1967) and *Nostoc muscorum* (Stewart and Singh, 1975; Padhy and Singh, 1978b; Vaishampayan and Prasad, 1984). In the experiments of Stewart and Singh (1975) a *nif⁻ het⁻* Sm^r mutant was mixed with the wild-type and clones *nif⁺ het⁺* Sm^r were isolated, at a frequency of up to 5×10^{-6}, 100 times higher than the spontaneous mutation and reversion frequences. They concluded that the "hybrid" phenotype resulted from the transfer of *nif* gene(s), although other explanations of the results obtained exist (see Herdman, 1982). Ladha and Kumar (1978) suggested that gene-transfer in mixed cultures was mediated by conjugation. Delaney et al. (1976) attributed these results to transformation by extracellular DNA released into the medium. However, the mechanism of the phenotypic changes discussed above remain very obscure and questionable. Kumar and Ueda (1984) carried out a scanning electron microscope study of mixed cultures of two mutant strains of *A. nidulans* Tx20 (*Synechococcus* 6301) and found physical pair-formation between cells at a frequency of 10^{-5} to 10^{-6} by "conjugation tubes." However, from the general conception of bacterial genetics it seems very unlikely to develop an efficient system of conjugation from co-culturing experiments with clonally-related mutants, even if sexuality is a real characteristic of cyanobacteria. It may be better to attempt crosses between mutants derived from distinct natural isolates and to search for a fertility factor. The creation of fertile strains of cyanobacteria, possessing the ability of high-frequency chromosome mobilization (Hfr-like), based on the modern methods of genetic engineering is obviously the most promising way to achieve this. Several plasmids can mobilize the chromosome in a wide range of bacteria which do not have their own gene-transfer systems. Mutant strains with a high chromosome mobilization ability can be obtained by genetic manipulation and transposition (Holloway, 1979; Pemberton and Buwen, 1981).

Delaney and Reichelt (1982) attempted to establish an R-factor mediated conjugation system in *Synechococcus* 6301. They reported on the stable integration of plasmid R68.45 into chromosomal DNA and suggested the possibility to create an Hfr-like strain. The cyanobacterial transconjugants recovered possessed the Km^r determinant, homologous to that of R68.45. The insertion, of most of R68.45 into the chromosome of the transconjugants was shown by restriction enzyme analysis. The construction of Hfr-donor strains, by chromosome integration of broad-host range plasmids harbouring Tn-elements, may allow the development of detailed genetic analysis of cyanobacteria but this useful system has not yet been developed.

Another important breakthrough in the use of a conjugation system in cyanobacteria was concerned with the transfer of plasmid vectors from *E. coli* into cyanobacteria via mobilization. Wolk et al. (1984) described a highly efficient and reproducible system of conjugative transfer of cloning vectors into *Anabaena* PCC7120, PCC7118 and M-131 and also for the unicell *A. nidulans* R2, with the re-engineered shuttle vector pSG111. The system, for filamentous

cyanobacteria, consisted of three elements: a) a shuttle vector based
on plasmid pBR322, carrying the *bom* (basis of mobility) and
antibiotic resistance regions and the cryptic cyanobacterial plasmid
replicon pDU1 ; b) a helper plasmid (colicin K or D) to mobilize the
shuttle vector in *trans*; c) plasmid RP-4 to provide a wide host-range
pilus. Triparental mating was performed on a nitrocellulose filter
with the cyanobacterial recipient strain, *E. coli* with RP-4 and *E.
coli* containing the shuttle vector and a helper plasmid (pGJ28 or
pDS4101). After one day of non-selective incubation the filter was
transferred to antibiotic containing agar for the selection of
transconjugant clones. By using arsenate or trimethoprim containing
plates or UV-treatment (to kill the *E. coli*) it is possible to obtain
axenic cyanobacterial clones without re-streaking (Thiel; Lammers,
cited from Haselkorn, 1985a). This conjugation technique has been
applied to many other filamentous cyanobacteria, shuttle vectors were
transferred to and maintained in some facultative heterotrophic
organisms of the *Nostoc* genus PCC6310, PCC7107, PCC73102 (Flores and
Wolk, 1985). It appears that plasmid pDU1, will function in several
(but not all, eg. *Anabaena* PCC7937) filamentous cyanobacteria after
mobilization in shuttle vectors (Wolk, personal communication). In
principle a similar technique for promoting gene-transfer may be
employed with other strains of cyanobacteria after a suitable shuttle
vector has been constructed. It should be mentioned that this system
also appears to work with vectors (lacking a cyanobacterial
replication origin) which cannot be maintained autonomously in
cyanobacterial cells but capable of integration or transposition into
the chromosome (Elhai, cited from Haselkorn, 1985a). The conjugal
transfer of a plasmid containing the ColE1 replicon, RP-4 *tra*
functions and transposon Tn*501* (encoding mercury resistance) into
Synechocystis PCC6714 has been demonstrated (Bullerjahn, cited from
Haselkorn, 1985a). It must be noted that transformation with
integrative DNA has two components, DNA uptake and recombination
within the host genome. Recombination in cyanobacteria has been
detected to occur by homologous and non-homologous recombination.
This system of transposition following conjugation may be extremely
useful for mutagenesis and for the isolation of genes in
cyanobacteria, similar to the work performed by transformation in *A.
nidulans* R2 (Tandeau de Marsac et al., 1982). Cobley (1985) reported
on the construction and transfer of mobilizable shuttle vectors from
E. coli into the filamentous cyanobacterium *Fremyella diplosiphon*
(*Calothrix* PCC7601), a strain capable of chromatic adaptation. It is
likely that the further application of broad-host range plasmids and
mobilizable cloning vehicles will allow optimisation of conjugal
gene-transfer systems in cyanobacteria (eg. McFarlane et al., 1987).
The work of Wolk's group has solved one of the most vital
requirements for the genetic analysis of filamentous cyanobacteria
and must be given full credit.

Some cyanobacterial viruses can enter into a lysogenic state. The
phenomenon of lysogeny by cyanophages in the LPP-group has been
described by several authors (Cannon et al., 1971; Cannon and Shane,
1972; Padan et al., 1972; Singh and Singh, 1972; Rimon and Oppenheim,
1975). Lysogeny in *Anabaenopsis raciborskii* by phage AR-1 (Singh and
Singh, 1972), *Anabaena variabilis* by the temperate phage A-4L
(Khudiakov and Gromov, 1973), *Plectonema boryanium* by a long-tailed

virus (Singh, 1975) and *Nostoc muscorum* by N-1 (Padhy and Singh, 1978a) has been reported. However, reproducible cyanophage mediated transduction has not yet been conclusively shown for any strain of cyanobacteria. Transfer, at low frequence, of Sm^r from a mutant strain of *Plectonema boryanum* to the wild-type with cyanophages LPP-1 and P-2 has been suggested by Singh and Singh (1972) but this observation awaits confirmation. A transduction system would certainly be a helpful additional tool for the genetic analysis of cyanobacteria but the reasons for the apparent difficulty in developing such a system remain unclear.

PLASMIDS

Plasmids are normally extrachromosomal elements composed of circular (no linear plasmid has been described for cyanobacteria) double-stranded DNA which have been of great use in the development of gene-cloning systems. Their presence in cyanobacteria was first observed by Asato and Ginoza (1973) for *Anacystis nidulans*, confirmed by Restaino and Frampton (1975). The detailed investigations by Roberts and Koths (1976) with *Agmenellum quadruplicatum* PR-6 using electron microscopy and gel electrophoresis analysis revealed the presence of six different covalently closed circular (CCC) plasmids. Subsequently, other unicellular cyanobacteria have been shown to contain CCC plasmid DNA (Van den Hondel et al., 1979; Lau and Doolittle, 1979; Friedberg and Seijffers, 1979; Lau et al., 1980; Buzby et al., 1983; Elanskaya et al., 1985a; Engwall and Gendel, 1985; Vakeria et al., 1985; Kohl et al., 1987). A survey of extrachromosomal DNA from a range of filamentous organisms was first performed by Simon (1978). Seven out of 13 different strains tested contained detectable plasmid DNA and all but one had multiple elements. Plasmids have been detected in other filamentous cyanobacteria (Friedberg and Seiffers, 1979; Reaston et al ., 1980; Lambert and Carr, 1982; Bogorad et al., 1983; Potts, 1984; Engwall and Gendel, 1985; Kumar and Tripathi, 1985; Franche and Reynaud, 1986). Cyanobacterial plasmids appear to range in molecular weight from 0.9 to 150 Md (1.38-230 kb) and most strains possess more than one plasmid molecule. Further information on cyanobacterial plasmids can be found in the recent comprehensive reveiw by Tandeau de Marsac and Houmard, (1987), who also discuss differences found for plasmid number and size by previous workers. One filamentous strain, *Nostoc* PCC7524, contains three different sizes of plasmid DNA (designated pDU1-3) and one, pDU2 is a dimer of the smallest plasmid pDU1, this was the first report of multimeric plasmid DNA in cyanobacteria (Reaston et al., 1980). Analysis of plasmid preparations from *Synechocystis* 6803 revealed a plasmid designated pSS5, twice the size of pSS2, which may be a dimer (Elanskaya, personal communication). However, Castets et al. (1986) found that pSY1 (pSS2) and pSY2 (pSS5) share no homologous DNA sequences.

The absence of extrachromosomal DNA in some cyanobacteria (see Tandeau de Marsac and Houmard, 1987) is notable and this may be due to limitations of the isolation methods used or low copy number. Compared with *E. coli*, plasmids are not as easy to isolate from cyanobacteria, the large scale method developed by Van den Hondel et al. (1979) with slight modification (Reaston et al., 1980) works well and is recommended. Lambert and Carr (1982) tested several known plasmid isolation techniques for cyanobacteria, some of which may be useful for small scale isolations. A rapid method of plasmid isolation was described by Engwall and Gendel (1985), based on the "in-well lysis" procedure of Eckhardt (1978). A modified "in-well lysis" technique was used by Franche and Reynaud (1986) to screen several strains of *Anabaena* and *Nostoc*, the largest plasmid found was 150 Md in size. A similar method was successfully used much earlier for *Nostoc* PCC7524 (Reaston, unpublished observations). This plasmid isolation technique may be a good gentle method for the screening of very large plasmids in cyanobacteria. Recently, Rebiâre et al. (1986) characterised a mega-plasmid (1000-1500 kb) from *Synechococcus* PCC7942, using this technique. Their modified method appears to be successful for many cyanobacteria and it will be a useful plasmid screening technique for other strains, after efficient lysis conditions have been found, a critical point in this method.

The lack of detectable plasmid DNA in some cyanobacteria implies that they may not be essential in these microorganisms. Comparable functions would be expected to be encoded by cyanobacterial plasmids as found in other bacteria (Hardy, 1986). However, no genetic property has yet been definitely assigned to any cyanobacterial plasmid, which are therefore termed "cryptic". The involvement of plasmid functions in photosynthesis was postulated by Asato and Ginoza (1973) but no plasmid DNA was found in *Synechococcus* PCC6715 (Van den Hondel et al., 1979), so their presence may not be an absolute requirement for photosynthesis. Simon (1978) concluded that no correlation could be made between the presence or absence of plasmid DNA with either metabolic activity or cell differentiation in the filamentous strains he tested. The plasmid content and size did not vary in two non-heterocystous, non-revertable mutants, with those found in the wild-type of *Nostoc* PCC7524 (Reaston et al., 1980). Plasmids from *Anabaena* PCC7120 did not hybridise to cloned *nif* genes (Mazur et al., 1980). Carboxysomes from three cyanobacteria did not contain extrachromosomal DNA (Vakeria et al., 1984). Also, RuBisCo genes have only been located on chromosomal DNA in *Microcystis* PCC7820 (see Codd and Vakeria, 1987). Kumar et al. (1967) suggested that the loss of Sm^r from *A. nidulans* Tx20 after treatment with the curing agent proflavine was due to the loss of plasmid DNA but drug-resistance was never established to be plasmid encoded. The possibility of plasmid involvement in the toxicity of *Microcystis aeruginosa* WR70 was proposed by Hauman (1981) but the cultures tested were "non-axenic." Recent work by Kohl et al. (1987) has also suggested this plasmid function in strains of *Microcystis aeruginosa*. Vakeria et al. (1985) "cured" axenic cultures of *Microcystis aeruginosa* (PCC7820) of plasmid DNA and found no significant change in toxicity and it is "unlikely that plasmids are involved in the toxicity" of this organism. Plasmid sequence homology was found in the chromosomal DNA of toxin producing strains of *M. aeruginosa* (Kohl et al., 1987). To resolve these anomalies it would be useful if plasmids from *Microcystis* species were probed with synthetic DNA sequences derived from the sequence of proteins involved with hepatotoxic peptide formation. The possibility that bacteriocin production in cyanobacteria is plasmid encoded has been suggested by Flores and Wolk (1986) but to our knowledge this has not been proven. Castets et al. (1986) observed with *Synechocystis* 6803 that the loss of plasmid pSY2 and that of cell motility occurred at a high frequency but no direct correlation between these two phenomena was found. A plasmid from *Fremyella diplosiphon* was found to hybridise differentially to RNA from green and red light grown cultures, green light appeared to induce a plasmid encoded gene (Bogorad et al., 1983). It has been suggested by Kumar and Tripathi (1985) that extrachromosomal genetic elements are involved in heterocyst differentiation and polymixin B tolerance in a *Nostoc sp.*, "probably *N. linckia*". However, the obvious, crucial experiments to prove this proposal using the transformation system reported by the same authors (Tripathi and Kumar, 1985) have not been reported. Tandeau de Marsac and Houmard (1987) discuss other plasmid encoded possibilities in cyanobacteria, which range from toxin production to the encoding of restriction-modification systems. It is to be hoped that the cryptic plasmids of cyanobacteria will eventually be found to have some true function.

Six strains of unicellular cyanobacteria have been found to contain two plasmids of similar size and restriction patterns (Van den Hondel and Van Arkel, 1980). These homologous or identical plasmids were found in five closely related strains of *Synechococcus* 6301, 6908 (Van den Hondel et al., 1979; Lau and Doolittle, 1979), 6311, *A. nidulans* R2 and 602 (Van den Hondel and Van Arkel, 1980). The sixth strain was *Synechococcus* 6707 (Van den Hondel et al., 1979) whose DNA base composition (Herdman et al., 1979; Wilmotte and Stam, 1984) differs from the others by 13 mol% GC. The presence of homologous plasmids in different strains of unicellular cyanobacteria may indicate that these plasmids are naturally transferable or have a common history. Similar plasmid homology has been shown for similar strains in the LPP group of filamentous cyanobacteria (Potts, 1984). Regions of sequence homology have been observed between plasmids in the same and different species of cyanobacteria (Lau et al., 1980). It was suggested that these regions may represent transposable genetic elements (insertion sequences or transposons).

From the range of native cyanobacterial plasmids characterised, five have been used to construct shuttle vectors (see Pouwels et al., 1985; Tandeau de Marsac and Houmard, 1987). *A. nidulans* R2 contains two easily detectable plasmids pUH24 and pUH25 (Van den Hondel et al., 1980), which were redesignated pANS and pANL, respectively by Gendel et al. (1983a). The other plasmids used for vector construction in unicells are pAQ1 from *Agmenellum quadruplicatum* PR-6 (Buzby et al., 1983) and pSS2 (designated pUG1 by Chauvat et al., 1986) from *Synechocystis* 6803 (Elanskaya et al., 1985a). One plasmid, pDU1, from *Nostoc* 7524 (Reaston et al., 1982b) has been used for some filamentous cyanobacteria. Details on the use of these plasmids to construct shuttle vectors is given below in the "Host-Vector Systems " section of this review. The complete nucleotide sequence of plasmid pUH24 (7845 bp) has been reported (Weisbeek et al., 1985), this sequence will help to improve pUH24 as a useful cloning vector and may indicate this plasmid's coding capacity and the exact location of its replication origin.

RESTRICTION AND MODIFICATION

Cyanobacteria have proved to be rich sources of sequence-specific endonucleases (restriction enzymes), with some strains possessing enzymes with unique specificities. See Tandeau de Marsac and Houmard (1987) for the current list of cyanobacterial restriction enzymes. Three types of restriction enzyme (type I-III) have been characterised in other prokaryotes but only type II (sequence-specific) restriction enzymes have been described for cyanobacteria. The occurrence of restriction-modification systems in cyanobacteria was initially suggested by Doolittle (1979) based on the observations of methylated bases in the DNA of cyanobacteria (Kaye et al.. 1967; Pakhomova et al., 1968; Pakhomova, 1974).

The resistance of cyanobacterial chromosomal DNA to cleavage by some type II restriction enzymes has been reported by several workers (Van den Hondel et al., 1983; Herrero et al., 1984; Lambert and Carr, 1984; Cangelosi et al., 1986). This resistance to restriction has been attributed to DNA modification and/or lack of restriction sites. Future investigators wishing to study the cleavage of cloned cyanobacterial DNA would be recommended to use the truly modification deficient strain (*E. coli* GM48) which was previously used by Wolk et al. (1984). A DNA modification system similar to the *E. coli dam* methylation system has been found in some cyanobacteria (Brooks et al., 1983; Lambert and Carr, 1984). This *dam*-like system was described for *Anabaena variabilis* (PCC7118), the strain which possesses *Ava*I, II and III. Chromosomal DNA was resistant to cleavage by *Mbo*I, which will only cut if the adenine residue within the *dam* sequence (GATC) is unmethylated. However, the cyanobacterial DNA showed no sequence homology when probed with the *Eco dam* gene (Brooks et al., 1983) and the *Anabaena* methylase may represent a new class of adenine methylase acting at the sequence GATC. The possession of a similar modification system in *Nostoc* 7524 was previously postulated by Reaston (1981) and later found in this strain (Reaston, unpublished observations).

A possible role for restriction enzymes in cyanobacteria was suggested by Currier and Wolk (1979) who studied cyanophage infection in two filamentous cyanobacteria. This work was expanded by Duyvesteyn et al. (1983), who showed that a correlation could be made between cyanophage infection and the content of restriction enzymes in several cyanobacteria. Cyanobacterial restriction enzymes can also play an important role in the host-restriction of cloning vectors transferred into cyanobacteria (discussed later). By contrast, the involvement of a phage encoded restriction enzyme produced by cyanophage AS-1 in the breakdown of host DNA after infection of *Synechococcus* 6301 has been described by Szekeres (1981). This enzyme (AS-1 endonuclease) acts in a sequence-specific manner and recognises and cleaves at the sequence PuG.CPy (Szekeres et al., 1983). An enzyme with similar properties has not been reported for other bacteria infected with virulent bacteriophages. The DNA modification which protects AS-1 DNA against the action of the AS-1 endonuclease may also be unusual (Szekeres, personal communication). Cyanobacteria possess restriction-modification systems analogous to other prokaryote systems. The precise role(s) of these enzymes in cyanobacteria is not fully understood and requires further investigation. It may be possible that DNA methylation is important in the control of gene expression in cyanobacteria. Recently, the isolation and characterisation of the type II modification enzyme, M.*AquI*, from *Agmenellum quadruplicatum* PR-6 has been described by Karreman et al. (1986), this enzyme modifies DNA against the action of *AquI* and its isoschizomer *AvaI*. The methylation site within the *AvaI* recognition sequence modified by M.*AquI* was found at the first cytosine residue in the *AvaI* recognition sequence CPyCGPuG. The same modification site was postulated as the protection site for an isoschizomer of *AvaI*, *NspI* (redesignated *Nsp*7524III by Reaston et al., 1982a) in the DNA of *Nostoc* 7524 (Reaston, 1981).

HOST-VECTOR SYSTEMS

    The ideal gene-cloning system for cyanobacteria would have two major essential features, both having several optimum requirements. First, the cyanobacterial host must be able to accept DNA by an efficient gene-transfer system; have a high plating efficiency; have low mutation frequency for the markers used on the cloning vector; have no or low host-restriction; be cured of the native plasmid used to construct the cloning vector and be Rec⁻ to stabilise the autonomous vector and allow the generation of merodiploids. The second prerequisite is the availability of a gene-cloning vector being as small as possible which contains expressable genetic markers and unique cloning sites. The plasmid cloning vector can be either shuttle, able to replicate in both *E. coli* and the cyanobacterial host or integrative, capable of insertion into the host genome. Using *E. coli* permits a straightforward approach making use of all the powerful genetic techniques developed for this model organism. As yet no completely optimised host-vector system has been developed for one cyanobacterium. However, many of the above requirements have been solved for different cyanobacterial host-vector systems, which are described below.

    No *E. coli* vector or broad-host range plasmid has yet been found capable of autonomous replication in unicellular cyanobacteria (Van den Hondel et al., 1980; Kuhlemeier et al., 1981; Gendel et al., 1983a; Shestakov et al., 1985b; Dzelzkalns and Bogorad, 1986; Lightfoot and Wootton, 1987). The inability of *E. coli* plasmids,

lacking a cyanobacterial replication origin (pDU1), to be
autonomously maintained in filamentous cyanobacteria has also been
found (Wolk et al., 1984). Recently, Daniell et al. (1986) reported
on the transformation of the *E. coli* vector pBR322 into permeaplasts
of *Synechococcus* 6301, after prolonged contact of donor DNA with
recipient permeaplasts. These workers were able to isolate intact
molecules of pBR322 (based on gel size) from the transformed
cyanobacterial host. However, Lightfoot and Wootton (1987) were
unable to obtain or detect successful transfer of pBR322 in the same
strain. They constructed integrative vectors with plasmid pBR322 and
randomly isolated chromosomal DNA fragments from 6301 before reliable
and reproducible plasmid transfer was achieved. *E. coli* vector
transfer via transformation was found in *Synechocystis* 6803 after UV
irradiation (Dzelzkalns and Bogorad, 1986). They discovered that *E.
coli* plasmids, without a cyanobacterial origin of replication, were
inserted into the *Synechocystis* chromosome and did not replicate
autonomously in the host. This UV-induced integrative transformation
was independent of homologous recombination. A similar integration
mechanism may have occurred in the experiments of Daniell et al.
(1986) but this awaits confirmation. Thus, the construction of
suitable cloning vectors has relied on the introduction of selectable
genetic markers into the cryptic plasmids of cyanobacteria.
Elaboration of various shuttle vectors that can transfer and
replicate in both *E. coli* and cyanobacteria allowed the great
potential for genetic engineering in the latter microorganisms to be
developed. Lists of *E. coli*::Cyanobacteria shuttle vectors, with
their relevant characteristics (markers, restriction sites, etc.) are
available elsewhere (Pouwels et al., 1985; Tandeau de Marsac and
Houmard, 1987).

The most widely used cyanobacterium for gene-transfer and cloning
studies has been *Anacystis nidulans* R2. This strain is very
attractive for these purposes because it is highly transformable,
contains plasmid DNA, has a high plating efficiency and appears to
lack *in vivo* restriction enzyme activity (but see later). Several
hybrid plasmids capable of replication in *A. nidulans* R2 and *E. coli*
have been constructed from plasmid pUH24. Van den Hondel et al.
(1980) first described the *in vivo* labelling of this small plasmid
with a genetic marker from *E. coli* expressed in *A. nidulans* R2.
Plasmid pRI46 harbouring the transposon Tn*901* (Ap^r) was introduced by
transformation into R2 and the Tn*901* element transposed into plasmid
pUH24, this was an important breakthrough. From the resulting
drug-resistant plasmid (pCH1) a deletion derivative (pUC1) was
constructed *in vitro* which lacked one of the inverted repeats of
Tn*901*, to stabilise the marker and prevent further transposition.
Plasmid pUC1 could be transformed in *A. nidulans* R2 but not in *E.
coli* with a high frequency (10^{-4} - 10^{-5}). The presence of single
cleavage sites for *Bam*HI, *Xho*I, two for *Bgl*II and the selective
Ap^r-marker made this plasmid a good candidate for the construction of
gene-cloning vectors. A family of shuttle vectors have been prepared
on the basis of plasmid pUC1 by Van Arkel's group. This plasmid was
cloned into the well characterised *E. coli* vector pACYC184, carrying
a Cm^r determinant which expressed well in the cyanobacterial cells
(Kuhlemeier et al., 1981). Plasmid pUC104 (Ap^r, Cm^r) obtained by the
ligation of *Bam*HI cleaved pUC1 and pACYC184 was stably maintained in

150

A. nidulans R2. The frequency of transformation was higher after chloramphenicol selection compared with ampicillin selection. In general, the Ap^r-marker is less suitable for cloning studies with cyanobacteria than resistance to chloramphenicol, kanamycin/neomycin (Km/Nm) or streptomycin. Another shuttle vector, pUC105, also consisting of pACYC184 but inserted at a different location (a *BgI*II site) and orientation of pUC1 was less stable in *A. nidulans* R2, possibly because of the interaction with the resident plasmid pUH24 (Kuhlemeier et al., 1981).

Vector pUC104 isolated from *A. nidulans* R2 transformed the cyanobacterium more efficiently than from *E.coli*. This phenomenon was explained by assuming that *A. nidulans* R2 and *E. coli* differ with respect to their post–replicational DNA modification (Kuhlemeier et al., 1981). This proposal appears to be justified as Gallagher and Burke (1985a) presented data for this cyanobacterium possessing a sequence–specific endonuclease which could be expected to restrict transformation with foreign DNA in *A. nidulans* R2. The restriction enzyme, *Ani*I was tested with plasmid DNA from *Bacillus* strains instead of the normally used DNA substrate of bacteriophage Lambda, unexplained by Gallagher and Burke (1985a). It is possible that the *Bacillus* DNA was *dam⁻* (Brooks et al., 1983) and such unmodified DNA was a good substrate for *Ani*I. The results of Weisbeek et al. (1985) expanded the observations of Gallager and Burke (1985a) about the existence of restriction enzyme activity in *A. nidulans* R2. This enzyme appears to recognise a sequence which contains the *dam* sequence and is sensitive to *dam* methylation. It is possible that *Ani*I is the same enzyme as *Ssp*27144I from *Synechococcus* ATCC27144 (PCC6301) isolated by Dainippon–Ink. (1985), which is an isoschizomer of *Cla*I, sensitive to *dam* methylation. It was fortuitous that all the shuttle vectors for *A. nidulans* R2 were constructed in *dam⁻* strains of *E. coli.*

Plasmid pUC104 was successfully used to construct a cosmid vector (pPUC29) by ligating the cohesive end DNA sequences of λ DNA. This shuttle vector can be packaged *in vitro* into λ phage particles and introduced by infection into *E. coli* cells (Tandeau de Marsac et al., 1982).

To optimise the host–vector system for gene–cloning in *A. nidulans* R2, Kuhlemeier et al. (1983) isolated a host strain (R2-SPc) cured of the indigenous plasmid pUH24 and also constructed vectors with more suitable markers. The convenient hybrid plasmid constructed, pUC303, has two markers (Cm^r, Sm^r) and three unique restriction sites. A single *Eco*RI site is located within the Cm^r determinant, allowing screening of recombinants by insertional inactivation. Strains of *A. nidulans* R2 lacking the small plasmid pUH24 have been isolated in two ways, after treatment with SDS (Kuhlemeier et al., 1983) or as a result of spontaneous disappearance under stress conditions (Chauvat et al., 1983). In such plasmid cured hosts no sequences homologous with vector plasmids exist and hence segregation cannot occur. The host–vector systems created by Kuhlemeier et al. (1981; 1983) were efficiently used to isolate and reintroduce several genes into *A. nidulans* R2, two genes involved with nitrate reductase activity (Kuhlemeier et al., 1984a) and a methionine gene clone (Kuhlemeier et al., 1985). A third gene associated with nitrate reductase was cloned from this strain using a shuttle gene library (Kuhlemeier et al., 1984b).

Another limitation for gene-cloning in autonomously replicating vectors is associated with possible recombination between a plasmid-borne gene and its chromosomally located homologue. This recombinational instability can be minimised by dual selection for both the vector and for the chromosomal insert (Kuhlemeier et al., 1984a; 1985). Isolation of Rec⁻ strains of R2 would be of great advantage for performing complementation analysis and studies of dominance. One approach suggested by Borrias et al. (1985) to attempt the isolation of a Rec⁻ mutant is by cloning the cyanobacterial Rec gene in *E. coli* and subsequently using a derivative of it to replace the wild-type gene in R2. Borrias et al., 1987 have cloned a *rec*A gene from *A. nidulans* R2. Murphy et al. (1987) have cloned and partially sequenced an *E. coli* recA gene homologue from *Synechococcus* 7002, which complements Rec⁻ *E. coli* strains in recombinational and UV–survival assays. A *rec*A-like gene has recently been cloned and partially characterised from the filamentous cyanobacterium *Anabaena variabilis* (PCC7937) by Owttrim and Coleman (1987), which complements the recombinational deficiences of *E. coli* recA mutants. These gene clones may possibly help to isolate recombination deficient strains of these cyanobacteria, after suitable manipulation.

Other hybrid plasmids serving as shuttle vectors in *A. nidulans* R2 have been constructed with pUH24 and *E. coli* plasmids of the pBR-family. The hybrid shuttle vector pLS103 (Cmʳ), constructed by Sherman and Van de Putte (1982) was obtained via cointegrate formation between plasmids pUH24 and pBR322::Cm, *in vivo*, after transforming *A. nidulans* R2 with the *E. coli* plasmid. Golden and Sherman (1983) constructed vector pSG111 on the basis of plasmid pBR328, carrying Apʳ and Cmʳ determinants which can be transformed and a derivative mobilized into *A. nidulans* R2 (Wolk et al., 1984). The recombinant plasmid pDF30 (Apʳ, Cmʳ), described by Friedberg and Seijffers (1983) is composed of vector pBR325 and the replicon of the indigenous plasmid (pDF3), of *Synechococcus* 6311, which is homologous to pUH24 (Friedberg and Seijffers, 1983). The hybrid plasmid pECAN1 (Apʳ, Cmʳ) based on pBR325 was constructed by Gendel et al. (1983a). Most of the shuttle vectors mentioned above contain more than one selectable marker but only 3–4 unique cloning sites which limits the possibilities for gene-cloning in *A. nidulans* R2. To overcome this problem a series of hybrid plasmids were constructed with pUH24 containing *E. coli* plasmids with a multi-site polylinker (Gendel et al., 1983a,b). Plasmid pPLANB2 derived from the *E. coli* plasmid pDL13 provides seven single restriction sites and its shortened derivative (pCB4) has five suitable unique sites, which significantly increases the flexibility for the cloning of a variety of DNA restriction fragments. These vectors, which only contain the Apʳ marker, transform both *E. coli* and *A. nidulans* R2 with a relatively high frequency (up to 10⁻⁶). A similar strategy was used by Lau and Stauss (1985) who constructed several small, versatile shuttle vectors. Plasmid pXB7 (Apʳ) has 10 unique cloning sites; pECAN8 (Apʳ) contains 4 sites within the α *lacZ* gene, which permits easy detection of DNA inserts in the presence of Xgal in some *E. coli* strains; pKBX has 10 single cloning sites and was genetically labelled with Km^r as the favoured selectable marker. Friedberg and Seijffers (1986) have constructed a series of cloning vectors (pPL-class) which utilise regulatory signals of bacteriophage Lambda and the expression of cloned genes in *A. nidulans* R2 was initiated and controlled by these phage signals.

The large plasmid pUH25 (pANL) of *A. nidulans* R2 has also been of interest for the construction of shuttle vectors. Its origin of replication is compatible with that of pUH24. The possibility to use two distinct cloning vectors based on independent replicons would allow the simultaneous cloning of different fragments of foreign DNA into *A. nidulans* R2. A physical map of pANL for five restriction enzymes has been constructed by Laudenbach et al. (1983), who showed that a 11.7 kb *Bam*HI fragment contains a functional origin of replication. A series of compact shuttle vectors derived from this *Bam*HI fragment have been constructed by ligation with an *E. coli* polylinker in plasmid pDPL13 labelled with Ap^r. The creation of these vectors (pANL01a and pANL01b) by deletion of pPLANBa1 led to the discovery that pANL has two distinct origins of replication supporting transformation in *A. nidulans* R2. These putative *ori*-regions do not overlap and show no DNA homology to the replication origin of the small plasmid pANS of R2 (Laudenbach et al., 1985). Vectors of the pANL0-class can exist autonomously in *A. nidulans* R2 but the possibility of recombination between these hybrids and endogenous pANL might certainly effect the vector stability in recombination proficient hosts as described above for pUH24 (pANS). Elanskaya et al. (1985c) have constructed several hybrid plasmids consisting of large DNA fragments of *Bam*HI partially digested pANL into plasmid pUB110 from *Staphylococcus aureus*, encoding Km^r and inserted at different *Bam*HI sites in pANL, these hybrids could transform *A. nidulans* R2 with a high frequency. They replaced the endogenous plasmid in selective conditions and were stably maintained in the cyanobacterium. Plasmid pUB110 has also been inserted into the small plasmid (pANS) resulting in a vector which can replicate and confer Km^r to *A. nidulans* R2 but is unstable in *Bacillus subtilis* (Vilchinskas and Elanskaya, personal communication).

Hybrid plasmids have been constructed with the small, cryptic plasmid pAQ1 from *Agmenellum quadruplicatum* PR-6 by Buzby et al. (1983). The unique *Hind*III site in pAQ1 was used to ligate it into *Hind*III cleaved plasmids pBR322 and pBR325, these hybrids (pAQE2 and pAQE10) could transform either *E. coli* or PR-6. The *in vivo* action of the restriction enzyme *Aqu*I (*Ava*I) produced by PR-6 (Lau and Doolittle, 1980) on the transfer ability of these shuttle vectors was clearly shown (Buzby et al., 1983). Plasmids isolated from the transformed cyanobacterium could transform PR-6 much more efficiently than plasmid DNA from *E. coli*. It is likely that the DNA was modified by the cyanobacterium and such modified DNA was resistant to the *in vivo* action of *Aqu*I. When the single *Ava*I site in pBR322 and pBR325 was deleted from the shuttle vectors better transformation occurred with such deleted vectors (pAQE5 and pAQE11). Thus, *Aqu*I may hamper gene-cloning in *A. quadruplicatum* PR-6. It may be possible to protect cloned DNA against the action of *Aqu*I by *in vitro* modification with

153

the companion modification enzyme M.*AquI*, isolated from this strain
(Karreman et al., 1986). However, an *in vivo* modification step may be
possible as the restriction-modification genes for *AvaI* and *AvaII*
from *Anabaena variabilis* (PCC7118) have now been cloned into *E. coli*
(Lunnen and Wilson, cited from New England Biolabs, Catalogue
1986/87). After suitable manipulation, these clones may modify *AvaI*
and *AvaII* sites by an easy *in vivo* methylation step in *E. coli*. To
prevent the host-restriction caused by *AquI*, Buzby et al. (1985) have
isolated two derivatives of PR-6, designated G23 and G38, which still
possess *AquI* activity but transform better than the wild-type strain.
These strains appear to have some change in the *AquI* restriction-
modification system and possibly may indicate a duplication of the
methylase gene (Porter, personal communication) but the actual
mechanism(s) which improves the transformation ability in these
strains is unclear.

Another important characteristic of plasmid transformation in PR-6
is that dimeric and trimeric forms of biphasic plasmids gave
significantly higher transformation frequences than did analogous
monomeric forms (Buzby et al., 1983). In this characteristic PR-6
contrasts with R2, which shows more efficient transformation with
monomeric plasmids in the absence of the resident helper plasmid
(Chauvat et al., 1983). The shuttle vector pAQE17 (Buzby et al.,1985)
has been used to express allophycocyanin genes, from the cyanelle
Cyanophora parapoxa, in *Synechococcus* 7002 (De Lorimer et al., 1987).
In an article by Porter et al. (1986), gene-cloning by complementation and the problems which exist in producing merodiploids in
A. quadruplicatum PR-6 are discussed.

The elaboration of host-vector systems for the facultative
heterotrophic cyanobacterium *Synechocystis* 6803 are especially
important for studies of plant-like photosynthesis. Shuttle vectors
from *Synechococcus* strains are unable to replicate in 6803. Therefore
the small plasmid of *Synechocystis* 6803 has been used independently
for the creation of shuttle cloning vectors for this cyanobacterium
(Shestakov et al., 1985b; Chauvat et al., 1986). Shestakov et al.
(1985b) and Elanskaya et al. (1987) have cloned the small plasmid,
designated pSS2, cleaved with *Sau*3AI into the *Bam*HI site of plasmid
pACYC184. The hybrid plasmid pSE7 obtained from a Cm^r Tc^s
transformant of *E. coli* could transform *Synechocystis* 6803 but most
transformants were unstable. In a stable transformant a recombinant
plasmid pSE76 was found which appeared as a result of *in vivo*
recombination between pSE7 and the resident plasmid pSS2. Plasmid
pSE76 capable of autonomous replication in both *E. coli* and
Synechocystis 6803 has a Cm^r-marker and a unique site for *Sal*I. When
isolated from 6803, pSE76 transformed the wild-type strain at a
frequency of 10^{-6} to 10^{-5}. . To optimise this vector system the DNA
fragment of plasmid pUC-4K, carrying the Km^r determinant was inserted
in the *Sal*I site of plasmid pSE76. The resulting vector pSE176
contains two selectable markers, a number of single restric- tion
sites within the Km^r gene and a unique *Eco*RI site in the Cm^r-region,
which permits gene-cloning by insertional inactivation (Elanskaya, et
al., 1987). Using the same small plasmid (designated pUG1) and
pACYC177, Chauvat et al. (1986) constructed the shuttle vector
pUF311, encoding Km^r in *E. coli* and *Synechocystis* 6803. Cloning of

the Cm^r determinant from pACYC184 into pUF311 resulted in vector plasmid pFCLV7 (Cm^r, Km^r) which transformed the wild-type strain with a low frequency (3×10^{-7}). However, recipient cells containing pUF311 as a resident plasmid were transformed more efficiently by pFC LV7, indicating again the importance of recombin- ation between vector and resident plasmids for the stable maintenance of hybrid plasmids in this cyanobacterium. For the optimisation of a gene-cloning system in *Synechocystis* 6803 a Rec$^-$ strain would be very helpful. Such mutants deficient in homologous recombination were found among mutants sensitive to UV-light and mitomycin C. There was no transformation with either chromosomal markers or with integrative vectors (see below) in the Rec$^-$ recipient, while autonomously replicative vector pSE76 transformed such cells at a high frequency (Shestakov et al., 1985b).

The construction of cloning vehicles and their use in the genetic analysis of unicellular cyanobacteria proceeded by some years the development of reliable and reproducible gene-transfer with filamentous cyanobacteria. One reason for this discrepancy with the filamentous organisms is the presence of restriction enzymes, which many contain. To our knowledge only one heterocystous cyanobacterium (*Nostoc* PCC73102) lacks detectable type II restriction enzyme activity (Duyvesteyn and De Waard , personal communication). The host-restriction caused by the endonucleases in some filamentous cyanobacteria was elegantly prevented by the construction of suitable shuttle vectors by the reduction or elimination of the recognition sequences for the most common cyanobacterial restriction enzymes, *Ava*I and *Ava*II (Wolk et al., 1984). The cyanobacterial plasmid replicon used to construct the hybrid vectors was pDU1 (Reaston et al., 1982b) from *Nostoc* 7524, a strain which produces five type II restriction enzymes, *Nsp*7524I-V (Reaston et al., 1982a), lacks the recognition sites for these enzymes (Reaston, unpublished observ- ations) and therefore lacks the sites for *Ava*I and II (Wolk et al., 1984). A similar paucity of cyanobacterial restriction enzyme sites has been observed in the DNA of *Anabaena variabilis* (Herrero et al., 1984) and has been suggested for another cyanobacterium *Synechocystis* 6701 (Calléja et al., 1985). One of the vectors constructed by Wolk et al. (1984), pVW1C (Ap^r, Cm^r), contained 1 *Ava*I and 8 *Ava*II sites, was not transferred efficiently, unlike pRL1 (Cm^r) which had none of these sites. Three other shuttle vectors constructed from pRL1 are pRL5 (Cm^r, Sm^r), pRL6 (Cm^r, Km^r/Nm^r) and pRL8 (Cm^r, Em^r) have either 1 *Ava*I or 1 *Ava*II site but they could be stably transferred. A possible explanation for the ability of these plasmids (and pVW1C) to be transferred into *Ava*I and *Ava*II containing cyanobacteria is the mechanism used for the gene-transfer event. It is possible that during mobilization the vectors are transferred in a single-stranded form, which might escape efficient restriction by *Ava*I and II and the DNA was modified before conversion back into the double-stranded form. To further optimise a system for the genetic analysis of filamentous cyanobacteria, Schmetterer et al. (1986) have constructed mobilizable plasmid vectors which can express luciferase in *Anabaena* species. These *lux* genes may be good candidates for use as promoter probes for studies on cell differentiation in filamentous cyano- bacteria.

Chromosome integration appears to be a useful technique for stabilising the inheritance of heterologous genes that are not easily established on autonomously replicating vectors. Williams and Szalay (1983) suggested a method for integrating foreign DNA stably into the chromosome of *A. nidulans*   R2. They constructed chimeric DNA consisting of three distinct elements:-

1) fragment of cyanobacterial chromosomal DNA (insertional DNA) which mediates the integration of the chimeric molecule into the recipient chromosome;
2) fragment of foreign DNA (interrupting DNA) inserted within the chromosomal fragment (in this case plasmid pACYC184 encoding Cm^r was used);
3) fragment of foreign DNA (flanking DNA) which flanks but does not interrupt the insertional DNA (plasmid pBR322 encoding Ap^r in this case).

The hybrid circular molecule (pKW1061) produced in this way replicated autonomously as a plasmid in *E. coli* but not in the cyanobacterium. Such a chimeric molecule could transform recipient cells of *A. nidulans* R2 and served as an integrative vector. Three types of transformants were obtained: type I, resistant to Cm only; type II, resistant to Ap only; type III, resistant to both antibiotics. The predominant transformant was of type I which indicates the more efficient integration of "interrupting" than "flanking" DNA. Integration into the chromosome occurred at the site of homology between chromosomal and "insertional" DNA. Stable transformants of type I seem to be a result of non-reciprocal recombination while transformants of types II and III contain duplication of "insertional" DNA formed by a reciprocal crossover.

Using derivative vectors in which the lengths of flanking or interrupting DNA vary, Kolowsky et al. (1984) and Kolowsky and Szalay (1986) showed several properties of the integration system of *A. nidulans* R2. The length of foreign element rather than its position in the donor molecule determined the ratio of the three transformant types obtained. The stability and the efficiency of integration appeared to depend on the position of a foreign DNA element within the donor molecule; the most stable DNA integrate was found in type I transformants only when the foreign DNA was within the cyanobacterial "insertional" DNA. Foreign DNA linked to the ends of the cyanocterial fragment can integrate only in type II and III configurations which are more unstable. Foreign DNA of 20 kb in length was found to be completely stable in type I transformants. Thus, this integration system shows the means of stably inserting large pieces of heterologous DNA into the chromosome of *A. nidulans* R2. Using a similar approach, Elanskaya et al. (1985b) prepared a series of integrative vectors for *A. nidulans* R2 on the basis of plasmid pACYC184 carrying the Cm^r marker. These chimeric plasmids contain either *Bam*HI (pIAB-class) or *Hind*III (pIAH-class) fragments from chromosomal DNA of R2 (with sizes of 2.4–5.2 kb), transformed the cyanobacterium with frequencies of up to 10^{-4}. One of these integrative vectors (pIAH4) has been used by Scanlan (personal communication) to insert the gene for pectate lyase (an extracellular

enzyme) from *Erwinia carotovora* into chromosomal DNA of *A. nidulans*
R2. Plasmid pIAHPEL was detected by Southern hybridization but the
gene was not expressed. However, using an integration system a
foreign gene (*lacZ*) was expressed in R2 (Scanlan, personal
communication). On the basis of the chromosome integration approach
Elanskaya et al. (1985c) transferred the gene for α-amylase of
Bacillus amyloliquefaciens into *A. nidulans* R2 with the help of the
integrative vector pIAH4. The transformants obtained produced
α-amylase which was secreted into the periplasm. Vectors carrying the
α-amylase gene and its regulatory elements will be useful for
constructing plasmid molecules providing secretion of proteins from
cyanobacterial cells. Another example of a heterologous gene
expressed in *A. nidulans* R2 from Gram-positive bacteria is Km^r
encoded by the *Staphylococcus* plasmid pUB110 (Elanskaya et al.,
1985c; Gallagher and Burke, 1985b).

Integrative vectors (Cm^r) of two classes have been constructed for
Synechocystis 6803 (Shestakov et al., 1985a,b). Vectors of the
pSIS-class contain *Sau*3AI fragments (0.6–1.2 kb) of chromosomal DNA,
vectors of the pSIB-class possess *Bam*HI fragments (2.7–5.2 kb) cloned
into the *Bam*HI site of pACYC184. The presence of integrated plasmids
in the chromosome of Cm^r transformants was demonstrated by Southern
hybridization. Some of the chimeric plasmids have unique sites for
restriction endonucleases and will enable the integration of foreign
genes into 6803 chromosomal DNA. By constructing a series of chimeric
integrative vectors using a variety of chromosomal DNA inserts it
will be possible to stably introduce a multitude of foreign genes at
different locations into the chromosome.

The system of chromosomally targeted transformation developed by
Williams and Szalay (1983), for the introduction of genes cloned into
insertional vehicles, will be extremely useful for the genetic
analysis of native and foreign genes in cyanobacteria. Recently,
Vermaas et al. (1987) isolated the complete *psb*B gene, encoding the
Photosystem II protein CP-47, from *Synechocystis* 6803. Using a
"Km-resistance cartridge" inserted into the incomplete *psb*B gene
(previously isolated by heterologous Southern hybridization using the
*psb*B fragment from *Cyanophora paradoxa*), which transformed 6803. The
Km^r transformants obtained carried the Km-resistance gene solely in
the cyanobacterial *psb*B gene region. This work shows how powerful
homologous chromosome integration can be to both isolate and mutate
cyanobacterial genes. Homologous chromosome integration will allow
the study of gene interaction, the regulation of gene expression and
create cyanobacterial strains of potential use for biotechnology. The
development of reliable gene-transfer and host-vector systems in
cyanobacteria has allowed modern genetic approaches to be started
with this important group of microorganisms. These systems will be
extremely useful for future studies on the molecular biology of
cyanobacteria.

ACKNOWLEDGEMENTS

We would like to thank all the friends and colleagues, too numerous to mention separately, whose research and help made this review possible.

REFERENCES

ASATO, Y and GINOZA, H.S. (1973). Separation of small DNA molecules from the blue-green alga *Anacystis nidulans*. Nature New Biol. 244:132–133.

ASTIER, C. and ESPARDELLIER, F. (1976). Mise en évidence d'un syesteme de transfert génétique chez une cyanophycée du genre *Aphanocapsa*. C.R. Acad. Sci. Paris 282:795–797.

BAZIN, M.J. (1968). Sexuality in a blue-green alga: genetic recombination in *Anacystis nidulans*. Nature 218:282–283.

BOGORAD, L., GENDEL, S.M., HAURY, J.H. and KOLLER, K.P. (1983). Photomorphogenesis and complementary chromatic adaptation in *Fremyella diplosiphon*. In: Papageorgioö, G.C. and Parker, L. (eds.) Photosynthetic Prokaryotes: Cell Differentiation and Function. Elsevier, Amsterdam and New York, pp.119–126.

BORRIAS, W.E., DE VRIEZE, G. and VAN ARKEL, G. A. (1985). In search of the "recA" gene of the cyanobacterium *Anacystis nidulans* R2. In: Abst. V Inter. Symp. Photosyn. Prok. Grindelwald, Switzerland, p.329.

BORRIAS, W.E., VAN DER PLAS, J., VAN ARKEL, G.A. and WEISBEEK, P.J. (1987) Recombination as a genetic tool in the analysis of cyanobacterial genes. In: Abst. EMBO Eorkshop on Oxygenic and Anoxygenic Electron Systems in Cyanobacteria. Cape Souniou, Greece p.12.

BROOKS, J.E., BLUMENTHAL, R.M. and GINGERAS, T.R. (1983). The isolation and characterization of the *Escherichia coli* DNA adenine methylase (*dam*) gene. Nucleic Acids Res. 11:837–851.

BUZBY, J.S., PORTER, R.D. and STEVENS, S.E. (1983). Plasmid transformation in *Agmenellum quadruplicatum* PR-6: construction of biphasic plasmids and characterization of their transformation properties. J. Bacteriol. 154:1446–1450.

BUZBY, J.S., PORTER, R.D. and STEVENS, S.E. (1985). Expression of the *Escherichia coli lacZ* gene on a plasmid vector in a cyanobacterium. Science 230:805–807.

CALLÉJA, F., TANDEAU DE MARSAC, N., COURSIN, T., VAN ORMONDT, H. and DE WAARD, A. (1985). A new endonuclease recognizing the deoxynucleotide sequence CCNNGG from the cyanobacterium *Synechococcus* 6701. Nucleic Acids Res. 13:6745–6751.

CANGELOSI, G.A., JOSEPH, C.M., ROSEN , J.J. and MEEKS, J.C. (1986). Cloning and expression of a *Nostoc* sp. leucine biosynthetic gene in *Escherichia coli*. Arch. Microbiol. 145:315–321.

CANNON, R.E. and SHANE, M.S. (1972). The effect of antibiotic stress on protein synthesis in the establishment of lysogeny of *Plectonema boryanum.* Virology 49:130-133.

CANNON, R.E., SHANE, M.S. and BUSH, V.N. (1971). Lysogeny of a blue-green alga *Plectonema boryanum.* Virology 45:149-153.

CASTETS, A.M., HOUMARD, J. and TANDEAU DE MARSAC, N. (1986). Is cell motility a plasmid encoded function in the cyanobacterium *Synechocystis* 6803? FEMS Microbiol. Lett. 37:277-281.

CHAUVAT, F., ASTIER, C., VEDEL, F. and JOSET-ESPARDELLIER, F. (1983). Transformation in the cyanobacterium *Synechococcus* R2: improvement of efficiency; role of the plasmid pUH24 plasmid. Mol. Gen. Genet. 191:39-45.

CHAUVAT, F., DE VRIES, L., VAN DER ENDE, A. and VAN ARKEL, G.A. (1986). A host-vector system for gene cloning in the cyanobacterium *Synechocystis* PCC6803. Mol. Gen. Genet. 204:185-191.

COBLEY, J. (1985). Chromatic adaptation in *Fremyella diplosiphon.* I, mutants hypersensitive to green light, II, construction of mobilizable vectors lacking sites for *Fdi*I and II. In: Abst. V Inter. Symp. Photosyn. Prok. Grindelwald, Switzerland, p.105.

CODD, G.A. and VAKERIA, D. (1987). Enzymes and genes of microbial autotrophy. Microbiol. Sci. 4:154-159. CRAIG, R. and REICHELT, B.Y. (1986). Genetic engineering in algal biotechnology. Trends Biotechnol. Nov. 1986:280-285.

CURRIER, T.C. and WOLK, C.P. (1979). Characteristics of *Anabaena variabilis* influencing plaque formation by cyanophage N-1. J. Bacteriol. 139:88-92.

DAINIPPON-INK. (1985). Preparation method for a restriction endonuclease from a *Synechococcus* species; isolation and characterization. Derwent Biotechnol. Abst. 4:98.

DANIELL, H. and McFADDEN, B.A. (1986). Characterization of DNA uptake by the cyanobacterium *Anacystis nidulans*. Mol. Gen. Genet. 204:243-248.

DANIELL, H., SAROJINI, G. and McFADDEN, B.A. (1986). Transformation of the cyanobacterium *Anacystis nidulans* 6301 with the *Escherichia coli* plasmid pBR322. Proc. Natl. Acad. Sci. USA 83:2546-2550.

DELANEY, S.F. and REICHELT, B.Y. (1982). Integration of the R plasmid, R68.45, into the genome of *Synechococcus* PCC6301. In: Abst. IV Inter. Symp. Photosyn. Prok. Bombannes, France, D5.

DELANEY, S.F., HERDMAN, M. and CARR, N.G. (1976). Genetics of blue-green algae. In: Lewin, R.A. (ed.) Genetics of the Algae. Blackwell, Oxford, pp.7-28.

DE LORIMIER, R., GUGLIELMI, G., BRYANT, D.A. and STEVENS, S.E. (1987). Functional expression of plasmid allophycocyanin genes in a cyanobacterium. J. Bacteriol. 169:1830-1835.

DEVILLY, C.I. and HOUGHTON, J.A. (1977). A study of genetic transformation in *Gleocapsa alpicola*. J. Gen. Microbiol. 98:277-280.

DOOLITTLE, W.F. (1979). The cyanobacterial genome, its expression, and the control of that expression. Adv. Microbiol. Physiol. 20:1-102.

DUYVESTEYN, M.G.C., KORSUIZE, J., DE WAARD, A., VONSHAK, A. and WOLK, C.P. (1983). Sequence-specific endonucleases in strains of *Anabaena* and *Nostoc*. Arch. Microbiol. 134:276-281.

DZELZKALNS, V.A. and BOGORAD, L. (1986). Stable transformation of the cyanobacterium *Synechocystis* sp. PCC6803 induced by UV irradiation. J. Bacteriol. 165:964-971.

ECKHARDT, T. (1978). A rapid method for the identification of plasmid desoxyribonucleic acid in bacteria. Plasmid 1:584-588.

ELANSKAYA, I.V., BIBIKOVA, M.V., BOGDANOVA, S.L., BABYKIN, M.M. and SHESTAKOV, S.V. (1987). Bireplonic vector plasmids for cyanobacterium *Synechocystis* 6803 and *Escherichia coli*. Dokl. Akad. Nauk. SSSR 293:716-719.
ELANSKAYA, I.V., BIBIKOVA, M.V., BOGDANOVA, S.L., KOKSHAROVA, T.A., AGAMALOVA, S.R., NIKITINA, E.I. and SHESTAKOV, S.V. (1985a). Plasmids from cyanobacterium *Synechocystis* sp. 6803. Mol. Genet. Mikrobiol. Virusol. (8):19-21.
ELANSKAYA, I.V., MORZUNOVA, I.B. and SHESTAKOV, S.V. (1985b). Recombinant plasmids capable of integration with the chromosome of cyanobacterium *Anacystis nidulans* R2. Mol. Genet. Mikrobiol. Virusol. (11):20-24.
ELANSKAYA, I.V., MORZUNOVA, I.B. and SHESTAKOV, S.V. (1985c). Expression of Gram-positive bacteria genes in cyanobacterium *Anacystis nidulans* R2. In: Abst. V Inter. Symp. Photosyn. Prok. Grindelwald, Switzerland, p.372.
ENGWALL, K.S. and GENDEL, S.M. (1985). A rapid method for identifying plasmids in cyanobacteria. FEMS Microbiol. Lett. 26:337-340.
FLORES, E. and WOLK, C.P. (1985). Identification of facultatively heterotrophic N2-fixing cyanobacteria able to receive plasmid vectors from *Escherichia coli* by conjugation. J. Bacteriol. 162:1339-1341.
FLORES, E. and WOLK, C.P. (1986). Production, by filamentous, nitrogen-fixing cyanobacteria, of a bacteriocin and of other antibiotics that kill related strains. Arch. Microbiol. 145:215-219.
FRANCHE, C. and REYNAUD, P.A. (1986). Characterization of several tropical strains of *Anabaena* and *Nostoc*: morphological and physiological properties and plasmid content. Ann. Inst. Pasteur/Microbiol. 137A:179-197.
FRIEDBERG, D. and SEIJFFERS, J. (1979). Plasmids in two cyanobacterial strains. FEBS Lett. 107:165-168.
FRIEDBERG, D. and SEIJFFERS, J. (1983). A new hybrid plasmid capable of transforming *Escherichia coli* and *Anacystis nidulans*. Gene 22:267-275.
FREIDBERG, D. and SEIJFFERS, J. (1986). Controlled gene expression utilising Lambda phage regulatory signals in a cyanobacterium host. Mol. Gen. Genet. 203:505-510.
GALLAGHER, M.L. and BURKE, W.F. (1985a). Sequence-specific endonuclease from the transformable cyanobacterium *Anacystis nidulans* R2. FEMS Microbiol . Lett. 26:317-321.
GALLAGHER, M.L. and BURKE, W.F. (1985b). Expression by *Anacystis nidulans* R2 of an antibiotic resistance gene originating from a Gram-positive organism. In: Abst. 85th. Am. Soc. Microbiol., annual meeting 1985:h155.
GENDEL, S., STRAUS, N., PULLEYBLANK, D. and WILLIAMS, J. (1983a). A novel shuttle cloning vector for the cyanobacterium *Anacystis nidulans*. FEMS Microbiol. Lett. 19:291-294.
GENDEL, S., STRAUS, N., PULLEYBLANK, D. and WILLIAMS, J. (1983b). Shuttle cloning vectors for the cyanobacterium *Anacystis nidulans*. J. Bacteriol. 156:148-154.
GOLDEN, S.S. and SHERMAN, L.A. (1983). A hybrid plasmid is a stable cloning vector for the cyanobacterium *Anacystis nidulans* R2. J. Bacteriol. 155:966-972.
GOLDEN, S.S. and SHERMAN , L.A. (1984). Optimal conditions for genetic transformation of the cyanobacterium *Anacystis nidulans* R2. J. Bacteriol. 158:36-42.

GRAY, M.W. and DOOLITTLE, W.F. (1982). Has the endosymbiont hypothesis been proven? Microbiol. Rev. 46:1-42.
GRIGORIEVA , G.A. (1985). Heterospecific transformation in the genus Synechococcus. Biol. Nauki 9:91-94.
GRIGORIEVA, G.A. and SHESTAKOV, S.V. (1976). The application of the genetic transformation method for the taxonomic analysis of unicellular blue-green algae . In: Abst. II Inter. Symp. Photosyn, Prok. Dundee, Scotland. pp.220-221.
GRIGORIEVA, G.A. and SHESTAKOV, S.V. (1982). Transformation in the cyanobacterium Synechocystis sp. 6803. FEMS Microbiol. Lett. 13:367-370.
HARDY, K. (1986). Bacterial Plasmids, Aspects of Microbiology 4. Van Nostrand, Reinhold, UK.
HASELKORN, R. (1985a). Genes and gene transfer in cyanobacteria. Plant Mol. Biol. Reporter 3:24-32.
HASELKORN, R. (1985b). Gene transfer systems in cyanobacteria. In: Workshop Proc. New Delhi, Sept. 1985:in press.
HAUMAN, J.H. (1981). Is a plasmid(s) involved in the toxicity of Microcystis aeruginosa? In: Carmichael, W.W. (ed.) The Water Environment Algal Toxins and Health. Plenum Press, New York and London, pp.97-102.
HERDMAN, M. (1973). Mutations arising during transformation in blue-green algae Anacystis nidulans. Mol Gen. Genet. 120:369-378.
HERDMAN, M. (1982). Evolution and genetic properties of cyanobacterial genomes. In: Carr, N.G. and Whitton, B.A. (eds.) The Biology of Cyanobacteria. Blackwell, Oxford, pp.263-305.
HERDMAN, M. and CARR, N.G. (1971). Recombination in Anacystis nidulans mediated by an extracellular DNA/RNA complex. J. Gen. Microbiol. 68:XIV.
HERDMAN, M., JANVIER, M., WATERBURY, J.D., RIPPKA, R., STANIER, R.Y. and MANDEL, M. (1979). Deoxyriboncleic acid base composition of cyanobacteria. J. Gen. Microbiol. 111:63-71.
HERRERO, A. and WOLK, C.P. (1986). Genetic mapping of the chromosome of the cyanobacterium Anabaena variabilis. J. Biol. Chem. 261:7748-7754 .
HERRERO, A., ELHAI, J., HOHN, B. and WOLK, C.P. (1984). Infrequent cleavage of cloned Anabaena variabilis DNA by restriction endonucleases from A. variabilis. J. Bacteriol. 160:781-784.
HOLLOWAY, B.A. (1979). Plasmids that mobilize bacterial chromosome. Plasmid 2:1-19.
KARREMAN, C., TANDEAU DE MARSAC, N. and DE WAARD, A. (1986). Isolation of a deoxycytidylate methyl transferase capable of protecting DNA uniquely against cleavage by endonuclease R.AquI (isoschizomer of AvaI). Nucleic Acids Res. 14:5199-5205.
KAYE, A.M., SALOMON, R. and FRIDLENDER, B. (1967). Base composition and presence of methylated bases in DNA from a blue-green alga Plectonema boryanum. J. Mol. Biol. 24:479-483.
KHUDYAKOV, I.Y. and GROMOV, B.V. (1973). Temperate cyanophage A-4(L) of blue-green alga Anabaena variabilis. Mikrobiologiya 42:904-907.
KIRBY, N.W., MUSGRAVE, S.C., ROWELL, P., SHESTAKOV, S.V. and STEWART, W.D.P. (1986). Photoproduction of ammonium by immobilized mutant strains of Anabaena variabilis. Appl. Microbiol. Biotechnol. 24:42-46.
KOHL, J.-G., BÖRNER, Th., HENNING, M., SCHWABE, W. and WEIHE, A. (1987). Plasmid content and differences in ecologically important

characteristics of different strains of *Microcystis aeruginosa*. Arch. Hydrobiol. Suppl.:in press.
KOLOWSKY , K.S. and SZALAY, A.A. (1986). Double-stranded gap repair in the photosynthetic prokaryote *Synechococcus* R2. Proc. Natl. Acad. Sci. USA 83:5578-5582.
KOLOWSKY, K.S., WILLIAMS, J.G.K. and SZALAY, A.A. (1984). Length of foreign DNA in chimeric plasmids determines the efficiency of its integration into the chromosome of the cyanobacterium *Synechococcus* R2. Gene 27:289-299.
KUHLEMEIER, C.J., BORRIAS,W.E., VAN DEN HONDEL, C.A.M.J.J. and VAN ARKEL, G.A. (1981). Construction and characterization of two hybrid plasmids capable of transformation to *Anacystis nidulans* R2 and *Escherichia coli* K12. Mol. Gen. Genet. 184:249-254.
KUHLEMEIER, C.J., THOMAS, A.A.M., VAN DER ENDE, A., VAN LEEN, R.W., BORRIAS, W.E., VAN DEN HONDEL, C.A.M.J.J. and VAN ARKEL, G.A. (1983). A host-vector system for gene cloning in the cyanobacterium *Anacystis nidulans* R2. Plasmid 10:156-163.
KUHLEMEIER, C.J., LOGTENBERG, T., STOORVOGEL, W., VAN HEUGTEN, H.A.A., BORRIAS, W.E. and VAN ARKEL, G.A. (1984a). Cloning of two nitrate reductase genes from the cyanobacterium *Anacystis nidulans* R2. J. Bacteriol. 159:36-41.
KUHLEMEIER, C.J., TEEUWSEN, V.J.P., JANSSEN, M.J.T. and VAN ARKEL, G.A. (1984b). Cloning of a third nitrate reductase gene from the cyanobacterium *Anacystis nidulans* R2 using a shuttle gene library. Gene 31:109-116.
KUHLEMEIER, C.J., HARDON, E.M., VAN ARKEL, G.A. and VAN DE VATE, C. (1985). Self-cloning in the cyanobacterium *Anacystis nidulans* R2: fate of a cloned gene after reintroduction. Plasmid 14:200-208.
KUMAR, H.D. (1962). Apparent genetic recombination in a blue-green algae. Nature 196:1121-1122.
KUMAR, H.D. and UEDA, K. (1984). Conjugation in the cyanobacterium *Anacystis nidulans*. Mol. Gen. Genet. 195:356-357.
KUMAR, H.D. and TRIPATHI, A.K. (1985). Mutagenic and plasmid-eliminating action of acridine orange in the cyanobacterium *Nostoc*. Curr. Sci. 54:845-852.
KUMAR, H.D., SINGH, H.N. and PRAKASH, G. (1967). The effect of proflavine on different strains of blue-green alga, *Anacystis nidulans*. Plant Cell Physiol. 8:171-179.
LADHA, J.K. and KUMAR, H.D. (1978). Genetics of blue-green algae. Biol. Rev. 53:355-386.
LAMBERT, G.R. and CARR, N.G. (1982). Rapid small-scale plasmid isolation by several methods from filamentous cyanobacteria. Arch. Microbiol. 133:122-125.
LAMBERT, G.R. and CARR, N.G. (1984). Resistance of DNA from filamentous and unicellular cyanobacteria to restriction endonuclease cleavage. Biochim. Biophys. Acta 781:45-55.
LAU, R.H. and DOOLITTLE, W.F. (1979). Covalently closed circular DNAs in closely related unicellular cyanobacteria. J. Bacteriol. 137:648-652.
LAU, R.H. and DOOLITTLE, W.F. (1980). *Aqu*I: a more easily purified isoschizomer of *Ava*I. FEBS Lett. 121:200-202.
LAU, R.H. and STRAUS, N.A. (1985). Versatile shuttle cloning vectors for the unicellular cyanobacterium *Anacystis nidulans* R2. FEMS Microbiol. Lett. 27:253-256.

LAU, R.H., SAPIENZA, C. and DOOLITTLE, W.F. (1980). Cyanobacterial plasmids: their widespread occurence and the existence of regions of homology between plasmids in the same and different species. Mol. Gen. Genet. 178:203–211.

LAUDENBACH, D.E., STRAUS, N.A., GENDEL, S. and WILLIAMS, J.P. (1983). The large endogenous plasmid of Anacystis nidulans: mapping, cloning and localization of the origin of replication. Mol. Gen. Genet. 192:402–407.

LAUDENBACH, D.E., STRAUS, N.A. and WILLIAMS, J.P. (1985). Evidence for two distinct origins of replication in the large endogenous plasmid of Anacystis nidulans R2. Mol. Gen. Genet. 199:300–305.

LIGHTFOOT, D.A. and WOOTTON, J.C. (1987). Stable transformation of the cyanobacterium Synechococcus PCC6301 using cloned DNA. J. Gen. Microbiol.:in press.

McFARLANE, G.J.B., MACHRAY, G.C. and STEWART, W.D.P. (1987). A simplified method for conjugal gene transfer into the filamentous cyanobacterium Anabaena sp. ATCC27893. J. Microbiol. Meth. 6:301–305.

MAZUR, B.J., RICE, D. and HASELKORN, R. (1980). Identification of blue-green algal nitrogen fixation genes by using heterologous hybridization probes. Proc. Natl. Acad. Sci. USA 77:186–190.

MITRONOVA, T.N., SHESTAKOV, S.V. and ZHEVNER, V.D. (1973). Properties of filamentous mutant of blue-green alga Anacystis nidulans sensitive to irradiation. Microbiologiya 42:519–524.

MURPHY, R.C., BRYANT, D.A. and PORTER, R.D. (1986). Molecular cloning and preliminary characterization of a recA gene from the cyanobacterium Synechococcus PCC7002. In: Biggins, J. (ed.) Progress in Photosynthesis Research, vol.4. Martinus Nijhoff, Dordrecht. The Netherlands, pp. 769–772. ORKWISZEWSKI, K.G. and KANEY, A.R. (1974). Genetic transformation of the blue-green bacterium Anacystis nidulans. Arch. Microbiol. 98:31–37.

OWTTRIM, G.W. and COLEMAN, J.R. (1987). Molecular cloning of a recA like gene from the cyanobacterium Anabaena variabilis. J. Bacteriol. 169:1824–1829.

PADAN, E., SHILO, M. and OPPENHEIM, A.B. (1972). Lysogeny of the blue-green alga Plectonema boryanum by LPP-2SPI cyanophage. Virology 47:525–526.

PADHY, R.N. and SINGH, P.K. (1978a). Lysogeny in the blue-green alga Nostoc muscorum. Arch. Microbiol. 117:265–268.

PADHY, R.N. and SINGH, P.K. (1978b). Genetical studies on the heterocyst and nitrogen fixation of the blue-green alga Nostoc muscorum. Mol. Gen. Genet. 162:203–211.

PAKHOMOVA, M.V . (1974). N⁶-Dimethylaminopurine in the DNA of certain species of algae. Dokl. Biochem. 214:71–73.

PAKHOMOVA, M.V., ZAITSEVA, G.N. and BELOZERSKII, A.N. (1968). Presence of 5-Methylcytosine and 6-Methylaminopurine in the DNA of some algae. Dokl. Biochem. 182:227–229.

PEMBERTON, J.M. and BOWEN, A.R.S. (1981). High-frequency chromosome transfer in Rhodopseudomonas sphaeroides promoted by broad-host range plasmid RP1 carrying mercury transposon Tn501. J. Bacteriol. 147:110–117.

PIKALEK, P. (1967). Attempt to find genetic recombination in Anacystis nidulans. Nature 215:666.

POLUKHINA, L.E. and SHESTAKOV, S.V. (1977). UV-sensitive and UV-resistant mutants of blue-green alga Anacystis nidulans. Genetika 13:689–695.

PORTER, R.D. (1986). Transformation in cyanobacteria. CRC Crit. Rev. Microbiol. 13:111-132.
PORTER, R.D., BUZBY, J.S., PILON, A., FIELDS, P. I., DUBBS, J.M. and STEVENS, S.E. (1986). Genes from the cyanobacterium *Agmenellum quadruplicatum* isolated by complementation: characterization and production of merodiploids. Gene 41:249-260.
POTTS, M. (1984). Distribution of plasmids in cyanobacteria of the LPP group. FEMS Microbiol. Lett. 24:351-354.
POUWELS, P.H., ENGER-VALK, B.A. and BRAMMAR, W.J. (1985). Vectors for cyanobacteria. In: Cloning Vectors, A Laboratory Manual. Elsevier, Amsterdam, New York, Oxford, sections IX-1 - IX5.
REASTON, J. (1981). Studies on the Molecular Biology of the Filamentous Cyanobacterium *Nostoc* PCC7524. Ph.D. Thesis. The University, Dundee, Scotland.
REASTON, J., VAN DEN HONDEL, C.A.M.J.J., VAN DER ENDE, A., VAN ARKEL, G.A., STEWART, W.D.P. and HERDMAN, M. (1980). Comparison of plasmids from the cyanobacterium *Nostoc* PCC7524 with two mutant strains unable to form heterocysts. FEMS Microbiol Lett. 9:185-188.
REASTON, J., DUYVESTEYN, M.G.C. and DE WAARD, A. (1982a). *Nostoc* PCC7524, a cyanobacterium which contains five sequence-specific deoxyribonucleases. Gene 20:103-110.
REASTON, J., VAN DEN HONDEL, C.A.M.J.J., VAN ARKEL, G.A. and STEWART, W.D.P. (1982b). A physical map of plasmid pDU1 from the cyanobacterium *Nostoc* PCC7524. Plasmid 7:101-104.
REBIERE, M.C., CASTETS, A.M., HOUMARD, J. and TANDEAU DE MARSAC, N. (1986). Plasmid distribution among unicellular and filamentous cyanobacteria : occurence of large and mega-plasmids. FEMS Microbiol. Lett. 37:269-275.
RESTAINO, L. and FRAMPTON, E.W. (1975). Labelling deoxyribonucleic acid of *Anacystis nidulans.* J. Bacteriol. 124:155-160.
RIMON, A. and OPPENHEIM, A.B. (1975). Heat - induction of the blue-green alga *Plectonema boryanum* lysogenic for the cyanophage SPIcts1. Virology 64:454-463.
RIPPKA, R., DERUELLES, J., WATERBURY, J.B., HERDMAN, M. and STANIER, R.Y. (1979). Generic assignments, strain histories and properties of pure cultures of cyanobacteria. J. Gen. Microbiol. 111:1-61.
ROBERTS, T.M. and KOTHS, K.E. (1976). The blue-green alga *Agmenellum quadruplicatum* contains covalently closed DNA circles. Cell 9:551-557.
ROMANOVA, N.I. and SHESTAKOV, S.V. (1976). Isolation of auxotrophic mutants in blue-green alga *Anacystis nidulans.* Dokl. Acad. Nauk SSSR 226:692-694.
SCHMETTERER, G., WOLK, C.P. and ELHAI, J. (1986). Expression of luciferases from *Vibrio harveyi* and *Vibrio fisheri* in filamentous cyanobacteria. J. Bacteriol. 167:411-414.
SHERMAN, L.A. and VAN DE PUTTE, P. (1982). Construction of a hybrid plasmid capable of replication in the bacterium *Escherichia coli* and the cyanobacterium *Anacystis nidulans.* J. Bacteriol. 150:410-413.
SHESTAKOV, S.V. and KHYEN, N.T. (1970). Evidence for genetic transformation in blue-green alga *Anacystis nidulans.* Mol. Gen. Genet. 107:372-375.
SHESTAKOV, S.V. and MITRONOVA, T.N. (1971). Mutants of blue-green alga *Anacystis nidulans* resistant to erythromycin and streptomycin. Biol. Nauki (14):98-102.
SHESTAKOV, S.V., ELANSKAYA, I.V., GLAZER, V.M. and ZHEVNER, V.D.

(1975). Repair of breaks induced by X-rays in DNA of blue-green algae *Anacystis nidulans*. Dokl. Akad. Nauk SSSR 222:1217-1219.

SHESTAKOV, S.V., KARBISHEVA, E.A. and ELANSKAYA, I.V. (1982). The nature of damage in mutants of *Anacystis nidulans* deficient in genetic transformation. Genetika 18:1271-1275.

SHESTAKOV, S.V., ELANSKAYA, I.V. and BIBIKOVA, M.V. (1985a). Integrative vectors for the cyanobacterium *Synechocystis sp.* 6803. Dokl. Akad. Nauk SSSR 282:176-179.

SHESTAKOV, S., ELANSKAYA, I. and BIBIKOVA, M. (1985b). Vectors for gene cloning in *Synechocystis* sp. 6803. In: Abst. V Inter. Symp. Photosyn. Prok. Grindelwald, Switzerland, p.109.

SIMON, R.D. (1978). Survey of extrachromosomal DNA found in the filamentous cyanobacteria. J. Bacteriol. 136:414-418.

SINGH, H.N. (1967). Genetic control of sporulation in the blue-green alga *Anabaena doliolum*. Planta 75:33-38.

SINGH, P.K. (1975). Lysogeny of blue-green alga *Plectonema boryanum* by long-tailed virus. Mol. Gen. Genet. 137:181-183.

SINGH, R.N. and SINHA, R. (1965). Genetic recombination in a blue-green alga, *Cylindrospermum majus* Kuetz. Nature 207:782-783.

SINGH, R.N. and SINGH, P.K. (1972). Transduction and lysogeny in Blue-Green Algae. Univ. Madras Press, India, pp.258-261.

SINGH, D.T., NIRMALA, K., MODI, D.R., and SINGH, H.N. (1987). Genetic transfer of herbicide resistance gene(s) from *Gloeocapsa* spp. to *Nostoc muscorum*. Mol. Gen. Genet. 208: 436-438. STANIER, R.Y. and COHEN-BAZIRE, G. (1977). Phototrophic prokaryotes: the cyanobacteria. Ann. Rev. Microbiol. 31:225-274.

STEVENS, S.E. and PORTER, R.D. (1980). Transformation in *Agmenellum quadruplicatum*. Proc. Natl. Acad. Sci. USA 77:6052-6056.

STEVENS, S.E. and PORTER, R.D. (1986). Heterospecific transformation among cyanobacteria. J. Bacteriol. 167:1074-1076.

STEWART, W.D.P. and SINGH, H.N. (1975). Transfer of nitrogen fixing (nif) genes in the blue-green alga *Nostoc muscorum*. Biochem. Biophys. Res. Comm. 62:62-69.

STOLETOV, V.N., ZHEVNER, V.D., GARIBYAN, D.V. and SHESTAKOV, S.V. (1965). Pigment mutations induced by nitrosomethylurea in *Anacystis nidulans*. Genetika 6:61-66.

SZEKERES, M. (1981). Phage-induced development of a site-specific endonuclease in *Anacystis nidulans*, a cyanobacterium. Virology 111:1-10.

SZEKERES, M., SZMIDT, A.E. and TÖRÖK, I. (1983). Evidence for a restriction/modification-like system in *Anacystis nidulans* infected by cyanophage AS-1. Eur. J. Biochem. 131:137-141.

SZYBALSKI, W. and BRYSON, V. (1952). Genetic studies on microbial cross resistance to toxic agents. I, cross resistance to Escherichia coli to fifteen antibiotics. J. Bacteriol. 64:489-499.

TANDEAU DE MARSAC, N. and HOUMARD, J. (1987). Advances in cyanobacterial molecular genetics. In: Van Baalen, C. and Fay, P. (eds.) The Cyanobacteria: A Comprehensive Review. Elsevier. Amsterdam. pp. 251-295.

TANDEAU DE MARSAC, N., BORRIAS, W.E., KUHLEMEIER, C.J., CASTETS, A.M., VAN ARKEL, G.A. and VAN DEN HONDEL, C.A.M.J.J. (1982). A new approach for molecular cloning in cyanobacteria: cloning of an *Anacystis nidulans met* gene using a Tn*901*-induced mutant. Gene 20:111-119.

TREHAN, K. and SINHA, U. (1981). Genetic transfer in a nitrogen

fixing filamentous cyanobacterium. J. Gen. Microbiol. 124:349-352.
TREHAN, K. and SINHA, U. (1982). DNA-mediated transformation in
Nostoc muscorum, a nitrogen-fixing cyanobacterium. Aust. J. Biol.
Sci. 35:573-577.
TRIPATHI, A.K. and KUMAR, H.D. (1985). Transformation in the
cyanobacterium *Nostoc*. Biol. Zbl. 104:239-243.
VAISHAMPAYAN, A. and PRASAD, A.B. (1984). Inter-strain transfer of a
pesticide-resistant marker in the N2-fixing cyanobacterium *Nostoc
muscorum*. Mol. Gen. Genet. 193:195-197.
VAKERIA, D., CODD, G.A. MARSDEN, W.J.N. and STEWART, W.D.P. (1984).
No evidence for DNA in cyanobacterial carboxysomes. FEMS Microbiol.
Lett. 25:14 9-152.
VAKERIA, D., CODD, G.A., BELL, S.G., BEATTIE, K.A. and PRIESTLY, I.M.
(1985). Toxicity and extrachromosomal DNA in strains of the cyano-
bacterium *Microcystis aeruginosa*. FEMS Microbiol. Lett. 29:69-72.
VAN BAALEN, C. (1965). Quantitative surface plating of coccoid
blue-green algae. J. Phycol. 1:19-22.
VAN DEN HONDEL, C.A.M.J.J. and VAN ARKEL, G.A. (1980). Development of
a cloning system in cyanobacteria. Antonie Leeuwenhoek 46:228-229.
VAN DEN HONDEL, C.A.M.J.J., KEEGSTRA, W., BORRIAS, W.E . and VAN
ARKEL, G.A. (1979). Homology of plasmids in strains of unicellular
cyanobacteria. Plasmid 2:323-333.
VAN DEN HONDEL, C.A.M.J.J., VERBEEK, S., VAN DER ENDE, A., WEISBEEK,
P.J., BORRIAS, W.E. and VAN ARKEL, G.A. (1980). Introduction of
transposon Tn901 into a plasmid of *Anacystis nidulans*: preparation
for cloning in cyanobacteria. Proc. Natl. Acad. Sci. USA
77:1570-1574.
VAN DEN HONDEL, C.A.M.J.J., VAN LEEN, R.W., VAN ARKEL, G.A.,
DUYVESTEYN, M.G.C. and DE WAARD, A. (1983). Sequence-specific
nucleases from the cyanobacterium *Fremyella diplosiphon*, and a
peculiar resistance of its chromosomal DNA towards cleavage by other
restriction enzymes. FEMS Microbiol. Lett. 16:7-12.
VERMAAS, W.F.J., WILLIAMS, J.G.K. and ARNTZEN, C.J. (1987).
Sequencing and modification of *psb*B, the gene encoding the CP-47
protein of Photosystem II, in the cyanobacterium *Synechocystis* 6803.
Plant Mol. Biol. 8:317-326.
WEISBEEK, P.J., TEERSTRA, R., VAN DIJK, M., BLOEMHEUVEL, G., DE BOER,
D., VAN DER PLAS, J ., BORRIAS, W.E. and VAN ARKEL, G.A. (1985).
Modification and sequence analysis of the small plasmid pUH24 of
Anacystis nidulans R2. In: Abst. V Inter. Symp. Photosyn. Prok.
Grindelwald, Switzerland, p.328.
WILLIAMS, J.G.K. and SZALAY, A.A. (1983). Stable integration of
foreign DNA into the chromosome of the cyanobacterium *Synechococcus*
R2. Gene 24:37-51.
WILMOTTE, A.M.R. and STAM, W.T. (1984). Genetic relationships among
cyanobacterial strains originally designated as "*Anacystis nidulans*"
and some other *Synechococcus* strains. J. Gen. Microbiol.
130:2737-2740.
WOLK, C.P. and KRAUS, J. (1982) Two approaches to obtaining low,
extracellular deoxyribonclease activity in cultures of
heterocyst-forming cyanobacteria. Arch. Microbiol. 131:302-307.
WOLK, C.P., VONSHAK, A., KEHOE,P. and ELHAI, J. (1984). Construction
of shuttle vectors capable of conjugative transfer from *Escherichia
coli* to nitrogen-fixing filamentous cyanobacteria. Proc. Natl. Acad.
Sci. USA 81:1561-1565.

Oxford Surveys of Plant Molecular and Cell Biology Vol.4 (1987) 167–220

VIRULENCE OF AGROBACTERIUM

L.S. Melchers and P.J.J. Hooykaas

*Dept. of Plant Molecular Biology
Leiden University, Biochemistry Lab,
Wassenaarseweg 64, 2333 AL. Leiden,
The Netherlands.*

INTRODUCTION

The gram–negative soil bacterium *Agrobacterium tumefaciens* is the causative agent of the plant disease crown gall (Smith and Townsend 1907), while the related *Agrobacterium rhizogenes* provokes hairy root in plants (Riker, 1930). Crown gall is the cause of important losses in agriculture and horticulture in Europe (Kerr and Panagopoulos 1977; Sule, 1978), North America (Dhavantari, 1978; Kennedy and Alcorn, 1980), Australia (New and Kerr, 1972) and Japan (Ohta and Nishiyama, 1984). Infections leading to hairy root formation do not result in crop losses, and may instead be useful for the controlled induction of rooting in plants (Jaynes and Strobel, 1981). The crown gall disease is characterized by the formation of tumours called crown galls at wound sites of infected plants (Fig.1). In the case of hairy root, abundant root proliferations occur in infected wound areas (Fig.1). In fact both crown gall and hairy root are neoplastic diseases. Crown gall cells and hairy root cells are tumorous and are characterized by an ability to proliferate unlimited and autonomously i.e. in the absence of added phytohormones, which are needed for the growth of normal plant cells (Braun, 1958). In the absence of phytohormones (in *in vitro* culture) crown gall cells form (amorphous) calli, whereas hairy root cells differentiate into roots (Tepfer and Tempé, 1981). A feature shared by crown gall and hairy root cells is the production and excretion of compounds (called opines) which are not formed in normal plant cells (reviewed by Tempé and Goldmann 1982).

Now it is well established that during crown gall and hairy root induction part of a large plasmid, which is present in virulent *Agrobacterium* strains, becomes integrated and expressed in the nuclear genome of plant cells at the infection sites (for a review see Bevan and Chilton, 1982). Expression of this transferred(T)–DNA in plant cells results in their transformation into opine producing

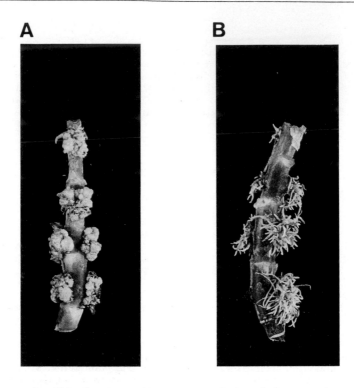

Figure.1. (A) Crown gall tumours induced by Agrobacterium
tumefaciens *and (B) Hairy roots induced by* Agrobacterium
rhizogenes *on a kalanchoe plant.*

Table 1. Characteristics of strains belonging to different
          Agrobacterium biotypes.

	Biotype 1	Biotype 2	Biotype 3
Ketolactose production	+	−	−
Maximum growth temperature	37	29	35
Growth on erythritol as a carbon source	−	+	−
Growth in the presence of 2% NaCl	+	−	+

Data from Kerr and Panagopoulos (1977) .

tumour cells. The opines that are produced can be catabolized by the virulent agrobacteria which induced tumour formation, but not by most other soil microorganisms. Thus *Agrobacterium* applies the genetic manipulation of plant cells in nature in order to create a favourable niche for itself. Therefore, this type of parasitism has been called "genetic colonization" (Schell et al., 1979). In this review we shall focus on the molecular mechanism responsible for crown gall and hairy root formation with emphasis on the bacterial virulence genes involved.

THE GENUS *AGROBACTERIUM*

Bacteria belonging to the genus *Agrobacterium* are gram-negative rods, which are peritrichously flagellated and thus motile (Kersters and De Ley, 1984). Agrobacteria are closely related to the root-nodule inducing bacteria of the genus *Rhizobium* and the leaf-nodule inducing phyllobacteria (Kersters and De Ley, 1984). Agrobacteria are well known as common soil inhabitants, but they have also been isolated from patients in hospitals (Riley and Weaver, 1977; Plotkin 1980; Swann et al., 1985). The taxonomy of *Agrobacterium* is based on the phytopathogenic properties of the bacteria. Strains able to induce crown gall are referred to as *Agrobacterium tumefaciens*, while hairy root inducing strains are called *A. rhizogenes*. Avirulent strains are designated *A. radiobacter*. For strains isolated from aerial galls on *Rubus* the name *A. rubi* has been used. In fact the phytopathogenic character of the bacteria is not a suitable criterion for taxonomic purposes, since it is determined by a mobile genetic element, *viz.* the Ti (tumour inducing)-plasmid in A. *tumefaciens* and *A. rubi* and the Ri (root inducing) plasmid of A. *rhizogenes*. Numerical taxonomic analyses have shown that in the genus *Agrobacterium* three biotypes can be distinguished, but strains of intermediate biotype have also been isolated (Keane et al., 1970; White, 1972; Kersters et al., 1973; Panagopoulos and Psallidas, 1973; Kerr and Panagopoulos, 1977). A number of prominent characteristics, which are used to classify *Agrobacterium* strains are shown in Table 1. Biotype 2 *Agrobacterium* strains are hardly distinguishable from strains belonging to the *Rhizobium leguminosarum* cluster. Besides their phytopathogenic (or symbiotic) properties and their sensitivity (resistance) to rhizobiophages only subtle differences have been found. The Nile Blue test is a plate assay which can be used to discriminate between rhizobia and agrobacteria (Skinner, 1977). The close relatedness between agrobacteria and rhizobia becomes apparent also from hybrids that were constructed. *Rhizobium* strains of the *R. leguminosarum* cluster (*R. leguminosarum+*, *R. trifolii*, R. *phaseoli*) became capable of inducing crown gall or hairy root upon receipt of a Ti plasmid or an Ri plasmid, respectively (Hooykaas et al., 1977; Hooykaas and Schilperoort, 1984). However, the tumours induced were much smaller in size and appeared later than those induced by the *Agrobacterium* parental strains. The latter is especially true for *R. meliloti* strains harbouring a Ti plasmid. Only tiny tumours were infrequently induced by such strains on a number of plant species (Hooykaas and Schilperoort, 1984; Hooykaas unpublished). Reversely, the transfer of Sym plasmids, which carry the genes for root nodulation, host specificity and nitrogen fixation, from *Rhizobium* to *Agrobacterium* strains of biotype 1 or 2 resulted in strains that were able to nodulate specific host plants. However, no nitrogen was fixed in the nodules (Hooykaas et al., 1981; 1982; 1985; Van Brussel et al., 1982; Banfalvi et al., 1983; Broughton et al., 1984; Hirsch et al., 1985). Transfer of the Ti, Ri or Sym plasmids to *E. coli* or other bacteria of the Enterobacteriaceae family did not result in bacteria that were able to induce tumors, hairy roots or root nodules, respectively. The reason for this is unclear at present, but:-

(1) certain chromosomal genes may be absent in *E. coli*,
(2) certain functions determined by *E. coli* may interfere with the expression of the Ti, Ri and Sym-determined gene products, or
(3) the virulence and nodulation genes may not be transcribed properly by *E. coli*.

DIFFERENT TYPES OF TUMOUR AND ROOT INDUCING PLASMIDS

Tumours and hairy roots induced by *Agrobacterium* produce a characteristic set of opines, which can be degraded by the virulent agrobacteria (Petit et al., 1970). The structural formulae of a number of opines are shown in Fig.2. Production of opines in the tumours and hairy roots as well as the catabolism of these compounds is determined by genes located in the Ti (Ri) plasmid (Bomhoff et al. 1976; Kerr and Roberts, 1976; Hooykaas et al., 1977; Guyon et al., 1980; Ellis and Murphy, 1981; Chang and Chen, 1983; Chang et al., 1983; Petit et al., 1983; Chilton and Chilton ,1984; Chilton et al., 1985a; 1985b). Based on the opines produced in the tumours (Table 2), Ti plasmids are divided in four groups, while the Ri plasmids fall in three groups. Octopine (N^2(D-1-carboxyethyl)-l-arginine) is the characteristic compound found in tumours induced by strains with an octopine Ti plasmid such as pTiB6, pTiAch5, pTiA6, pTiAg57, or pTi15955. Nopaline (N^2(D-1,3-dicarboxypropyl)-l-arginine) is found in tumours formed by strains with a nopaline Ti plasmid (examples: pTiC58, pTiT37, pTi1D135, pTi223). Strains which induce tumours in which neither octopine nor nopaline is present were originally called null-type. Now it is known that these strains fall into two different types; those, e.g. strains with pTi542, pTiAT4, which provoke tumors with the opine leucinopine (N^2(L-1, 3-dicarboxypropyl)-l-leucine), and those, e.g. strains with pTiEU6 or pTiAT181, giving tumours with the opine D,L-succinamopine (asparaginopine) (N^2(D-1, 3-dicarboxy propyl) -l-asparagine). It has to be noted here that an epimer (L,L-succinamopine) is a typical compound of tumours formed after infection with leucinopine strains (Chilton et al., 1985b). There are three types of Ri plasmids (Petit et al., 1983; Brevet et al., 1986). Hairy roots induced by strains with a mannopine Ri plasmid (e.g.

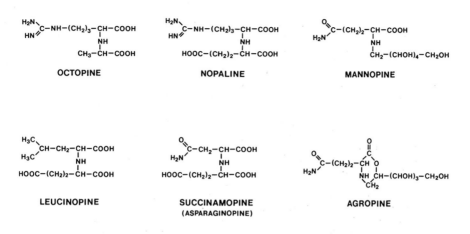

Figure.2. Structural formulae of opines.

pRi8196, pRiTR7) contain mannopine (N^2(1-D-mannityl)-1-glutamine). Agropine Ri plasmids (e.g. pRi1855, pRi15834, pRiA4, pRiHRI) are responsible for the formation of agropine (1-2'-δ-lactone of mannopine) besides mannopine in hairy roots. Strains with a third class of Ri plasmid (e.g. pRi2659) provoke hairy roots with cucumopine (structure unknown). Agropine and mannopine are also produced in tumours induced by octopine and leucinopine strains of *A. tumefaciens*. Another class of opines, the agrocinopines, which are phosphorylated sugar derivatives (Ellis and Murphy, 1981; Ryder et al., 1984; Messens et al., 1986), are present in tumours and hairy roots induced by various types of *Agrobacterium* strains (Table 2).

Most of the Ti and Ri plasmids isolated so far have a size of between 190-240 kbp. (Zaenen et al., 1974; Sciaky et al., 1978; Costantino et al., 1981). However, recently a completely different class of Ti plasmids was isolated from strains that came from tumors on *Lippia* species. These had an unusually large size of 500 kbp (Unger et al., 1985). The restriction patterns of Ti and Ri plasmids belonging to each of the different "opine types" are different, but within certain types the plasmids can be quite similar or even identical. Wide host range (WHR) octopine Ti plasmids form a very homogenous group (Sciaky et al., 1978). The same is true for the mannopine Ri and agropine Ri plasmids (Costantino et al., 1981). The nopaline Ti plasmids form a heterogenous group (Sciaky et al., 1978), although certain nopaline Ti plasmids (pTiT37, pTiH100) are very similar to the succinamopine Ti plasmids (pTiEU6, pTi181, pTiT10/73).

In order to be able to localize genes on the various Ti and Ri plasmids, physical maps have been constructed for these plasmids. For the WHR octopine Ti plasmids pTiAch5, pTiA6 and pTiB6, which are almost identical, maps were made for the restriction enzymes *Sma*I, *Hpa*I (Chilton et al., 1978), *Kpn*I, *Xba*I (Ooms et al., 1980), *Bam*H1,

Table 2. Opine profiles of different plasmid types.

Opine	Plasmid type					
	octTi	nopTi	lopTi	sapTi	agrRi	mopRi
octopine	+	−	−	−	−	−
nopaline	−	+	−	D	−	−
leucinopine	−	−	+	D	−	−
dl-succinamopine	−	−	−	+	−	−
ll-succinamopine	−	−	+	D	−	−
mannopine	+	−	+	−	S	+
agropine	+	−	+	−	+	−
agrocinopine A	−	+	−	+	+*	−
agrocinopine C	−	−	+	−	−	+*

Data from Bomhoff et al. (1976), Chang and Chen (1983), Chang et
al. (1983), Chilton and Chilton (1984), Chilton et al. (1985a),
Chilton et al. (1985b), Dahl et al. (1983), Ellis and Murphy
(1981), Firmin and Fenwick (1978), Guyon et al. (1980), Petit et al.
(1970), Petit et al. (1983), Tepfer and Tempé (1981).

* = degradation was not tested.
D = degradation, but no synthesis in the tumours
S = synthesis in the hairy roots, but degradation not encoded by
    the Ri plasmid; a catabolic plasmid present in the same
    A. rhizogenes strain may determine catabolism (Petit et al.
    1983).
oct = octopine ; nop = nopaline ; lop = leucinopine ;
sap = succinamopine ; agr = agropine ; mop = mannopine.

*Eco*RI, *Hind*III (De Vos et al., 1981), *Xho*I, *Sst*I and
*Sal*I (Knauf and Nester, 1982). For the limited host range (LHR)
octopine Ti plasmid pTiAg162, a map was constructed for the enzymes
*Eco*RI, *Kpn*I, *Sal*I and *Xho*I (Knauf et al., 1984), while for the
nopaline Ti plasmid pTiC58 this was done for *Bam*HI, *Eco*RI, *Hind*III,
*Hpa*I, *Kpn*I, *Sma*I and *Xba*I by Depicker et al. (1980) and for the
leucinopine Ti plasmid pTi542 for *Sal*I and *Bam*HI by Hood et al.
(1984) and Komari et al. (1986). For the agropine Ri plasmid pRi1855,
a physical map for *Eco*RI fragments was constructed by Pomponi et al.
(1983) and for the closely related plasmid pRiHRI for *Bam*HI, *Eco*RI,
*Kpn*I *Sma*I and *Xba*I by Jouanin (1984). A *Bam*HI restriction map of the
mannopine Ri plasmid pRi 8196 was worked out by Koplow et al. (1984).

FUNCTIONS DETERMINED BY Ti AND Ri PLASMIDS

The Ti and Ri plasmids confer a number of properties to *Agrobacterium*. First of all they contain two regions (Fig.3) which are involved in tumorigenesis, *viz.* a T-region, which is transferred to the plant nucleus during tumour induction, and secondly a Vir (virulence)–region, which determines the system by which the T–DNA is transported (Holsters et al., 1980; Garfinkel and Nester, 1980; Ooms et al., 1980). The combination of T-region and Vir–region determines host range properties for tumour induction (Loper and Kado, 1979; Thomashow et al., 1980a; Hoekema et al., 1984a; Buchholz and Thomashow, 1984a; Yanofsky et al., 1985a; 1985b). Thus, besides wide host range (WHR) Ti plasmids also limited host range (LHR) Ti plasmids are known. The LHR Ti plasmids are prevalent especially in biotype 3 strains isolated from *Vitis* species (Panagopoulos and Psallidas, 1973), while WHR Ti plasmids have been found in strains of all three biotypes (Perry and Kado, 1982). Tumour morphology is a Ti plasmid determined trait (Hooykaas et al., 1977; Hooykaas and Schilperoort, 1984). On *Kalanchoe daigremontiana* octopine strains induce tumours with a 'rough' surface which are surrounded by adventitious roots. Leucinopine strains induce 'rough' tumours too, but these are not accompanied by root formation. Both nopaline and succinamopine strains form 'smooth' tumours from which shoots emerge occasionally. This underscores the relatedness between nopaline and succinamopine Ti plasmids. Strains with mannopine Ri plasmids and agropine Ri plasmids differ slightly in virulence. Root formation by agropine Ri strains usually is more abundant than that by mannopine Ri strains (Petit et al., 1983).

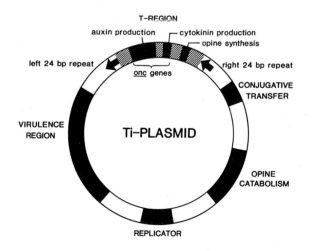

Figure.3.   *Genetic map of an octopine Ti plasmid.*

173

Another difference is that in contrast with mannopine Ri strains agropine Ri strains are capable of inducing tumours on several plant species (e.g. pea, sunflower, kalanchoe). Roots may emerge from such tumors. Mannopine and cucumopine Ri strains are polar i.e. they induce hairy roots on the apical side of carrot slices, but not on the basal side (Ryder et al., 1985). Agropine Ri strains form hairy roots on both sides of carrot slices.

Besides virulence and opine synthesis Ti and Ri plasmids determine a number of other functions. As can be seen in Table 2 they have genes for the catabolism of the opines that are formed in the tumors which they induce (Bomhoff et al., 1976). These genes embrace not only catabolic genes, but also genes for the specific uptake of certains opines (Klapwijk et al., 1977). The primary degradation products of the opines are amino acids and small sugars. Interestingly, octopine and nopaline Ti plasmids were found to have genes for the catabolism of amino acids such as arginine and ornithine (Dessaux et al., 1986) via proline (Farrand and Dessaux, 1986). Ti plasmids are conjugative via inducible *tra*-genes. Initially transfer of Ti and Ri plasmids was observed *in planta* only (Kerr and Roberts, 1986; Albinger and Beiderbeck, 1977). Later it was found that opines functioned as the inducers of the *tra*-genes as well as of the opine catabolism genes (Klapwijk et al., 1978; Petit and Tempé, 1978). For octopine Ti plasmids octopine was found to act as an inducer (Petit et al., 1978; Hooykaas et al., 1979), while for nopaline Ti plasmids agrocinopine A and for leucinopine Ti plasmids agrocinopine C is the inducer of the *tra*-genes (Ellis et al., 1982). Which opines function as inducers of the Ri plasmids is unknown. Another character determined by octopine Ti plasmids and some nopaline Ti plasmids is interference with phage AP1 multiplication (Hooykaas et al., 1977; Van Larebeke et al., 1977). This phenomenon has been called phage AP1 exclusion. Octopine Ti plasmids inhibit the multiplication of another phage, called Ψ, as well (Expert and Tourneur, 1982; Expert et al., 1982). Bacteriocins are produced by certain *Agrobacterium* strains such as the biotype 2 strain K84 (Kerr and Htay, 1974). Sensitivity to agrocin 84 is determined by a locus on nopaline and succinamopine Ti plasmids (Engler et al., 1975; Kerr and Robert, 1976). Nopaline and succinamopine Ti plasmids code for an uptake system, which is meant for the transport of agrocinopine A but via which also agrocin 84 is carried into the bacterial cell (Murphy and Roberts, 1979; Murphy et al, 1981). Agrocin 84 is a phosphorylated sugar derivative (Tate et al., 1979; Murphy et al., 1981) and therefore has some structural homology with agrocinopine A (Ryder et al., 1984; Messens et al. 1986). Sensitivity to the agrocin of strain D286 is determined by a locus that is present on octopine, nopaline and perhaps also leucinopine Ti plasmids (Hendson et al., 1983). Ti plasmids as well as *Agrobacterium* chromosomal genes may specify the production of phytohormones. A gene on the nopaline Ti plasmid called *tzs* (*trans*-zeatin secretion) codes for an enzyme involved in the production of *trans*-zeatin (Regier and Morris, 1982; Akiyoshi et al. 1985; Beaty et al., 1986). The same plasmid also has a locus for the production of the auxin indole acetic acid (IAA) (Liu and Kado, 1979; Liu et al., 1982).

The region responsible for replication and incompatibility has been localized and cloned for the WHR octopine Ti plasmid (Koekman et al., 1979; 1980; 1982; Knauf and Nester, 1982), the LHR octopine Ti plasmid (Knauf et al., 1984), the nopaline Ti plasmid (Gallie et al. 1985), the leucinopine Ti plasmid (Komari et al., 1986) and the agropine Ri plasmid (Jouanin et al., 1985; Nishiguchi et al., 1987). For the agropine Ri plasmid the replication region was sequenced, which showed the presence of three large open reading frames (Nishiguchi et al., 1987). Incompatibility studies have shown that octopine Ti plasmids, nopaline Ti plasmids and succinamopine Ti plasmids belong to the same incompatibility group *inc*Rh-1 (Hooykaas, 1979; Hooykaas et al., 1980; Knauf and Nester, 1982). Also the LHR octopine Ti plasmids belong to this group (Knauf and Nester, 1982; Knauf et al., 1984; Hooykaas and Schilperoort, 1984). However, the leucinopine Ti plasmids belong to a different group (*inc*Rh-2) together with certain non-Ti nopaline catabolic plasmids (Hooykaas 1979; Hooykaas and Schilperoort, 1984; Komari et al., 1986). The agropine Ri plasmids form a third group called *inc*Rh-3 (Hooykaas 1979; Costantino et al., 1980; White and Nester, 1980). Remarkably these Ri plasmids are incompatible with certain *Rhizobium trifolii* and *R. phaseoli* Sym plasmids (O'Connell and Pühler, pers. commun; our unpublished results). A locus linked to the replicator of Ti and Ri plasmids is *ein* (entry inhibition) (Hooykaas et al., 1980; Koekman et al., 1982). Entry inhibition operates between heterologous Ti plasmids (Hooykaas et al., 1980; Knauf et al., 1984; Komari et al., 1986) or between wild-type Ti plasmids and cloned mini-replicons (Koekman et al., 1982; Komari et al., 1986; Nishiguchi et al., 1987), but not between homologous Ti plasmids (Hooykaas et al., 1980). All octopine Ti plasmids belong to the same *ein*-group, but nopaline Ti plasmids fall in at least two different *ein*-groups (Hooykaas, unpublished). Neither the molecular mechanism underlying entry inhibition, nor the gene specifying this activity are known at the moment.

STRUCTURE OF THE T-DNA

During tumour and hairy root induction by *Agrobacterium* part of the Ti plasmid (or Ri plasmid, respectively) is transferred to plant cells, where it becomes covalently linked to the nuclear DNA of these plant cells (for a review see Bevan and Chilton, 1982). There are some indications that T-DNA may enter the chloroplast as well, but cannot be stabilized there (De Block et al., 1985). The transferred DNA originates from a fixed part of the Ti (Ri) plasmid, which is surrounded by an imperfect direct repeat of 24 bp. The sequences of these repeats, which play an essential role in T-DNA transfer, have been determined for the nopaline Ti plasmid (Yadav et al., 1982; Zambryski et al., 1982), the octopine Ti plasmid (Simpson et al., 1982; Barker et al., 1983; Gielen et al., 1984) and the agropine Ri plasmid (Slightom et al., 1985;1986). The consensus sequence of the repeat is shown in Table 3. Within the T-region there may be sequences which have some similarity to the 24 bp border repeats

Table 3. T-region border repeats

Consensus:	G	g	C	A	G	G	A	T	A	T	A	T	N	N	N	N	t	g	T	A	A	a/t	t/c
A	0	2	0	8	0	0	8	0	8	0	8	0	1	0	2	1	1	0	1	0	7	8 5	1
T	0	0	0	0	0	0	0	8	0	8	0	8	0	8	4	1	3	0	3	7	0	8 0	3 3
C	0	0	8	0	0	0	0	0	0	0	0	0	0	1	3	1	2	1	0	0	0	1 0	0 3
G	8	6	0	0	8	8	0	0	0	0	0	0	2	4	2	5	3	1	7	0	0	0 0	1

A compilation of the sequences of eight different T-region border repeats, taken from pTi15955, pTiC58 and pRiA4 is shown. The derived 24bp border consensus sequence comprises two conserved regions of 12bp and 7bp with, in between, a variable region of 5bp.

(Hepburn and White 1985; Peerbolte, 1986; Slightom et al., 1986; Van Haaren et al., 1986).

Whether these pseudo-repeats play any role in T-DNA transfer from wild-type Ti plasmids is not clear. Border fragments in which Ti plasmid DNA is linked to plant DNA have been isolated from transformed plant cells and mapped (Zambryski et al., 1982; Holsters et al., 1983; Jorgensen et al., 1987). In general there was more variation at left end junctions than at right end junctions. Jorgensen et al. (1987) found that two thirds of the right end junctions were lying within 11 bp of the right border repeat, while two thirds of the left end junctions were within 217 bp of the left repeat. From the right repeat, at most, the left 3 bp are integrated into the plant genome, while for the left repeat this can be as much as the right 12 bp (Yadav et al., 1982; Zambryski et al., 1982; Holsters et al., 1983).

   The T-DNA structure is different for each tumour line. Some lines contain only one copy of the Ti T-DNA, but other may harbour up to a dozen or more copies, which can either be complete or truncated (Lemmers et al., 1980; Thomashow et al., 1980b; De Beuckelaer et al. 1981; Ooms et al., 1982a; Hepburn et al., 1983a; Holsters et al., 1983; Ursic et al., 1983; Kwok et al., 1985; Peerbolte et al., 1986a; Van Lijsebettens et al., 1986). If more than one copy of T-DNA is present, this may be at one locus or at several loci in the plant genome (Kwok et al., 1985; Budar et al., 1986; Spielmann and Simpson, 1986). Multiple copies present at one locus may be arranged as a tandem repeat (Lemmers et al., 1980; Ooms et al., 1982; Holsters et al., 1983) or in an inverted repeat structure (Kwok et al., 1985; Spielman and Simpson 1986; Jorgensen et al., 1987). Although less work has been done to study the T-DNA structure in lines transformed with Ri T-DNA, the picture which begins to emerge is similar to what has been described for Ti T-DNA (Chilton et al., 1982; Spano et al., 1982; White et al. 1982; Tepfer, 1984; Durand-Tardif et al., 1985; Jouanin et al., 1987). The key factors which are of primary importance in determining the transformation efficiency by *Agrobacterium* and the structure of the integrated T-DNA are unknown. It will be clear, however, that many factors may play a role in this, *viz Agrobacterium* and Ti plasmid type (T-region; Vir-region), T-region copy numbers in the bacterium (*in cis* vs. *in trans* towards the Vir-region), the multiplicity of infecting agrobacteria, the amount of *vir*-inducers present (see following sections), the plant species used, its growth conditions and the method of infection applied, and eventually the selection regimes applied.

T-DNA may integrate at different sites in the nuclear genome of the plant cell. However, it is unknown whether there are preferred sites for integration. The selection for expression of T-DNA will of course limit the potential number of insertion sites to chromosomal DNA regions which are transcriptionally active. T-DNA can be present both in single copy DNA and in repeated DNA (Zambryski et al., 1982). For *Crepis capillaris*, it was shown that mannopine Ri T-DNA can integrate into any of its three chromosomes (Ambros et al., 1986). For tomato, seven Ti T-DNA inserts were mapped on five different chromosomes (Chyi et al., 1986), while for petunia, mapping of six Ti T-DNA inserts showed that three were located at different sites in chromosome I, two in chromosome III and one in chromosome IV (Wallroth et al., 1986).

Nopaline and succinamopine Ti plasmids as well as mannopine Ri plasmids have only one T-region. However, octopine Ti plasmids – both WHR and LHR types –, leucinopine Ti plasmids and agropine Ri plasmids have two T-regions called T_L- and T_R- region respectively, which are separated from each other by a segment of DNA, which is called T_C-region. The T_L- as well as the different T_R-regions are surrounded by the 24 bp direct repeats. Therefore, during transformation plant cells may receive either the T_L-DNA or the T_R-DNA or both at the same time (Thomashow et al., 1980b; Ooms et al., 1982a; Czako and Marton 1986). In some transformed lines the T_L- and T_R-DNA are at the same locus (Peerbolte et al., 1986a), where they may be linked in an inverted orientation (Jouanin et al., 1987). The T_C-region has never been observed as an independent unit in transformed plant cells. However, in an exceptional sunflower tumour line the T_C-DNA is present as part of a large insert comprising the T_L-, T_C- and T_R-DNA (Ursic et al., 1983). Strains have been constructed containing more than one type of Ti or Ri plasmid. Upon infection of plant cells with such strains (in a multiplicity of infection of 100 per plant cell) cotransfer of both T-regions to the same plant cell was observed with a frequency which could be as high as 80% (An et al., 1985; Depicker et al., 1985; Petit et al., 1986). However, when two strains with different Ti or Ri plasmids were coinoculated, the frequency with which cells were found containing both T-regions was much lower (Depicker et al., 1985; Petit et al., 1986). This was interpreted to indicate that transformation is rather limited by the establishment of an bacteria–plant cell interaction than by the process of DNA integration or the number of plant cells that are sensitive for transformation (Depicker et al., 1985).

EXPRESSION OF THE T-DNA

The T-DNA contains a number of genes which are expressed in the plant cell. Transcript maps have been made for the octopine Ti T_L-DNA (Gelvin et al., 1982; Murai and Kemp, 1982a; Willmitzer et al., 1982a), the octopine Ti T_R-DNA (Karcher et al., 1984; Winter et al., 1984), the nopaline T-DNA (Willmitzer et al., 1983) and the agropine Ri T_L- and T_R-DNA (Taylor et al., 1985). Expression of the T-DNA genes is independent of the surrounding plant sequences. Apparently, these bacterial genes contain expression signals that are recognized by the plant system. For some T-DNA genes the 5' and 3' regulatory sequences have partly been defined. It was found that the T-DNA genes depend for expression on typical eukaryotic expression signals such as a TATA-box, an upstream element and in some cases an enhancer in the 5' part and a AATAAA-box and a GT-rich box in the 3' part of the gene (Shaw et al., 1984; An et al., 1986; Bruce and Gurley, 1987; De Pater 1987). Expression of T-DNA genes has also been studied in plants regenerated from transformed cells. A remarkable finding was that, although most T-DNA genes were expressed more or less constitutively in all the different tissues, some showed a distinct tissue specificity for expression (Durand-Tardif et al., 1985; Koncz and Schell, 1986; Ooms et al., 1986).

Although T-DNA genes do not depend on plant sequences for expression this is no guarantee that T-DNA will be expressed in the transformed cell. Even if multiple copies of T-DNA are present in the genome, expression may not occur in particular transformed lines (Van Lijsebettens et al., 1986). The lack of expression may be due to the DNA-structure at the insertion sites, which may be influenced for instance by methylation. Indeed treatment with the de-methylating agent 5-azacytidine can induce expression of T-DNA genes in some transformed lines (Hepburn et al., 1983b; Amasino et al., 1984; Van Slogteren et al., 1984).

In *Agrobacterium* the octopine Ti T-region genes are not transcribed (Stachel et al., 1985a), but some regions in the nopaline Ti T-region (Janssens et al., 1984) and the agropine Ri T_L-region (Plessis et al. 1985) are transcribed in the bacterium. Whether expression of T-region genes in the bacterium plays any role in the tumour induction process is uncertain.

FUNCTIONS DETERMINED BY T-DNA GENES

Genes located in the T-DNA determine the two characteristic properties of crown gall and hairy root cells:-

1) phytohormone independent growth
2) production and excretion of opines.

Mutagenesis of the T_L–region of the octopine Ti plasmid revealed that mutations at three loci led to changes in host range for tumor induction and/or in the morphology of the tumours induced on plants such as kalanchoe and tobacco (Garfinkel et al., 1981; Ooms et al. 1981). Such mutants were avirulent on tomato, but supplementation of either auxin or cytokinin (depending on the particular mutant studied) restored tumorigenicity on this plant (Ooms et al., 1981). Apparently the mutations had inactivated loci which cause either an auxin or a cytokinin effect in plants.   Therefore the Ti loci involved were called *aux* locus and *cyt* locus respectively. (Fig. 4). The *aux* mutants formed shooty tumours on tobacco, whereas the *cyt* mutants provoked rooty tumors. This agrees with the response of tobacco callus *in vitro* towards relatively high concentrations of cytokinin and auxin, respectively (Skoog and Miller, 1957). Direct evidence for the presence of such levels of cytokinin in shooty tumours and relatively low levels in rooty tumours was provided by Akiyoshi et al. (1983). Because of the tumour phenotype on tobacco and kalanchoe the *aux* locus has also been called *tms* ("tumour morphology shoot") or *shi* ("shoot inhibition") and the *cyt* locus has also been named *tmr* ("tumour morphology root') or *roi* ("root inhibition") by the crown gall groups in Seattle (Garfinkel et al., 1981) and Ghent (Leemans et al., 1982) respectively.

Northern blots of tumour RNA revealed that the *cyt*–locus is expressed in one polyA–RNA of 1200 bases, while the *aux*–locus gives two polyA–RNAs of 2700 and 1600 bases respectively. (Willmitzer et al., 1982). In agreement herewith DNA sequence analysis of the octopine Ti T–region showed that the *cyt*–locus determines one open reading frame (orf) for a protein of 27 kDa (Barker et al., 1983; Heidekamp et al., 1983; Gielen et al., 1984; Lichtenstein et al. 1984), while the *aux*–locus codes for two proteins of 50 and 84 kDa (Barker et al., 1983; Gielen et al., 1984; Klee et al., 1984; Sciaky and Thomashow, 1984).

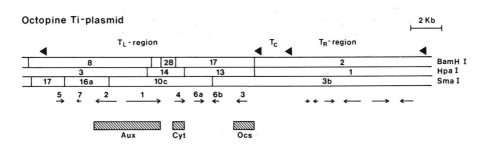

Figure.4.   Physical   map   of the T–region of the octopine Ti plasmid.   The T_L–region, T_C–region and T_R–region   segments are surrounded by 24 bp direct   border   repeats (arrow heads).   Open reading frames on T_L–DNA and T_R–DNA are indicated with black arrows according to Barker et al. (1984). The T–DNA loci aux, cyt and ocs are boxed.

In order to determine the precise function of the proteins encoded by the *cyt*-gene and the two *aux*-genes, these genes were expressed in *E. coli* after cloning in an expression vector. Three groups showed independently that the protein that was expressed from the *cyt*-gene was able to catalyze the reaction between isopentenyl-pyrophosphate and AMP to form isopentenyl-AMP, a compound with cytokinin activity (Akiyoshi et al., 1984; Barry et al., 1984; Buchmann et al., 1985). The enzyme is an isopentenyl transferase and therefore the *cyt*-gene is now called *ipt* as well. The enzymes determined by the *aux*-genes were found to catalyze the conversion of tryptophan into indole acetic acid (IAA), an auxin. The gene product of the *aux*-1 gene turned out to be a monooxygenase which could convert tryptophan into indole acetamide (IAM) – which in itself has no auxin-activity – (Thomashow et al., 1986) while the *aux*-2 encoded enzyme was found to be an amidohydrolase able to hydrolyze IAM with the formation of IAA (Schröder et al., 1984; Thomashow et al., 1984; Kemper et al., 1985) Therefore the genes *aux*-1 and *aux*-2 are now called *iaa*M and *iaa*H as well. The production of an auxin and a cytokinin via the octopine Ti T-DNA explains why T-DNA containing cells grow unlimited as tumor cells. In the T-region of the nopaline and succinamopine Ti plasmids similar *aux*- and *cyt*-genes are present (Joos et al., 1983; Goldberg et al., 1984; Blundy et al., 1986). None of the other genes located in the T-regions of octopine, nopaline or succinamopine Ti plasmids specifies a function which is essential for tumour induction (Garfinkel et al., 1981; Leemans et al., 1982; Ooms et al., 1982b; Joos et al., 1983). However, the genes called 5 and 6b which are expressed in plant cells transformed with the octopine T_L-DNA or the nopaline T-DNA can slightly modify the plant response to infection, which becomes visible only in infections with *cyt*- or *aux*-mutants (Joos et al., 1983; Ream et al., 1983). A number of genes present in the T-DNA code for enzymes involved in the biosynthesis of opines. The gene at the rightmost end of the T-DNA codes for octopine synthase in the case of the octopine T_L-DNA (Garfinkel et al., 1981; Schröder et al. 1981; Murai and Kemp, 1982b), nopaline synthase in the case of the nopaline T-DNA (Holsters et al., 1980; Joos et al., 1983) and succinamopine synthase in the case of the succinamopine T-DNA (Blundy et al., 1986). In the middle of the nopaline and succinamopine T-DNA a gene for agrocinopine synthesis is present (Joos et al., 1983; Blundy et al., 1986), while the T_R-DNA of the octopine Ti plasmid specifies three enzymes for agropine and mannopine synthesis via the intermediate fructopine (Ellis et al., 1984; Salomon et al., 1984; Komro et al., 1985). The gene 6a, which is present in the different Ti plasmids, confers the capacity to excrete opines from the transformed cells (Messens et al., 1985).

The genetic organization of other types of Ti plasmids and Ri plasmids has not been established in detail yet. The leucinopine Ti plasmid pTi542 has a split T-region like the octopine Ti plasmid (Hood et al., 1986a; Komari et al., 1986). The tumor-inducing *onc*-genes are located in the T_L-region, while genes for agropine and mannopine synthesis are located in the T_R-region (Hood et al., 1986a; Komari et al., 1986). Interestingly, the T_L-region of pTi542 shares DNA homology with the *aux*-genes of the octopine Ti plasmid, but not with the *cyt*-gene (Komari et al., 1986). The leucinopine Ti plasmid has a broad host range for tumour induction. From experiments with the octopine Ti plasmid, however, it is known that the absence of a functional *cyt*-gene leads to a reduction in the host range for tumor induction (Ooms et al., 1981). Therefore, the absence of *cyt*-gene homology in pTi542 indicates that either a *cyt*-gene is present in pTi542 which lacks enough homology to be detected in Southern blots, or that the *cyt*-gene in pTi542 is replaced by an other *onc*-gene with an unknown function. The LHR octopine Ti plasmid pTiAg162 (Fig.5) has two T-regions like the WHR octopine Ti plasmids (Buchholz and Thomashow, 1984b; Yanofsky et al., 1985a;1985b). The distance between the two T-regions, which are called T_A- and T_B-region, however, is much larger (32 kbp) than in the case of the WHR octopine Ti plasmid where the T_C-DNA is not more than 1,8 kbp. The T_A-region contains sequences with homology to the WHR octopine Ti plasmid *cyt* and *ocs* (octopine synthase) genes as well as the genes 5,6a and 6b, while the T_B-region has sequences homologous to the *aux*-1 and *aux*-2 genes (Buchholz and Thomashow, 1984b; Yanofsky et al., 1985a). Further results showed that in the LHR octopine Ti plasmid the *onc*-genes are indeed divided over the T_A- and the T_B-region. The T_A-region was

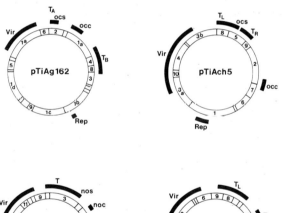

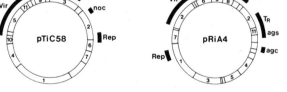

Figure.5. Comparison of a limited host range octopine Ti plasmid (pTiAg162), wide host range (WHR) octopine Ti plasmid (pTiAch5), WHR nopaline Ti plasmid (pTiC58) and a WHR agropine Ri plasmid (pRiA4).The circular KpnI-restriction map and the locations of different functional regions common to all four plasmids are indicated: Virulence region, T-region(s), opine synthase, opine catabolism, origin of replication (rep) and transfer functions. Abbreviations: ocs, octopine synthase; occ, octopine catabolism; nos, nopaline synthase; noc, nopaline catabolism; ags, agropine synthase; agc, agropine catabolism.

responsible for oncogenicity on *Vitis* species, while the T_B-region
was the cause of the formation of rooty tumours on *Nicotiana rustica*
(Yanofsky et al., 1985b). The T_A-region has a segment with homology
to the cyt gene, but this does not represent a functional *cyt*-gene.
(Yanofsky et al., 1985a). Addition of the *cyt*-gene from a WHR Ti
plasmid to LHR strains leads to an extension of their host range
(Buchholz and Thomashow, 1984a; Hoekema et al., 1984a; Yanofsky et
al. 1985a). It could therefore be concluded that the absence of a
functional *cyt*-gene is one of the reasons for the limited host range
of the LHR strains. Reversely, however, it seems that this lack of a
cyt-gene makes the LHR strains more adapted to induce tumours on
plant species such as *Vitis*, since the presence of a cyt gene in the
WHR Ti plasmid is the cause of a "hypersensitive" response to
infection with wild-type WHR strains (Hemsted and Reisch, 1985).

The mannopine Ri plasmids contain one T-region, whereas in the
agropine Ri plasmids the T-region is split in a T_L- and a T_R-region,
which are 25 kbp apart (De Paolis et al., 1985; White et al., 1985;
Offringa et al., 1986). The T_L-region of the agropine Ri plasmid is
largely homologous to the T-region of the mannopine Ri plasmid (Spano
et al., 1982). However, in the middle of T_L a region is found, called
C_T-DNA, which is absent from the mannopine Ri T-region (Spano et al.
1982). The C_T-DNA (cellular T-DNA) has been given this name, because
certain plant species naturally carry homologous DNA in their genome
(Spano et al., 1982; White et al., 1983). The C_T-DNA was cloned from
Nicotiana glauca and precisely defined by sequencing; many
restriction sites turned out to have been conserved (Furner et al.
1986). The distribution of the C_T-DNA over the plant kingdom led
Furner et al. (1986) to conclude that the C_T-DNA probably was a relic
from earlier transformation events by *A. rhizogenes*. Sequencing of
the entire T_L-region of an agropine Ri plasmid led to the
identification of a large number of open reading frames (Slightom et
al., 1986). The oncogenic properties of T_L-region mutants on
Kalanchoe daigremontiana leaves identified a number of loci involved
in root induction, namely *rol*A, *rol*B, *rol*C, *rol*D (White et al.,
1985). A segment of the Ri plasmid containing *rol*C was found to be
able to inhibit adventitious root formation around tumours induced by
strains containing a Ti plasmid with *rol*C in the T-region (our
unpublished results). The *rol*-mutants of the agropine Ri plasmid were
still oncogenic on plant species such as carrot, pea and sunflower.
This is due to the presence of the T_R-region, which has a segment
with homology to the Ti *aux*-1 and *aux*-2 genes (Huffman et al., 1984;
White et al., 1985). Complementation experiments with Ti *aux*-mutants
showed that the T_R-region indeed contains an *aux*-1 and an *aux*-2 gene,
which are functional in the transformed plant cells (Offringa et al.,
1986). Both the T_L-region and the T_R-region of the agropine Ri
plasmid can separately induce proliferations in plants (Offringa et
al., 1986; Vilaine and Casse-Delbart, 1987). The presence of the
T_R-region in agropine Ri strains makes these more strongly oncogenic
and rhizogenic than mannopine Ri strains (Petit et al., 1983;
Cardarelli et al., 1985; Ryder et al., 1985). Nevertheless, hairy
root lines induced by agropine strains contain the T_L-DNA, but
usually not the T_R-DNA or only a part of T_R which does not overlap
with both *aux*-genes (Spano et al., 1982; Peerbolte, 1986).

This suggests that:-

(1) the presence of functional *aux*-genes in some transformed cells at the wound site stimulates the growth of neighbouring plant cells with the T_L-region only
(2) the presence of functional *aux*-genes inhibits the growth or the differentiation into hairy roots of the cells containing the T_R-DNA or the T_L- and the T_R-DNA.

In the mannopine Ri T-region segments of homology were found with the Ti plasmid loci involved in mannopine and agrocinopine synthesis, which is in agreement with the production of these opines in hairy roots of the mannopine type (Lahners et al., 1984). In the agropine Ri T_L-region a segment with homology to the locus for agrocinopine synthesis was detected, while the T_R-region has the genes for agropine and mannopine synthesis (Willmitzer et al., 1982b; Huffman et al., 1984; Jouanin, 1984; De Paolis et al., 1985).

GENES INVOLVED IN THE T-REGION TRANSFER PROCESS

Crown gall tumorigenesis is a multi step process including attachment of *Agrobacterium* to plant cells, transfer of T-DNA to plant cells followed by integration and expression of T-DNA genes in the plant genome (see Fig. 6). The T-region genes are not involved in the T-DNA transfer process as all the DNA between the 24 bp T-region border repeats can be deleted without affecting its transfer to plant cells (Leemans et al., 1982; Zambryski et al., 1983). The genes responsible for T-DNA transfer are located in the Vir-region of the Ti plasmid and in the bacterial chromosome.

The Chromosomal Virulence Loci

The first step in tumour formation is the attachment of *Agrobacterium* to a specific host cell receptor site (Lippincott and Lippincott, 1969; Schilperoort, 1969) This initial event is mediated by genes located on the bacterial chromosome. Strains carrying mutations at a number of loci (*chv*A, *chv*B, *att* and *psc*A or *exo*C) are defective in plant cell attachment and therefore avirulent or strongly attenuated in virulence (Garfinkel et al., 1980; Matthysse, 1987; Thomashow et al., 1987; Cangelosi et al., 1987). Mutations at certain chromosomal loci such as *chv*C were found to lead to a restriction in the host range for tumour induction (Hooykaas et al., 1984). Garfinkel and Nester (1980) reported that *chv*A and *chv*B mutants were avirulent on plant species such as *Kalanchoe*, sunflower and tomato (Table 4); therefore it was assumed that *chv*A and *chv*B genes are essential for tumorigenicity on all plant species (Douglas et al., 1985a). Surprisingly, however, we found that such mutants were still virulent on plant species such as

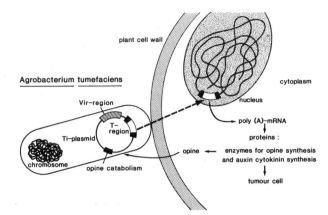

Figure.6. Schematic drawing of the tumor-induction process. The T-region of a Ti plasmid is transferred and integrated as T-DNA in chromosomal DNA of the plant cell nucleus. Expression of T-DNA genes leads to synthesis of opines and the phytohormones auxin and cytokinin; the hormones provoke tumour development. The opines are useful only for the bacteria as a source of nutrition and as inducers of bacterial conjugation.

Solanum tuberosum and *Nicotiana knightiana* (Hooykaas unpublished). The functions determined by *chv*A, *chv*B and *exo*C (*psc*A) are not yet known. The *chv*A and *chv*B genes are expressed constitutively (Douglas et al., 1985a). Mutations in these genes have a pleiotropic effect:–

(1) the mutants lack flagella and are therefore non–motile,
(2) *chv*B and *psc*A (*exo*C) mutants do not produce the exopolysaccharide (EPS) β–1,2–glucan,
(3) *psc*A (*exo*C) mutants are defective in the production of succinoglycan, another exopolysaccharide, and therefore form dark colonies on calcofluor plates (Bradley et al., 1984; Douglas et al. , 1985a; Puvenesarajah et al., 1985; Cangelosi et al., 1987; Thomashow et al., 1987).

Whether any of these exopolysaccharides directly plays a role in tumour induction – by mediating attachment for instance – is not known at the moment. *Agrobacterium* is able to produce cellulose fibrils (β–1,4–glucan). Mutants unable to produce cellulose were isolated (Matthyse et al., 1981; Matthysse, 1983). It was subsequently found that such mutants were normally virulent, albeit that washing of infected wound sites reduced the incidence of tumour formation (Deasey and Matthysse, 1984). This indicates that cellulose fibrils play an accessory role in tumour induction, namely by anchoring the bacteria firmly to the plant cells so that they cannot be removed by simple washing. Recently non–attaching *Agrobacterium* mutants were isolated by Matthysse (1987). These mutants were

Table 4.   Tumour induction on several plant species

| Mutant | Virulence on | | | | |
	Tomato	Sunflower	Pea	Nicotiana glauca	Kalanchoe tubiflora
*vir*A	−	−	−	−	−
*vir*B	−	−	−	−	−
*vir*G	−	−	−	−	−
*vir*D	−	−	−	−	−
*vir*C	−	+++	−	++	++
*vir*E	−	−	−	+	++
*vir*F	+	+++	++	+	+++
*chv*A	−	−	−	−	−
*chv*B	−	−	−	−	−
*chv*C	−	+++	−	+++	+
*psc*A (*exo*C)*	ND	−	ND	ND	ND
att°	ND	ND	ND	ND	ND
wild-type	+++	+++	+++	+++	+++

*   *psc*A (*exo*C) mutant is severely attenuated in virulence on carrot
discs (Thomashow et al. 1987).
o   *att* mutant is avirulent on *Kalanchoe daigremontiana* and carrot
discs (Matthysse 1987).
ND   not determined

virulent; they had a mutation in a 12 kb *Eco*RI fragment of the
Agrobacterium C58 chromosome, which defines a new chromosomal
virulence locus *att*. The *att* mutants had flagella and fimbriae and
were motile.   They normally produced β-1,2-glucan and did not have
alterations in their lipopolysaccharide structure. Closer examination
of these mutants revealed that the mutants lack two polypeptides (34
kDa and 38 kDa), which are found in the outer-membrane or periplasmic
space of the wild-type bacteria. Future studies with these
att-mutants may possibly determine the functional relationship
between these polypeptides and bacterial binding to plant cells.

Homologous sequences to *chv*A, *chv*B and *psc*A (*exo*C) also exist in
the genomes of fast growing rhizobia including *Rhizobium trifolii*, *R.
leguminosarum*, *R. meliloti* and *R. phaseolii* and are required for
nodule development on leguminous plants (Dylan et al., 1986; Finan et
al., 1986; Cangelosi et al., 1987; Thomashow et al., 1987; Van Veen
et al., 1987). Complementation experiments have indeed shown that
corresponding genes of *Agrobacterium* and *Rhizobium* are functionally

interchangeable (Dylan et al., 1986; Cangelosi et al., 1987). Therefore, these genes may play a role in the establishment of a general type of attachment, which is necessary for tumour formation by *Agrobacterium* as well as for root nodule formation by *Rhizobium*.

Virulence Region of the Ti Plasmid

The *vir*-genes of the Ti (and Ri) plasmid Vir-region direct the events which result in T-DNA transfer to the plant cell genome. The virulence regions of the WHR octopine Ti plasmid pTiAch5 and of the nopaline Ti plasmid pTiC58 were compared precisely by heteroduplex analysis (Engler et al., 1981). This study revealed that the Vir-regions of these plasmids have extensive DNA homology but nevertheless also contain small areas of non-homology. The disruptions in the homologous areas are probably due to insertions and deletions which occurred in these Ti plasmids during evolution. One area of non-homology between the octopine and nopaline Vir-regions was found to share extensive sequence homology with the transposable element IS66 (Waldron and Hepburn, 1983; Machida et al. 1984). Also it was shown that Ri plasmids (White and Nester, 1980; Risuleo et al., 1982; Huffman et al., 1984; Jouanin, 1984) and LHR Ti plasmids (Knauf et al., 1984) share DNA homology with the Vir-regions of WHR Ti plasmids. Indeed, complementation experiments showed that *vir*-genes of different Ti plasmids and Ri plasmids are interchangeable (Hooykaas et al., 1984; Hoekema et al., 1984b).

The Vir-region encodes *trans*-acting functions (Hille et al., 1982; Klee et al., 1982; Hoekema et al., 1983) required to promote T-region transfer. The *vir*-DNA itself is not integrated into the plant genome. Transfer of the T-region from the bacterium to the plant cell is still possible after separation of the T-region from the rest of the Ti plasmid. (De Framond et al., 1983; Hoekema et al., 1983). Even from a chromosomal location T-region transfer occurs normally if the *vir*-genes are present in *trans* (Hoekema et al., 1984c).

During the past decade, the Vir-regions of the octopine Ti plasmid and the nopaline Ti plasmid were characterized in detail by transposon mutagenesis, complementation analysis of *vir*-mutants, nucleotide sequence determination and by expression studies. The 40 kbp Vir-region of the Ti plasmid is located adjacent to the left border repeat of the T-region (Garfinkel and Nester, 1980; Holsters et al., 1980; Ooms et al., 1980). Extensive transposon mutagenesis of the octopine Ti plasmid combined with complementation analysis of isolated *vir*-mutants revealed that the octopine Ti Vir-region contains seven transcriptional units (Hille et al., 1982; Klee et al. 1983; Hille et al., 1984; Hooykaas et al., 1984; Stachel and Nester 1986). Some of these Vir-loci are very large which suggests that these embrace operons with several cistrons. Different names have been given to the seven different *vir*-operons, but at the First International Congress of Plant Molecular Biology (Savannah, USA, 1985) it was agreed to use the following nomenclature (reading clockwise towards the T-region) *vir*A, *vir*B, *vir*G, *vir*C (previously *vir*O, *bak*), *vir*D (previously *vir*C, *hdv*), *vir*E and *vir*F (Fig.7), whereby the individual open reading frames (cistrons) within each operon are given the operon name followed by a number, e.g. *vir*B1, *vir*B2, *vir*B3 etc.

Mutational analysis of these Vir–loci identified two distinct classes namely, essential Vir–loci and 'host range' Vir–loci. The essential vir operons (*vir*A, *vir*B, *vir*D and *vir*G) contain genes which are crucial for tumour induction on all plant species. In contrast, mutations in the operons *vir*C, *vir*E, *vir*F only lead to a reduction in the host range for tumour induction (Table 4). The organization of the Vir–region of the nopaline Ti plasmid pTiC58 is similar to that of the octopine Ti plasmid (Lundquist et al., 1984; Rogowsky et al. 1987). The Vir–region of pTiC58 consists of six Vir–loci corresponding with *vir*A, *vir*B, *vir*G, *vir*C, *vir*D, *vir*E of the octopine Ti plasmid (Kao et al., 1982; Lundquist et al., 1984; Hagiya et al. 1985; Hirooka and Kado, 1986; Rogowsky et al., 1987).

Specific features of the nopaline Ti plasmid Vir–region (in comparison with the octopine Ti plasmid) are that it lacks the *vir*F locus (Otten et al., 1985), but instead contains a *tzs*-gene which confers on *Agrobacterium* the ability to produce the phytohormone *trans*-zeatin (Regier and Morris, 1982; Akiyoshi et al., 1985; Beaty et al., 1986). The *tzs*-gene is located at the most lefthand end of the nopaline Ti Vir–region next to *vir*A.

Initially the VirB loci of both nopaline and octopine Ti plasmids were reported to contain six complementation groups (Iyer et al. 1982; Lundquist et al., 1984), one of which could not be *trans*-complemented (Klee et al., 1982; Iyer et al., 1982; Lundquist et al., 1984). However, further studies of the octopine Ti *vir*B region clearly established that this region consists of a single locus and that all mutations are *trans*-complementable (Hille et al., 1982; Hooykaas et al., 1984; Stachel and Nester, 1986). Therefore, it is likely that further analysis of the nopaline *vir*B region will likewise lead to the elimination of the hypothesis of a *cis* dominant locus.

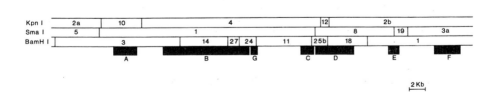

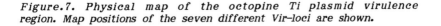

Figure.7. Physical map of the octopine Ti plasmid virulence region. Map positions of the seven different Vir–loci are shown.

187

NUCLEOTIDE SEQUENCE ANALYSIS OF THE Ti VIR-REGION

A more detailed picture of the structural organization of the virulence region has been obtained by DNA sequence analysis of the Vir-region. To date the nucleotide sequence determination of the octopine Ti plasmid (pTi15955, pTiA6) Vir-region has nearly been completed, while also for the nopaline Ti plasmid (pTiC58) Vir-region certain parts of the sequence have been determined (Table 5). In agreement with the structural and functional homology found between octopine and nopaline Ti Vir-regions, the nucleotide sequences of the nopaline *vir*-operons *vir*C, *vir*D and *vir*E showed considerable homology to the corresponding loci of the octopine Ti plasmid (Table 5). The number of open reading frames which are present in the different Vir-loci of the octopine Ti plasmid turned out to be one for *vir*A, eleven for *vir*B, one for *vir*G, two for *vir*C, four for *vir*D, and two for *vir*E. These data were in agreement with an operon structure of certain Vir-loci which was already hypothesized on the basis of the results from genetic studies. In total the octopine Ti Vir-region encodes at least 20 different Vir-proteins, the characteristics of which - number of amino acids, molecular weight and net charge - are listed in Table 5 and will be discussed in the following sections.

Das and coworkers (1986) determined the transcription initiation sites of six *vir*-promoters (*vir*A,B,C,D,E,G). A comparison of these promoter regions with the -10 and -35 consensus sequences of *E. coli* promoters is shown in Table 6. The -10 regions of *vir*B and *vir*D contain strong similarity to the *E. coli* consensus -10 sequence (5'-TATAAT-3'), but those of the other *vir*-promoters contain less homology. The -35 region of *vir*A, *vir*G and *vir*D share sequences with the consensus -35 region (5'-TTGACA-3') of *E. coli* promoters, while the -35 region of *vir*B, *vir*C and *vir*E lack strong similarity to the corresponding *E. coli* consensus region. Several *vir*-genes (*vir*B, *vir*C, *vir*D, *vir*E) contain in their upstream regions two common hexanucleotide sequences (CGAGTA and GCAATT) (Das et al., 1986). Besides, we found that in these upstream regions the consensus 5'-CAATTGAAA-3' is present in one or more copies. some of these sequences, might function as *cis*-acting regulatory elements for inducible expression of *vir*-genes (see section: Expression of *vir*-genes). The 3' ends of the 16S ribosomal RNAs of *Agrobacterium* (5'-CACCUCCUUU-3') and *E. coli* (5'-CACCUCCUUA-3') are almost identical (Weisburg et al., 1985). Therefore it is not surprising that all analyzed *vir*-operons except *vir*G show significant homology to the *E. coli* consensus ribosome binding site sequence (TAAGGAGGTG(5-9bp)ATG) (Shine and Dalgarno 1974).

A compilation of the codon usage of 21 *vir*-genes (*vir*A, *vir*B1, *vir*B2, *vir*B3, *vir*B4, *vir*B5, *vir*B6, *vir*B8, *vir*B9, *vir*B10, *vir*B11, *vir*G, *vir*C1, *vir*C2, *vir*D1, *vir*D2, *vir*D3, *vir*D4, *vir*E1 and *vir*E2) is shown in Table 7 next to a published compilation of that of 27 highly expressed *E. coli* genes (Sharp and Li, 1986). In these highly expressed *E. coli* genes certain codons (for example AUA (Ile) and AGG (Arg)) are used very infrequently. The *Agrobacterium* *vir*-genes

188

utilize all codons with uniform frequency. This is in agreement with findings that the *vir*-gene products are not formed abundantly.

Table 5. Characteristics of Vir-proteins of the octopine Ti plasmid.

Vir locus	Number of cistrons	Protein	Amino acids	M_r (000s)	Net- Charge	Literature citation
*vir*A	1	VirA	829	91.6	-2	Leroux et al. 1987; Melchers et al. 1987
*vir*B	11*	VirB$_1$	239	25.9	-3	Thompson & Melchers,
		VirB$_2$	121	12.3	4	unpublished
		VirB$_3$	108	11.8	2	
		VirB$_4$	587	64.4	9	
		VirB$_5$	191	21.6	-2	
		VirB$_6$	220	23.5	-5	
		VirB$_7$	295	31.8	-7	
		VirB$_8$	257	28.4	1	
		VirB$_9$	293	32.2	3	
		VirB$_{10}$	377	40.7	-3	
		VirB$_{11}$	343	38.0	-7	
*vir*C	2	VirC$_1$	231	25.7	-8	Yanofsky et al.,
		VirC$_2$	202	22.7	6	1986b
*vir*D	4	VirD$_1$	147	16.2	2	Thompson & Melchers,
		VirD$_2$	424	47.5	1	unpubl.; Yanofsky et
		VirD$_3$	201	21.4	15	al. 1986a.
		VirD$_4$	656	72.4	-1	
*vir*E	2	VirE$_1$	65	7.1	-4	Winans et al. 1987.
		VirE$_2$	533	60.5	-3	
*vir*F	N.D.					
*vir*G	1	VirG	267	29.6	1	Melchers et al. 1986 Winans et al. 1986.

N.D.   = not determined
Note: *vir*-loci sequenced of other Ti plasmids are: *vir*A of
        pTiAg162 (LHR) (Leroux et al. 1987) *tzs*, *vir*C, *vir*D and
        *vir*E of pTiC58 (Beaty et al 1986; Close et al. 1987; Hagiya
        et al. 1985; Hirooka et al. 1986)

Table 6. Homology of *vir*-promoter sequences to the -10 and -35 consensus sequence of *E. coli* promoters.

	E. coli consensus	
	-35 region	-10 region
Promoter	t c T T G A C a	t g - T A t A a T
*vir*A	C A • • T • • o	C A G C C o • o C
*vir*B	o T • C • • G T	C C C G • o • o •
*vir*G	o T • • • T • o	o A C • T G G T •
*vir*D	o G • • • • T T	G A C • T o • o •
*vir*C	o T A • A • • G	o A G • G o • o C
*vir*E	A o C C • • G T	A o A • • o G o G
Common	- - T T G A C -	- a - T - T A A t

Vir-promoter base pairs identical to highly conserved base pairs of *E. coli* are indicated with a black dot. The open circles represent base pairs which are identical to the weakly conserved base pairs of *E. coli* promoters (Hawley and McClure 1983) *Vir*-promoter regions were defined based on transcription initiation sites (S_1-mapping) of the different *vir*-genes determined by Das et al. (1986).

EXPRESSION OF *vir*-GENES

The transcriptional organization of the octopine Ti Vir-region has been analyzed by inserting a promoterless *lacZ* gene into each of the different *vir*-operons (Okker et al., 1984; Stachel et al., 1985a; Stachel and Nester, 1986). Expression of *vir*-genes could be monitored by assaying for β-galactosidase activity as *Agrobacterium* does not have interfering β-galactosidase activity of its own (Okker et al. 1984; Plessis et al., 1985). In agreement with sequence data the *vir*-genes turned out to be located clockwise towards the T-region with the exception of *vir*C, which is transcribed counterclockwise (Hille et al., 1984; Stachel and Nester, 1986).

In contrast to the chromosomal *chv*-genes the expression of most of the plasmidal *vir*-genes was found to be regulated. During growth on standard, rich or minimal media, only *vir*A and *vir*G are significantly expressed. However, when bacteria are exposed to plant cells or their

Table 7. Codon usage in 21 *A. tumefaciens vir*-genes compared with that in 27 *E. coli* genes.

Codon	21 *vir* -genes	27 *E. coli* genes[*]	Codon	21 *vir* -genes	27 *E. coli* genes[*]
	total	total		total	total
GGG-Gly	75	6	AUA-Ile	57	1
GGA-Gly	85	3	AUU-Ile	142	55
GGU-Gly	114	315	AUC-Ile	176	298
GGC-Gly	176	228			
			ACG-Thr	112	17
GAG-Glu	177	95	ACA-Thr	80	13
GAA-Glu	202	367	ACU-Thr	64	166
			ACC-Thr	127	172
GAU-Asp	185	118			
GAC-Asp	181	272	UGG-Trp	58	34
GUG-Val	104	67	UGU-Cys	21	10
GUA-Val	51	150	UGC-Cys	32	20
GUU-Val	139	303			
GUC-Val	130	20	UAU-Tyr	93	32
			UAC-Tyr	67	134
GCG-Ala	179	129			
GCA-Ala	151	178	UUG-Leu	132	8
GCU-Ala	150	304	UUA-Leu	52	8
GCC-Ala	167	37	CUG-Leu	128	403
			CUA-Leu	73	3
AGG-Arg	52	0	CUU-Leu	143	17
AGA-Arg	46	1	CUC-Leu	142	15
CGG-Arg	76	1			
CGA-Arg	71	1	UUU-Phe	136	47
CGU-Arg	108	261	UUC-Phe	104	159
CGC-Arg	144	93			
			CAG-Gln	138	202
AGU-Ser	75	10	CAA-Gln	141	25
AGC-Ser	129	48			
UCG-Ser	102	2	CAU-His	71	23
UCA-Ser	67	9	CAC-His	63	79
UCU-Ser	64	117			
UCC-Ser	89	87	CCG-Pro	101	171
			CCA-Pro	87	23
AAG-Lys	123	95	CCU-Pro	68	12
AAA-Lys	124	375	CCC-Pro	69	2
AAU-Asn	113	13			
AAC-Asn	105	256			
AUG-Met	155	130			

* Data from Sharp & Li (1986)

exudates the expression of the other *vir*-genes becomes induced to high levels (Stachel and Nester, 1986). Activation of the *vir*-genes is the direct result of the recognition by *Agrobacterium* of signal molecules produced by wounded plant cells, which trigger the vir regulatory system (Okker et al., 1984; Stachel et al., 1985a). The signal molecules present in *Nicotiana tabacum* exudates were purified and identified as the phenolic compounds acetosyringone (AS) and α-hydroxyacetosyringone (OH-AS) by Stachel et al.,(1985b). Structural formulae can be seen in Fig 8. A derivative of AS which lacks one methoxy-group (acetovanillone) has somewhat attenuated *vir*-inducing activity. Removal of both methoxy-groups (4-hydroxyacetophenone) or the hydroxyl-group (3,5-dimethoxyacetophenone) from AS leads to inactive compounds. These observations underscore the specificity of the plant signal molecules. Still other phenolic compounds such as syringaldehyde, syringic acid, sinapinic acid, vanillin, 2,6 –dimeth oxyphenol, furulic acid, 3,4-dihydroxybenzaldehyde, and coniferyl alcohol were found to have partial or full inducing activity (Stachel et al.,1985b; Melchers unpublished.) In contrast to Bolton et al. (1986) , in our experiments, the phenolic compounds 4-hydroxybenzoic acid, β-resorcyclic acid, protocatechuic acid, gallic acid, pyrogallic acid and catechol do not act as inducers of the *vir* genes. Data are accumulating that many monocotyledonous plant species do not produce *vir*-inducer molecules and this might be an important reason why these plants are inefficiently transformed by *Agrobacterium* (Okker et al., 1984; Usami et al., 1987).

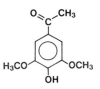 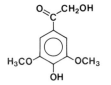

acetosyringone   α-hydroxyacetosyringone

Figure.8. Structural formulae of the vir*-inducing molecules.*

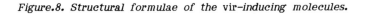

The regulatory system which controls the expression of the inducible *vir*-genes is likely to be essential in the tumour induction process. In fact it turned out that the essential *vir*-genes *vir*A and *vir*G are the regulators of the *vir*-regulon (Stachel and Zambryski 1986). The VirA and VirG proteins must carry out at least two distinct activities. Firstly, extracellular recognition of the plant signal molecule(s) and transduction of this environmental signal to inside the bacterium. Secondly, the activation of several separate *vir*-operons in response to the plant signal. Interestingly, the regulatory proteins VirA and VirG showed significant homology to several bacterial proteins which are already known to play a role in the transcriptional regulation of genes of K. pneumoniae or *E. coli*. The VirA protein has homology with the C-terminal region (about 250 amino acids) of four regulatory proteins *viz*. EnvZ, NtrB, PhoR and CpxA (Nixon et al., 1986; Leroux et al., 1987; Melchers et al., 1987). These proteins are part of a two component positive regulatory system (i.e. sensory protein/activator protein) which is present in a variety of prokaryotic species: EnvZ/OmpR, PhoR/PhoB, CpxA/Sfr of *E. coli* and NtrB/NtrC of *K. pneumoniae* and *Bradyrhizobium sp.* (Nixon et al., 1986). Genetic analysis has shown that *envZ* and *phoR* encode membrane proteins which are essential for the activation of the *omp-* and *pho-*genes respectively (Hall and Silhavy, 1981; Shinagawa et al. 1983). Recently, the 91.6 kDa VirA protein was localized in the *Agrobacterium* inner-membrane by immunodetection using antibodies raised against the VirA product (Leroux et al., 1987; Melchers et al. 1987). The VirA protein resembles *E. coli* EnvZ (Comeau et al., 1985; Liljeström, 1986) and various other transmembrane chemoreceptor proteins (Krikos et al., 1985) in that it contains two hydrophobic regions which can anchor the protein in the inner-membrane, a periplasmic domain and a cytoplasmic domain. Analogous to the two-component regulatory model, the VirA protein probably functions as an environmental sensor of plant derived inducer molecules and transmits the extracellular signal inside the bacterium (see Fig. 9).

The VirG protein showed significant homology with the positive regulatory proteins OmpR, PhoB, Sfr (=Dye) and NtrC (Melchers et al. 1986; Winans et al., 1986). Hence, it is likely that VirG is a positive regulator which activates *vir-*gene expression. It is yet unknown how the VirA protein regulates the activity of VirG. It may be that VirG is covalently modified – by phoshorylation for instance – by an activated VirA protein. Alternatively, it may be that VirA

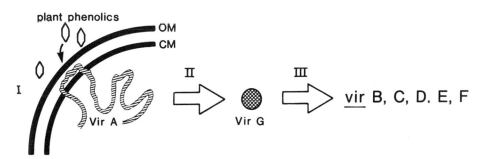

Figure.9. The Vir regulatory system of Agrobacterium *(I) Phenolic compounds (like acetosyringone) present in wound exudates of plants are recognized by the bacterium. (II) The regulatory protein VirA transduces a signal to VirG which in turn becomes activated (III). The positive regulatory protein VirG activates the transcription of the* vir-genes virB,C,D,E *and* F.
Abbreviations: OM, outer-membrane; CM, cytoplasmic membrane.

allows VirG to interact directly with the inducer molecules such as acetosyringone.

The kinetics of expression of the inducible *vir*-genes are similar. Within one hour after the addition of AS the expression of the inducible *vir*-genes is significantly elevated and becomes maximal in the following 6 - 12 hours, after which it returns to a pre-induced level in about 48 hours (Bolton et al., 1986; Winans et al., 1986; Melchers unpublished). The pH of the medium has a strong influence on *vir*-induction: induction is only optimal within a pH-range of 5.0 - 5.4 and does not occur at pH 6.3 or above (Stachel et al., 1986a; Veluthambi et al., 1987).

A second type of *vir*-gene regulation is mediated by a chromosomal locus called *ros*, which negatively regulates the expression of the *vir*C and *vir*D operons exclusively (Close et al., 1985). The expression of *vir*C and *vir*D of both the nopaline and the octopine Ti plasmids are elevated in the *Agrobacterium* ros mutant (Rogowsky et al., 1987). Identification of a physiological signal for Ros will probably provide more insight in this alternative type of *vir*-regulation which appears to be unlinked to the positive regulatory system involving *vir*A and *vir*G.

T-DNA INTERMEDIATE FORMATION: ROLE OF VirD PROTEINS AND BORDER REPEATS

The T-regions in the Ti and Ri plasmid are surrounded by a 24 bp imperfect direct repeat. These border repeats form the signals that are recognized by the transfer system of *Agrobacterium*. The right border (defined arbitrarily, see Table 3) is very important for T-DNA transfer (Ooms et al., 1982b; Shaw et al., 1984b; Wang et al., 1984; Hepburn and White, 1985; Gardner and Knauf, 1986; Rubin, 1986) but the left border can be deleted without any loss of tumorigenicity of the bacterium (Hille et al., 1983; Joos et al., 1983). Furthermore, the activity of the right border repeat is strongly dependent on its orientation (Wang et al., 1984; Peralta and Ream, 1985; Van Haaren et al., 1986). Inversion of the 24 bp repeat leads to a strong reduction in the transformation efficiency. Next to the right border repeats an enhancer sequence is present (Peralta et al., 1986; Van Haaren et al. 1986;1987) which contributes significantly to the efficiency of T-DNA transfer. This enhancer element comprises a 24 bp sequence located at a distance of 13 and 14 bp repectively from the right border repeats of the octopine T_L and T_R regions (Peralta et al., 1986). Although the border repeats are polar in their activity, the enhancer element functions independent of its position and orientation relative to the right border repeat and is active even at distances of more than 4000 bp from the border repeat (Van Haaren et al., 1986; Van Haaren unpublished; Ream pers. commun.).

Activation of the Vir–system leads to the nicking of the bottom strands of the 24 bp repeats (Yanofsky et al., 1986a; Albright et al. 1987; Wang et al., 1987). Border nicking occurs predominantly at the sequence 3'–CCG₊TCCT–5' of both the nopaline Ti (Wang et al., 1987) and the octopine Ti border repeats (Albright et al., 1987). At low frequency also top strand interruptions were detected which occurred within 5 bp of the bottom strand nick site (Yanofsky et al., 1986a; Albright et al., 1987). Other authors described double strand cleavages which were found at or around the border repeats during cocultivation of the bacteria with plant cells (Veluthambi et al. 1987).

Putative T–DNA intermediates can be detected in *Agrobacterium* during cocultivation with plant cells or incubation with *vir*-inducers. Circular T–DNA copies can be rescued from *Agrobacterium* (Koukolikova-Nicola et al., 1985; Machida et al., 1986; Yamamota et al. 1987) or from *E. coli* containing the *vir*-genes (Alt-Moerbe et al. 1986) by strongly selective methods. Such T–circles contain a "hybrid" 24 bp border repeat which originates from a fusion between the left and right border repeats originally surrounding the T–region (Koukolikova-Nicola et al., 1985; Alt-Moerbe et al., 1986; Machida et al., 1986). Although T–circles can be rescued from *Agrobacterium* by strong selection, such molecules cannot be detected in this bacterium by Southern blot analysis. Instead single stranded linear molecules, corresponding to the bottom strand of the T–region, can be detected in this way in AS–induced *Agrobacterium* strains (Stachel et al. 1986b;1987; Albright et al., 1987; Van Haaren unpublished). These molecules, which are called T–strands (Fig.10), are produced at

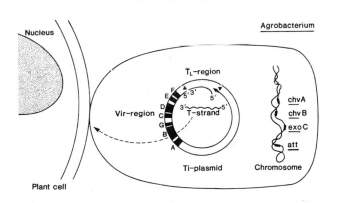

Figure.10. Generation of a single stranded T-molecule in Agrobacterium. *For schematic reasons only the TL–region of the* octopine Ti plasmid is shown. *VirD proteins (VirD1 and VirD2) are involved in site specific nicking at the bottom strand of the 24 bp border repeats (arrow heads). One model for T–strand formation suggests that the T–strand is released from the Ti plasmid by helicase and replacement strand synthesis of the bottom strand of the T–region. The final structure of the transferred intermediate is still obscure.*

approximately one copy per bacterium (Stachel et al., 1986b). Double stranded linear T-copies could be detected also, but only after treatment of the DNA with S1 nuclease, which suggests that S1 introduced a double strand cut at sites where already a nick was present (Stachel et al., 1986b).

Upon activation of the Vir-system border nicks can be detected within 4-12 hours (Wang et al., 1987), while T-strands can be found 8 hours after induction (Stachel et al., 1986b). In *Agrobacterium* strains with the octopine Ti plasmid pTiA6, which contains a T_L- and a T_R-region, nicks are formed independently at each of the four border repeats. Six different T-strands corresponding to T_L, T_R, T_C, T_L + T_C, T_C + T_R, and T_L + T_C + T_R - can be found in such strains after *vir*-induction (Stachel et al., 1987). These data suggest that border nicks function as initiation and termination sites of T-strand formation. It is not known how T-strands are formed, but two possibilities are obvious, firstly, excision in a process similar to what occurs during DNA-repair, and secondly new synthesis from the nick in the right repeat to that in the left repeat with release of the parental bottom strand.

Since the frequency of T-circle formation in *Agrobacterium* is very low (10^{-5}) in comparison with the transformation frequencies that are usually obtained with *Agrobacterium* (Machida et al., 1986), it is unlikely that the T-circles are true T-DNA intermediates. Linear T-DNA intermediates are more likely not only because of their prevalence, but also in view of the absence of permutations in the T-DNA in the transformed plant cell versus the T-region as present in the Ti plasmid. The linear T-molecules detected in *Agrobacterium* have end points at the nick sites in the 24 bp border repeats, which is consistent with the maximum extent of the T-DNA found in the transformed plant cells (See section: Structure of the T-DNA). Whether the single stranded T-strands, the double stranded T-molecules or both are the long-sought T-DNA intermediates is unclear. In this regard it is striking that in agrobacteria with a wild-type octopine Ti plasmid, six different T-strands are formed with equal frequency (Stachel et al., 1987), but that in tumour cells (i.e. after selection for T_L-DNA) T_L is found much more frequently than T_L + T_C or T_L + T_C + T_R. Apparently, there is a preference for the transfer and/or the integration of T_L alone in plant cells, although a more trivial explanation of the results might be that 'in real life' nicking occurs much more frequently so that combinations of T_L + T_C or T_L + T_C + T_R become extremely rare. The role of the enhancer in border nicking and T-strand formation is unclear, but data show that the enhancer is not absolutely essential for these processes (Wang et al., 1987; Van Haaren, unpublished). The enhancer may act as an entry site for the border-specific endonuclease or helper proteins, thereby increasing the frequency of border nicking and/or T-strand formation.

Proteins encoded by the VirD locus are the only Vir-proteins that are essential for border cleavage (Veluthambi et al., 1987), T-strand formation (Stachel et al., 1987) and T-circle production (Alt-Moerbe et al., 1986; Yamamoto et al., 1987). Although the VirD locus codes for four proteins, only two of these (VirD1 and VirD2) are necessary.

However, in view of the complexity of the processes it is likely that more proteins are involved. These must be encoded by the *Agrobacterium* chromosome.

Many DNA-binding proteins have a common amino acid sequence of Ala-$(N)_3$-Gly-$(N)_5$-(Ile/Val) in the domain which interacts with DNA (Pabo and Sauer, 1984). Crystallographic structures determined for the Cro and cI-repressor protein of bacteriophage λ and the CAP protein of *E. coli* revealed the existence of conserved α-helical structures which computer modelling suggests are the DNA-binding sites for these proteins. The DNA-binding structure (helix-turn-helix motif) comprises two α-helices which are called helix-2 and helix-3. Helix-3 fits into the major groove of DNA and helix 2 lies across helix-3 holding it in position. The amino acids which are important for the conformation of this structure are conserved in the studied proteins (Cro, cI-Rep, CAP) and appears to be also present in other DNA binding proteins (see Fig 11). Alanine is usually found in position 5, glycine in position 9, isoleucine or valine in position 15, and hydrophobic residues are found in positions 4,8,10 and 18. The side chains at positions 8-10 are small to permit a tight turn between the α-helices. The VirD1 proteins of the octopine and nopaline Ti plasmid turned out to contain these conserved amino acids with only one exception: a hydrophobic residue at position 4 is not present. The presence of a putitive DNA-binding domain in the $VirD_1$ protein was not unexpected in view of the role of the VirD1 protein in border nicking.

FUNCTIONS DETERMINED BY OTHER Vir-LOCI

Besides the regulatory genes *vir*A and *vir*G (See section: Expression of *vir*-genes) and the *vir*D1 and *vir*D2 genes, which are involved in T-DNA intermediate formation (See previous section), there are two other sets of *vir*-genes which are essential for tumorigenicity, namely the eleven genes of the VirB locus and the *vir*D3 and *vir*D4 genes of the VirD locus. Recently a technique called "agroinfection" was developed, which allowed to discriminate between the steps involved in T-DNA transfer and later steps leading to tumour formation (Gardner and Knauf, 1986; Grimsley et al., 1986). The technique employs an *Agrobacterium* strain with a T-region in which a (partial) tandem repeat of a plant virus, such as CaMV or a viroid such as PSTV, is present. The presence of tandemly repeated sequences of these viruses allows them to "escape" from the T-region in the infected plant cells and to give rise to systemic infection potentially even when tumour formation does not occur. Such constructs have been tested in combination with the different vir mutations (Gardner and Knauf,, 1986; Hille et al., 1986). It was found that no systemic infections occurred in combination with *vir*A, *vir*G, *vir*B or *vir*D mutations, which indicates that no T-DNA transfer occurs from such mutants. For *vir*A, *vir*G and the *vir*D genes this was expected, since they are involved in activation of the Vir-system and

T–DNA intermediate formation, respectively. The *vir*B genes are essential for tumorigenicity. The absence of T–DNA transfer from *vir*B mutants combined with the strong hydrophobicity of the proteins encoded by the VirB locus (our unpublished data) and the location of VirB proteins in the *Agrobacterium* membrane (Engström cited in Stachel et al., 1987), makes the VirB proteins candidates for being structural proteins mediating T–DNA transfer to plant cells. Systemic infection occurred after infection with the "host range" mutants *vir*C,

Putative DNA-binding domain within the VirD$_1$ protein .

		1	2	3	4	5	6	7	8	9	10	11	12	13	14	15	16	17	18	19	20	
VirD$_1$ pTi15955	40	Phe	Ser	Tyr	Gln	Ala	Arg	Leu	Leu	Gly	Leu	Ser	Asp	Ser	Met	Ala	Ile	Arg	Val	Ala	Val	...
VirD$_1$ pTiC58	40	Phe	Ser	His	Gln	Ala	Arg	Leu	Leu	Gly	Leu	Ser	Asp	Ser	Met	Ala	Met	Arg	Val	Pro	Val	...
λ cI Rep	33	Gln	Glu	Ser	Val	Ala	Asp	Lys	Met	Gly	Met	Gly	Gln	Ser	Gly	Val	Gly	Ala	Leu	Phe	Asn	...
λ Cro	16	Gln	Thr	Lys	Thr	Ala	Lys	Asp	Leu	Gly	Val	Tyr	Gln	Ser	Ala	Ile	Asn	Lys	Ala	Ile	His	...
P22 Rep	21	Gln	Ala	Ala	Leu	Gly	Lys	Met	Val	Gly	Val	Ser	Asn	Val	Ala	Ile	Ser	Gln	Trp	Gln	Arg	...
P22 Cro	13	Gln	Arg	Ala	Val	Ala	Lys	Ala	Leu	Gly	Ile	Ser	Asp	Ala	Ala	Val	Ser	Gln	Trp	Lys	Glu	...
CAP	169	Arg	Gln	Glu	Ile	Gly	Gln	Ile	Val	Gly	Cys	Ser	Arg	Glu	Thr	Val	Gly	Arg	Ile	Leu	Lys	...
Lac Rep	6	Leu	Tyr	Asp	Val	Ala	Glu	Tyr	Ala	Gly	Val	Ser	Tyr	Gln	Thr	Val	Ser	Arg	Val	Val	Asn	...
Gal Rep	4	Ile	Lys	Asp	Val	Ala	Arg	Leu	Ala	Gly	Val	Ser	Val	Ala	Thr	Val	Ser	Arg	Val	Ile	Asn	...
P22 cI	26	Gln	Arg	Lys	Val	Ala	Asp	Ala	Leu	Gly	Ile	Asn	Glu	Ser	Gln	Ile	Ser	Arg	Trp	Lys	Gly	...
Mat a$_1$	100	Lys	Glu	Glu	Val	Ala	Lys	Lys	Cys	Gly	Ile	Thr	Pro	Leu	Gln	Val	Arg	Val	Trp	Cys	Asn	...
Trp Rep	66	Gln	Arg	Glu	Leu	Lys	Asn	Glu	Leu	Gly	Ala	Gly	Ile	Ala	Thr	Ile	Thr	Arg	Gly	Ser	Asn	...
Tn3 Resolvase	161	Ala	Thr	Glu	Ile	Ala	His	Gln	Leu	Ser	Ile	Ala	Arg	Ser	Thr	Val	Tyr	Lys	Ile	Leu	Glu	...
Ara C	196	Ile	Ala	Ser	Val	Ala	Gln	His	Val	Cys	Leu	Ser	Pro	Ser	Arg	Leu	Ser	His	Leu	Phe	Arg	...
Lex Rep	28	Arg	Ala	Glu	Ile	Ala	Gln	Arg	Leu	Gly	Phe	Arg	Ser	Pro	Asn	Ala	Ala	Glu	Glu	His	Leu	...
					H	Ala			H	Gly	H					Ile/Val			H			

├──── Helix 2 ────┤   ├──── Helix 3 ────┤

Figure.11. Putative DNA–binding domain within the VirD1 proteins of the octopine (pTi15955) and nopaline (pTiC58) Ti plasmids are aligned with prokaryotic and potential yeast DNA–binding domains (Pabo and Sauer, 1984). The positions of the two α–helices of the DNA–binding domain (numbered 2 and 3 according to convention for the Cro protein) are shown at the bottom. Conserved residues within the domain, important for close aligment of the two helices, are boxed. The number to the left of each sequence is the position of the first amino acid shown within the complete protein. The positions of the amino acids are arbitrarily numbered at the top (1–20). The collection of bacterial sequences shown is from a review by Pabo and Sauer (1984). Agrobacterium vir D sequences are cited in Table 5.

virE and *virF*, although for *virE* mutants systemic infection became visible only in 10-20% of the infected plants (Hille et al., 1986). It can therefore be concluded that T-DNA transfer is possible from such mutants, although the efficiency may be significantly lower than from the wild-type. It can therefore be envisaged that the proteins encoded by these genes are involved in efficiency of intermediate formation, efficiency of transport, protection of the intermediate, integration of T-DNA into the plant genome or conditioning of the plant cells.

The efficiency of T-DNA transfer is one of the key factors which determine whether tumours are formed or not on a particular plant species. Indirect evidence suggests that the *virC* "host range" genes contribute to the efficiency of T-DNA transfer (Yanofsky et al. 1985a; Alt-Moerbe et al., 1986). *Agrobacterium* wide-host range (WHR) strains are unable to induce tumours on certain cultivars of *Vitis* (grapevine) plants. However, a derived *virC* mutant turned out to be tumorigenic. Avirulence of the WHR strains is due to a "hyper sensitive response" (Yanofsky et al., 1985a) in which the plant cells at the wound site turn dark brown and die. Apparently the WHR strains induce the hypersensitive response on *Vitis* because too many cells receive T-DNA perhaps leading to the overproduction of phytohormones. Certain plant species, like *Vitis*, may require the incorporation of T-DNA sequences into only a few cells at the site of inoculation to induce tumour proliferation. However, other plant species (e.g. kalanchoe), may require the incorporation of the T-DNA element into many cells to establish wild-type tumour induction. On such plant species *virC* mutants are attenuated. It has been observed that the VirC locus is not necessary for the efficient transfer of T-DNA from a small binary vector to plant cells, but is essential for an effective transfer from the Ti plasmid itself (Horsch et al. 1986). The reason for this is unclear at the moment.

Pairwise coinfection experiments of *Agrobacterium* mutants have shown that the virulence of *virE* and *virF* mutants can be restored by coinfection with mutants that are wild-type for the VirE or the VirF locus, respectively (Otten et al., 1984;1985). This suggests that the *virE* and *virF* genes determine diffusable factors that are required for efficient transformation. Recently, the nucleotide sequences of *virE* from both the octopine (Winans et al. , 1987) and nopaline (Hirooka et al., 1987) Ti plasmid were determined. The *virE* operon of the octopine Ti plasmid potentially encodes two proteins (7 kDa and 60.5 kDa), whereas that of the nopaline Ti plasmid possibly determines an extra protein (9 kDa, 7.1 kDa and 63.5 kDa). Analysis of the distribution of hydrophilic and hydrophobic amino acid residues showed that no obvious signal peptide is present in these proteins.

The VirF locus is present in the octopine and the leucinopine Ti plasmid as well as the agropine Ri plasmid, but is not encoded by the nopaline Ti plasmid (Otten et al., 1985). The introduction of the VirF locus into nopaline Ti carrying strains enhances the virulence of such strains on plants such as *Kalanchoe* (Otten et al., 1985). Reversely, also the nopaline Ti plasmid appeared to encode a specific host range function, namely via the plant-inducible *tzs*-gene. The

difference in oncogenicity of octopine and nopaline strains on plants
of the genus *Brassica* is due to the absence and presence of the
tzs-gene, respectively (R.Shields pers. commun.). For poplar this
might be the case too (Fillatti et al., 1987). Sequence studies
indicated that *tzs* has substantial homology (55%) with the T-region
cyt-gene (Beaty et al., 1986; Akiyoshi et al., 1985) which determines
an enzyme (isopentenyl-transferase) involved in the cytokinin
biosynthetic pathway in plant cells (Akiyoshi et al., 1984; Barry et
al., 1984). Although both *tzs* and *cyt* are involved in the production
of the phytohormone *trans*-zeatin these genes are expressed in
distinct types of organisms, i.e. *Agrobacterium* and plant cells
respectively. Furthermore, there are indications that a gene on the
nopaline Ti plasmid is involved in the production of the plant
hormone indole acetic acid by *Agrobacterium* (Liu et al., 1982). This
gene (called *iaa*P) was located in the middle of the nopaline Ti
Vir-region and turned out to be necessary for virulence (Liu et al.,
1982). It is not known whether the *iaa*P gene forms part of one of the
vir-operons.

PLANT GENETIC ENGINEERING

Agrobacterium is now frequently used as a vector for the
introduction of new genes into plant cells. Since the presence of
onc-genes in the T-DNA is incompatible with the regeneration of the
transformed cells into complete plants, preferably disarmed strains
are used as vectors. On the basis of the finding that a physical
separation of the T-region and the Vir-region does not lead to a
reduced potential for transformation (De Framond et al., 1983;
Hoekema et al., 1983), a binary vector system was developed
consisting of a T-region vector plasmid, into which genes can be
cloned directly (in *E. coli*) before transfer to *Agrobacterium*, and a
helper plasmid such as pAL4404 (Ooms et al., 1982b), which contains
the *vir*-genes. Versatile "binary vectors" have been constructed in
several groups (e.g. Bevan, 1984) and even special purpose vectors
such as promoter-probe vectors are available now (An 1986). Whether
derived cosmids (Simoens et al., 1986) shall become useful for plant
transformation has to await future experiments. This depends on the
precision with which *Agrobacterium* transfers large pieces of DNA to
plant cells. Indeed, it is not yet known what the length of DNA is
that can reproducibly be transferred faithfully to plant cells via
Agrobacterium.

The range of plant species for which *Agrobacterium* can be used as
a vector is very broad. Application is limited mainly by the lack of
cell-culture techniques for many plant species. Although this is true
in general, it has to be noted that *Agrobacterium* strains may differ
in their ability to transform a particular plant species. For *Populus*
it was found that a nopaline Ti plasmid served well as a helper
plasmid for the transfer of a binary vector, but an octopine Ti
plasmid would not do (Fillatti et al., 1987). The leucinopine pTi542

plasmid carrying strains are said to be "super-virulent": they form (large) tumours on an unusually wide range of different plant species. This is due to an enhanced ability to transfer T-DNA (An et al.1985). Recently, it was shown that the super-virulence trait resided in the Vir-region of pTi542 (Hood et al., 1986b). For use in a particular crop it is thus advisable to look for an optimal combination of *Agrobacterium* background (*chv*-genes), helper plasmid (*vir*-genes) and (binary) vector. Certain plants produce not enough of the *vir*-inducer molecules in their wound-sap to allow efficient transformation (Usami et al., 1987). The addition of AS may then lead to increased transformation frequencies, as was found for *Arabidopsis* to be the case indeed (Sheikholeslam et al., 1987).

For a long time it was thought that the host range for transformation by *Agrobacterium* was restricted to the dicotyledonous plants (De Cleene and De Ley, 1976). More recently, however, we could demonstrate that, although no tumour formation occurs, T-DNA transfer and expression does occur in certain monocots at least (Hooykaas-Van Slogteren et al., 1984). The evidence for this was that the opines octopine and nopaline were produced in tissues infected by virulent octopine and nopaline strains respectively, but not in tissues infected by strains lacking a Ti plasmid or with a mutation on an essential *vir*-gene. Our results were confirmed and extended by Hernalsteens et al. (1984) and Graves and Goldman (1986;1987) for a number of other monocotyledonous plant species. Recently, Christou et al. (1986) suggested that these data are not necessarily correct, because the opine and nopaline can be formed spontaneously in tissues that are fed with arginine for a long time. We are confident that we interpreted our data in 1984 correctly, since:-

(1) our infected tissues were not fed with arginine, which according to Christou et al. is necessary for nopaline production,
(2) in tissues infected with octopine strains octopine was found indeed, whereas Christou et al. did not detect octopine in any of the tissues in their investigation,
(3) we used an *in vitro* enzyme assay for opine production as described by Otten and Schilperoort (1978) and in fact measured enzyme activities *in vitro* using extracts from tissues in contrast to Christou et al. (1986) who monitored directly for the presence of the compound nopaline in the tissue.

In addition, recently it was demonstrated that the monocot maize can be transformed by *Agrobacterium* by a completely different approach, namely using agroinfection (Grimsley et al., 1987). These data all together show that monocots of the families Amaryllidaceae, Gramineae, Iridaceae and Liliaceae can be transformed by *Agrobacterium*. Although this is very likely, it must be noted that it has still to be proved that T-DNA is stably integrated into the genome of monocots.

The precise molecular mechanism of T-DNA transfer and integration are still largely unknown. Therefore, it is difficult to predict whether one day it will be possible to use the *Agrobacterium* system for instance for the introduction of new DNA into the chloroplasts or the mitochondria of plant cells, or for the site-directed mutagenesis

201

of the plant genome. However, we can be sure that more knowledge of the basics of the Agrobacterium system will pave the way for future applications in basic studies on the molecular biology of plants and in the genetic engineering of crops.

ACKNOWLEDGEMENTS

We thank our colleagues for sending preprints and communicating their results before publication. We are grateful to Dr. M.J.J. Van Haaren and Dr. C.W. Rodenburg for critical reading of the manuscript, to Mrs. M.J.G. Bergenhenegouwen for typing the manuscript, and to G.P.G. Hock and J. Simons for preparation of the figures.

REFERENCES

AKIYOSHI, D.E., MORRIS, R.O., HINZ, R., MISCHKE, B.S., KOSUGE, T., GARFINKEL, D.J., GORDON, M.P., and NESTER, E.W. (1983) Cytokinin–auxin balance in crown gall tumours is regulated by specific loci in the T–DNA. Proc. Natl. Acad. Sci. USA, 80:407–411.

AKIYOSHI, D.E., KLEE, H., AMASINO, R.M., NESTER, E.W. and GORDON, M.P. (1984) T–DNA of *Agrobacterium tumefaciens* encodes an enzyme of cytokinin biosynthesis. Proc. Natl. Acad. Sci. USA, 81:5994–5998.

AKIYOSHI, D.E., REGIER, D.A., JEN, G. and GORDON, M.P. (1985) Cloning and nucleotide sequence of the *tzs* gene from *Agrobacterium tumefaciens* strain T37. Nucleic Acids Res. 13:2773–2788.

ALBINGER, G. and BEIDERBECK, R. (1977) Ubertragung der Fahigkeit zur Wurzelinduktion von *Agrobacterium rhizogenes* auf *A. tumefaciens*. Phytopath. Z. 90:306–310.

ALBRIGHT, L. M., YANOFSKY, M.F., LEROUX, B., MA, D. and NESTER, E.W. (1987) Processing of the T–DNA of *Agrobacterium tumefaciens* generates border nicks and linear, single stranded T–DNA. J. Bacteriol. 169:1046–1055.

ALT–MOERBE, J., RAK, B. and SCHRÖDER, J. (1986) A 3.6–kbp segment from the Vir region of Ti plasmids contains genes responsible for border sequence–directed production of T–region circles in *E. coli*. EMBO J. 5: 1129–1135.

AMASINO, R.M., POWELL, A.L.T. and GORDON, M.P. (1984) Changes in T–DNA methylation and expression are associated with phenotypic variation and plant regeneration in a crown gall tumour line. Mol. Gen. Genet. 197: 437–446.

AMBROS, P.F., MATZKE, A.J.M. and MATZKE, M.A. (1986) Localization of *Agrobacterium rhizogenes* T–DNA in plant chromosomes by *in situ* hybridization. EMBO J. 5: 2073–2077.

AN, G., WATSON, B.D., STACHEL, S., GORDON, M.P. and NESTER, E.W. (1985) New cloning vehicles for transformation of higher plants. EMBO

J. 4:277–284.

AN, G. (1986) Development of plant promoter expression vectors and their use for analysis of differential activity of nopaline synthase promoter in transformed tobacco cells. Plant Physiol. 81:86–91.

AN. G., EBERT, P.R., YI, B-Y., and CHOI, C.-H. (1986) Both TATA box and upstream regions are required for the nopaline synthase promoter in transformed tobacco cells. Mol. Gen. Genet. 203:245–250.

BANFALVI, Z., RANDHAWA, G.S., KONDOROSI, E., KISS, A. and KONDOROSI, A. (1983) Construction and characterization of R-prime plasmids carrying symbiotic genes of *R. meliloti*. Mol. Gen. Genet. 189:129–135.

BARKER, R.F., IDLER, K.B., THOMPSON, D.V. and KEMP, J.D. (1983) Nucleotide sequence of the T-DNA region from *Agrobacterium tumefaciens* octopine Ti plasmid pTi15955. Plant Mol. Biol. 2:335–350.

BARRY, G.F., ROGERS, S.G., FRALEY, R.T. and BRAND, L. (1984) Identification of a cloned cytokinin biosynthetic gene. Proc. Natl. Acad. Sci. USA, 81:4776–4780.

BEATY, J.S., POWELL, G.K., LICA, L., REGIER, D.A., MACDONALD, E.M.S., HOMMES, N.G. and MORRIS, R.O. (1986) *Tzs*, a nopaline Ti plasmid gene from *Agrobacterium tumefaciens* associated with *trans*-zeatin biosynthesis. Mol. Gen. Genet. 203:274–280.

BEVAN, M.W. and CHILTON, M-D. (1982) T-DNA of the *Agrobacterium* Ti and Ri plasmids. Ann. Rev. Genet. 16:357–384.

BEVAN, M. (1984) Binary *Agrobacterium* vectors for plant transformation. Nucleic Acids Res. 12:8711–8721.

BLUNDY, K.S., WHITE, J., FIRMIN, J.L. and HEPBURN, A.G. (1986) Characterization of the T-region of the SAP-type Ti-plasmid pTiAT181: identification of a gene involved in SAP synthesis. Mol. Gen. Genet. 202:62–67.

BOLTON, G.W., NESTER, E.W. and GORDON, M.P. (1986) Plant phenolic compounds induce expression of the *Agrobacterium tumefaciens* loci needed for virulence. Science 232:983–985.

BOMHOFF, G., KLAPWIJK, P.M., KESTER, H.C.M., SCHILPEROORT, R.A., HERNALSTEENS, J.P. and SCHELL, J. (1976) Octopine and nopaline synthesis and breakdown genetically controlled by a plasmid of *Agrobacterium tumefaciens*. Mol. Gen. Genet. 145:177–181.

BRADLEY, D.E., DOUGLAS, C.J. and PESCHON, J. (1984) Flagella-specific bacteriophages of *Agrobacterium tumefaciens*: demonstration of virulence of nonmotile mutants. Can. J. Microbiol. 30:676–681.

BRAUN, A.C. (1958) A physiological basis for autonomous growth of crown gall tumour cell. Proc. Natl. Acad. Sci. USA. 44:344–349.

BREVET, J., PETIT, A., COMBARD, A., BOROWSKI, D. and TEMPÉ, J. (1986) Studies on hairy root – the cucumber strains of *Agrobacterium rhizogenes*. Abst. Fallen Leaf Lake Conf. on the genus *Agrobacterium* and Crown Gall, p33.

BROUGHTON, W.J., HEYCKE, N., MEYER, H. and PANKHURST, C.E. (1984) Plasmid-linked *nif* and "*nod*" genes in fast-growing rhizobia that nodulate *Glycine max*, *Psophocarpus tetragonolobus*, and *Vigna unguiculata*. Proc. Natl. Acad. Sci. USA 81:3093–3097.

BRUCE, W.B. and GURLEY, W.B. (1987) Functional domains of a T-DNA promoter active in crown gall tumors. Mol. Cell. Biol. 7:59–67.

BUCHHOLZ, W.G. and THOMASHOW, M.F. (1984a). Host range encoded by the *Agrobacterium tumefaciens* tumour inducing plasmid pTiAg63 can be expanded by modification of its T-DNA oncogene complement. J. Bacteriol. 160:327–332.

BUCHHOLZ, W.G. and THOMASHOW, M.F. (1984b). Comparison of T-DNA oncogene complements of *Agrobacterium tumefaciens* tumour-inducing plasmids with limited and wide host ranges. J. Bacteriol. 160:319–326. 3

BUCHMANN, I., MARNER, F.-J., SCHRÖDER, G., WAFFENSCHMIDT, S. and SCHRÖDER, J. (1985) Tumour genes in plants: T-DNA encoded cytokinin biosynthesis. EMBO J. 4:853–859.

BUDAR, F., THIA-TOONG, L., VAN MONTAGU M. and HERNALSTEENS, J.-P. (1986) *Agrobacterium*-mediated gene transfer results mainly in transgenic plants transmitting T-DNA as a single mendelian factor. Genetics 114:303–313.

CANGELOSI, G.A., HUNG, L., PUVANESARAJAH, V., STACEY, G., OZGA, D.A., LEIGH, J.A. and NESTER, E.W. (1987) Common loci for *Agrobacterium tumefaciens* and *Rhizobium meliloti* exopolysaccharide synthesis and their roles in plant interactions. J. Bacteriol 169:2086–2091.

CARDERELLI, M., SPANO, L., DE PAOLIS, A., MAURO, M.L., VITALI, G. and COSTANTINO, P. (1985) Identification of the genetic locus responsible for non–polar root induction by *Agrobacterium rhizogenes* 1855. Plant Mol. Biol. 5:385–391.

CHANG, C.-C. and CHEN, C.-H. (1983) Evidence for the presence of N^2–(1,3 dicarboxypropyl)–L–amino acids in crown gall tumours induced by *Agrobacterium tumefaciens* strains 181 and EU6. FEBS Lett. 162:432–435.

CHANG, C.-C., CHEN, C.-M., ADAMS, B.R. and TROST, B.M. (1983) Leucinopine, a characteristic compound of some crown–gall tumors. Proc. Natl. Acad. Sci. USA 80:3573–3576.

CHILTON, M.-D., MONTOYA, A.L., MERLO, D.J., DRUMMOND, M.H., NUTTER, R., GORDON, M.P. and NESTER, E.W. (1978) Restriction endonuclease mapping of a plasmid that confers oncogenicity upon *Agrobacterium tumefaciens* strain B6–806. Plasmid 1:254–269.

CHILTON, M.-D., TEPFER, D.A., PETIT, A., DAVID, C., CASSE–DELBART, F. and TEMPÉ, J. (1982) *Agrobacterium rhizogenes* inserts T-DNA into plant roots. Nature 295:432–434.

CHILTON, W.S. and CHILTON, M.-D. (1984) Mannityl opine analogs allow isolation of catabolic pathway regulatory mutants. J. Bacteriol. 158:650–658.

CHILTON, W.S., HOOD, E. and CHILTON, M.-D. (1985a) Absolute sterochemistry of leucinopine, a crown gall opine. Phytochemistry 24:221–224.

CHILTON, W.S., HOOD, E., RINEHART, K.L. and CHILTON, M.-D. (1985b) L.L–succinamopine: an epimeric crown gall opine. Phytochemistry 24:2945–2948.

CHRISTOU, P., PLATT, S.G. and ACKERMAN, M.C. (1986) Opine synthesis in wild–type plant tissue. Plant Physiol. 82:218–221.

CHYI, Y.-S., JORGENSEN, R.A., GOLDSTEIN, D., TANKSLEY, S.D. and LOAIZA-FIGUEROA, F. (1986) Locations and stability of *Agrobacterium*-mediated T-DNA insertions in the *Lycopersicon* genome. Mol. Gen. Genet. 204:64–69.

CLOSE, T.J., TAIT, R.C. and KADO, C.I. (1985) Regulation of Ti plasmid virulence genes by a chromosomal locus of *Agrobacterium tumefaciens.* J. Bacteriol. 164:774–781.

CLOSE, T.J., TAIT, R.C., REMPEL, H.C., HIROOKA, T., KIM, L. and KADO, C.I. (1987) Molecular characterization of the *vir*C genes of the Ti plasmid. J. Bacteriol. 169:2336–2344.

COMEAU, D.E., IKENAKA, K., TSUNG, K. and INOUYE, M. (1985) Primary characterization of the protein products of the *Escherichia coli* ompB locus: Structure and regulation of synthesis of the OmpR and EnvZ proteins. J. Bacteriol. 164:578-584.

COSTANTINO, P., HOOYKAAS, P.J.J., DEN DULK-RAS, H. and SCHILPEROORT, R.A. (1980) Tumour formation and rhizogenicity of *Agrobacterium rhizogenes* carrying Ti plasmids. Gene 11:79-87.

COSTANTINO, P., MAURO, M.L., MICHELI, G., RISULEO, G., HOOYKAAS, P.J.J. and SCHILPEROORT, R.A. (1981) Fingerprinting and sequence homology of plasmids from different virulent strains of *Agrobacterium rhizogenes*. Plasmid 5:170-182.

CZAKO, M. and MARTON, L. (1986) Independent integration and seed-transmission of the T_R-DNA of the octopine Ti plasmid pTiAch5 in *Nicotiana plumbaginifolia*. (1986) Plant Mol. Biol. 6:101-109.

DAHL, G.A., GUYON, P., PETIT, A. and TEMPÉ, J. (1983) Silver nitrate-positive opines in crown gall tumors. Plant Sc. Lett. 32:193-203.

DAS, A., STACHEL, S., EBERT, P., ALLENZA, P., MONTOYA, A. and NESTER, E. (1986) Promoters of *Agrobacterium tumefaciens* Ti-plasmid virulence genes. Nucleic Acids Res. 14:1355-1364.

DEASEY, M.C. and MATTHYSSE, A.G. (1984) Interactions of wild-type and a cellulose-minus mutant of *Agrobacterium tumefaciens* with tobacco mesophyll and tobacco tissue culture cells. Phytopathology 74:991-994.

DE BEUCKELEER, M., LEMMERS, M., DE VOS, G., WILLMITZER, L., VAN MONTAGU, M. and SCHELL J. (1981) Further insight on the transferred-DNA of octopine crown gall. Mol. Gen. Genet. 183:283-288.

DE BLOCK, M., SCHELL, J. and VAN MONTAGU, M. (1985) Chloroplast transformation by *Agrobacterium tumefaciens*. EMBO J. 4:1367-1372.

DE CLEENE, M. and DE LEY, J. (1976) The host range of crown gall. Botan. Rev. 42:389-466.

DE FRAMOND, A.J., BARTON, K.A. and CHILTON, M.-D. (1983) Mini-Ti: a new vector strategy for plant genetic engineering. BioTechnology 1:262-269.

DE PAOLIS, A., MAURO, M.L., POMPONI, M., CARDARELLI, M., SPANO, L. and COSTANTINO, P. (1985) Localization of agropine-synthesizing functions in the T_R region of the root-inducing plasmid of *Agrobacterium rhizogenes* 1855. Plasmid 13:1-7.

DE PATER, S. (1987) Plant expression signals of the *Agrobacterium* T-cyt gene. Thesis, Leiden University, 108 pp.

DEPICKER, A., DE WILDE, M., DE VOS, G., DE VOS, R., VAN MONTAGU, M. and SCHELL, J. (1980) Molecular cloning of overlapping segments of the nopaline Ti-plasmid pTiC58 as a means to restriction endonuclease mapping. Plasmid 3:193-211.

DEPICKER, A., HERMAN, L., JACOBS, A., SCHELL, J and VAN MONTAGU, M. (1985) Frequencies of simultaneous transformation with different T-DNAs and their relevance to the *Agrobacterium*/plant cell interaction. Mol. Gen. Genet. 201:477-484.

DESSAUX, Y., PETIT, A., TEMPÉ, J., DEMAREZ, M., LEGRAIN, C. and WIAME, J.-M. (1986) Arginine catabolism in *Agrobacterium* strains: role of the Ti plasmid. J. Bacteriol. 166:44-50.

DE VOS, G., DE BEUCKELEER, M., VAN MONTAGU, M. and SCHELL, J. (1981) Restriction endonuclease mapping of the octopine tumour-inducing plasmid pTiAch5 of *Agrobacterium tumefaciens*. Plasmid 6:249-253.

DHANVANTARI, B.N. (1978) Characterization of *Agrobacterium* isolates from stone fruits in Ontario. Can. J. Bot. 56:2309-2311.

DOUGLAS, C.J., STANELONI, R.J., RUBIN, R.A. and NESTER, E.W. (1985a)

Identification and genetic analysis of an *Agrobacterium tumefaciens* chromosomal virulence region. J. Bacteriol. 161:850–860.

DOUGLAS, C., HALPERIN, W., GORDON, M. and NESTER, E. (1985b) Specific attachment of *Agrobacterium tumefaciens* to bamboo cells in suspension cultures. J. Bacteriol. 161: 764–766.

DURAND–TARDIF, M., BROGLIE, R., SLIGHTOM, J. and TEPFER, D. (1985) Structure and expression of Ri T–DNA from *Agrobacterium rhizogenes* in *Nicotiana tabacum*. Organ and phenotypic specificity.   J. Mol. Biol. 186:557–564.

DYLAN, T., IELPI, L., STANFIELD, S., KASHYAP, L., DOUGLAS, C., YANOFSKY, M., NESTER, E., HELINSKY, D.R. and DITTA, G. (1986) *Rhizobium meliloti* genes required for nodule development are related to chromosomal virulence genes in *Agrobacterium tumefaciens*. Proc. Natl. Acad. Sci. USA 83:4403–4407.

ELLIS, J.G. and MURPHY, P.J. (1981) Four new opines from crown gall tumours – their detection and properties. Mol. Gen. Genet. 181:36–43.

ELLIS, J.G., KERR, A., PETIT, A. and TEMPÉ, J. (1982) Conjugal transfer of nopaline and agropine Ti–plasmids. The role of agrocinopines. Mol. Gen. Genet. 186:269–274.

ELLIS, J.G., RYDER, M.H. and TATE, M.E. (1984) *Agrobacterium tumefaciens* T_R–DNA encodes a pathway for agropine biosynthesis.   Mol. Gen. Genet. 195:466–473.

ENGLER, G., HOLSTERS, M., VAN MONTAGU, M., SCHELL, J., HERNALSTEENS, J.P. and SCHILPEROORT, R. (1975) Agrocin 84 sensitivity: a plasmid determined property in *Agrobacterium tumefaciens*. Mol. Gen. Genet. 138:345–349.

ENGLER, G., DEPICKER, A., MAENHAUT, R., VILLARROEL, R., VAN MONTAGU, M. and SCHELL, J. (1981) Physical mapping of DNA base sequence homologies between an octopine and a nopaline Ti plasmid of *Agrobacterium tumefaciens*. J. Mol. Biol. 152:183–208.

EXPERT, D. and TOURNEUR, J. (1982)Ψ, a temperate phage of *Agrobacterium tumefaciens*, is mutagenic. J. Virol. 42:283–291.

EXPERT, D., RIVIERE, F. and TOURNEUR, J. (1982) The state of phage DNA in lysogenic cells of *Agrobacterium tumefaciens*. Virology 121:82–94.

FARRAND, S.K. and DESSAUX, Y. (1986) Proline biosynthesis encoded by the *noc* and *occ* loci of *Agrobacterium* Ti plasmids. J. Bacteriol. 167:732–734.

FILLATTI, J.J., SELLMER, J., McCOWN, B., HAISSIG, B. and COMAI, L. (1987) *Agrobacterium* mediated transformation and regeneration of *Populus*. Mol. Gen. Genet. 206:192–199.

FINAN, T.M., KUNKEL, B., DE VOS, G.F. and SIGNER, E.R. (1986) Second symbiotic megaplasmid in *Rhizobium meliloti* carrying exopolysaccharide and thiamine synthesis gene. J. Bacteriol. 167:66–72.

FIRMIN, J.L. and FENWICK, G.R. (1978) Agropine – a major new plasmid–determined metabolite in crown gall tumors. Nature 276:842–844.

FURNER, I.J., HUFFMAN, G.A., AMASINO, R.M., GARFINKEL, D.J., GORDON, M.P. and NESTER, E.W. (1986) An *Agrobacterium* transformation in the evolution of the genus *Nicotiana*. Nature 319:422–427.

GALLIE, D.R., HAGIYA, M. and KADO, C.I. (1985) Analysis of the *Agrobacterium tumefaciens* plasmid pTiC58 replication region with a novel high–copy–number derivative. J. Bacteriol. 161:1034–1041.

GARDNER, R.C. and KNAUF, V.C. (1986) Transfer of *Agrobacterium* DNA to

plants requires a T-DNA border but not the *vir*E locus. Science 231:725–727.
GARFINKEL, D.J. and NESTER, E.W. (1980) *Agrobacterium tumefaciens* mutants affected in crown gall tumorigenesis and octopine catabolism. J. Bacteriol. 144:732–743.
GARFINKEL. D.J., SIMPSON, R.B., REAM, L.W., WHITE, F.F., GORDON, M.P. and NESTER, E.W. (1981) Genetic analysis of crown gall: fine structure map of the T-DNA by site-directed mutagenesis. Cell 27:143–153.
GELVIN, S.B., THOMASHOW, M.F., McPHERSON, J.C., GORDON, M.P. and NESTER, E.W. (1982) Sizes and map positions of several plasmid DNA-encoded transcripts in octopine-type crown gall tumours. Proc. Natl. Acad. Sci. USA 79:76–80.
GIELEN, J., DE BEUCKELEER, M., SEURINCK, J., DEBOECK, F., DE GREVE, H., LEMMERS, M., VAN MONTAGU, M. and SCHELL, J. (1984) The complete nucleotide sequence of the T-DNA of the *Agrobacterium tumefaciens* plasmid pTiAch5. EMBO J. 3:835–846.
GOLDBERG, S.B., FLICK, J.S. and ROGERS, S.G. (1984) Nucleotide sequence of the *tmr* locus of *Agrobacterium tumefaciens* pTiT37 T-DNA. Nucleic Acids Res. 12:4665–4677.
GRAVES, A.C.F. and GOLDMAN, S.L. (1986) The transformation of *Zea mays* seedlings with *Agrobacterium tumefaciens*. Plant Mol. Biol. 7:43–50.
GRAVES, A.C.F. and GOLDMAN, S.L. (1987) *Agrobacterium tumefaciens*-mediated transformation of the monocot genus *Gladiolus*: detection of expression of T-DNA encoded genes. J. Bacteriol. 169:1745–1746.
GRIMSLEY, N., HOHN, B., HOHN, T. and WALDEN, R. (1986) "Agroinfection", an alternative route for viral infection of plants by using the Ti plasmid. Proc. Natl. Acad. Sci. USA 83:3282–3286.
GRIMSLEY, N., HOHN, T., DAVIES, J.W. and HOHN, B. (1987) *Agrobacterium*-mediated delivery of infectious maize streak virus into maize plants. Nature 325:177–179.
GUYON, P., CHILTON, M.-D., PETIT, A. and TEMPÉ, J. (1980) Agropine in "null-type" crown gall tumours: evidence for generality of the opine concept. Proc. Natl. Acad. Sci. USA 77:2693–2697.
HAGIYA, M., CLOSE, T.J., TAIT, R.C. and KADO, C.I. (1985) Identification of pTiC58 plasmid-encoded proteins for virulence in *Agrobacterium tumefaciens*. Proc. Natl. Acad. Sci. USA 82:2669–2673.
HALL, M.N. and SILHAVY, T.J. (1981) Genetic analysis of the ompB locus in *Escherichia coli* K-12. J. Mol. Biol. 151:1–15.
HAWLEY, D.K. and McCLURE, W.R. (1983) Compilation and analysis of *Escherichia coli* promoter DNA sequences. Nucleic Acids Res. 11:2237–2255.
HEIDEKAMP, F., DIRKSE, W.G., HILLE, J. and VAN ORMONDT, H. (1983) Nucleotide sequence of the *Agrobacterium tumefaciens* octopine Ti-plasmid encoded *tmr* gene. Nucleic Acids Res. 11:6211–6223.
HEMSTAD, P.R. and REISCH, B.I. (1985) *In vitro* production of galls induced by *Agrobacterium tumefaciens* and *Agrobacterium rhizogenes* on *Vitis* and *Rubus*. J. Plant Physiol. 120:9–17.
HENDSON, M., ASKJAER, L., THOMSON, J.A. and VAN MONTAGU, M. (1983) A broad host-range agrocin of *Agrobacterium tumefaciens*. Appl. Env. Micr. 45:1526–1532.
HEPBURN, A.G., CLARKE, L.E., BLUNDY, K.S. and WHITE, J. (1983a) Nopaline Ti-plasmid, pTiT37, T-DNA insertions into a flax genome. J.

Mol. Appl. Genet. 2:211–224.
HEPBURN, A.G., CLARKE, L.E., PEARSON, L. and WHITE, J. (1983b) The role of cytosine methylation in the control of nopaline synthase gene expression in a plant tumour. J. Mol. Appl. Genet. 2:315–329.
HEPBURN, A.G. and WHITE, J. (1985) The effect of right terminal repeat deletion on the oncogenicity of the T–region of pTiT37. Plant Mol. Biol. 5:3–11.
HERNALSTEENS, J.-P., THIA-TOONG, L., SCHELL, J. and VAN MONTAGU, M. (1984) An *Agrobacterium*-transformed cell culture from the monocot *Asparagus officinalis*. EMBO J. 3:3039–3041.
HILLE, J., KLASEN, I. and SCHILPEROORT, R. (1982) Construction and application of R prime plasmids, carrying different segments of an octopine Ti plasmid from *Agrobacterium tumefaciens* for complementation of *vir* genes. Plasmid 7:107–118.
HILLE, J., WULLEMS, G. and SCHILPEROORT, R. (1983) Non–oncogenic T–region mutants of *Agrobacterium tumefaciens* do transfer T–DNA into plant cells. Plant Mol. Biol. 2:155–163.
HILLE, J., VAN KAN, J. and SCHILPEROORT, R. (1984) *Trans*-acting virulence functions of the octopine Ti plasmid from *Agrobacterium tumefaciens*. J. Bacteriol 158:754–756.
HILLE, J., DEKKER, M., OUDE LUTTIGHUIS, H., VAN KAMMEN, A. and ZABEL, P. (1986) Detection of T–DNA transfer to plant cells by *A. tumefaciens* virulence mutants using agroinfection. Mol. Gen. Genet. 205:411–416.
HIROOKA, T. and KADO, C.I. (1986) Location of the right boundary of the virulence region on *Agrobacterium tumefaciens* plasmid pTiC58 and a host–specifying gene next to the boundary. J. Bacteriol. 168:237–243.
HIROOKA, T., ROGOWSKY, P.M. and KADO, C.I. (1987) Characterization of the *virE* locus of *Agrobacterium tumefaciens* plasmid pTiC58. J. Bacteriol. 169:1529–1536.
HIRSCH, A.M., DRAKE, D., JACOBS, T.W. and LONG, S.R. (1985) Nodules are induced on alfalfa roots by *Agrobacterium tumefaciens* and *Rhizobium trifolii* containing small segments of the *Rhizobium meliloti* nodulation region. J. Bacteriol. 161:223–230.
HOEKEMA, A., HIRSCH, P.R., HOOYKAAS, P.J.J. and SCHILPEROORT, R.A. (1983) A binary plant vector strategy based on separation of *vir* and T–region of the *Agrobacterium tumefaciens* Ti-plasmid. Nature 303:179–180.
HOEKEMA, A., DE PATER, B.S., FELLINGER, A.J., HOOYKAAS, P.J.J. and SCHILPEROORT, R.A. (1984a) The limited host range of an *Agrobacterium tumefaciens* strain extended by a cytokinin gene from a wide host range T–region. EMBO J. 3:3043–3047.
HOEKEMA, A., HOOYKAAS, P.J.J. and SCHILPEROORT, R.A. (1984b) Transfer of the octopine T–DNA segment to plant cells mediated by different types of *Agrobacterium* tumour- or root-inducing plasmids: generality of virulence systems. J. Bacteriol. 158:383–385.
HOEKEMA, A., ROELVINK, P.W., HOOYKAAS, P.J.J. and SCHILPEROORT, R.A. (1984c) Delivery of T–DNA from the *Agrobacterium tumefaciens* chromosome into plant cells. EMBO J. 3:2485–2490.
HOLSTERS, M., SILVA, B., VAN VLIET, F., GENETELLO, C., DE BLOCK, M., DHAESE, P., DEPICKER, A., INZÉ, D., ENGLER, G., VILLARROEL, R., VAN MONTAGU, M. and SCHELL, J. (1980) The functional organization of the nopaline *Agrobacterium tumefaciens* plasmid pTiC58. Plasmid 3:212–230.
HOLSTERS, M., VILLARROEL, R., GIELEN, J., SEURINCK, J., DE GREVE, H.,

VAN MONTAGU, M. and SCHELL, J. (1983) An analysis of the boundaries of the octopine T_L-DNA in tumours induced by *Agrobacterium tumefaciens*. Mol. Gen. Genet. 190:35–41.

HOOD, E.E., JEN, G., KAYES, L., KRAMER, J., FRALEY, R.T. and CHILTON, M.-D. (1984) Restriction endonuclease map of pTiBo542, a potential Ti plasmid vector for genetic engineering of plants. Bio/Technology 2:702–709.

HOOD, E.E., CHILTON, W.S., CHILTON, M.-D. and FRALEY, R.T. (1986a) T-DNA and opine synthetic loci in tumours incited by *Agrobacterium tumefaciens* A281 on soybean and alfalfa plants. J. Bacteriol. 168:1283–1290.

HOOD, E.E., HELMER, G.L., FRALEY. R.T. and CHILTON, M.-D. (1986b) The hypervirulence of *Agrobacterium tumefaciens* A281 is encoded in a region of pTiBo542 outside of T-DNA. J. Bacteriol. 168:1291–1301.

HOOYKAAS-VAN SLOGTEREN, G.M.S., HOOYKAAS, P.J.J. and SCHILPEROORT, R.A. (1984) Expression of Ti plasmid genes in monocotyledonous plants infected with *Agrobacterium tumefaciens*. Nature 311:763–764.

HOOYKAAS, P.J.J. (1979) The role of plasmid determined functions in the interactions of *Rhizobiaceae* with plant cells. A genetic approach. Thesis, Leiden University, 168 pp.

HOOYKAAS, P.J.J., KLAPWIJK, P.M., NUTI, M.P., SCHILPEROORT, R.A. and RÖRSCH, A. (1977) Transfer of the *Agrobacterium tumefaciens* Ti plasmid to avirulent agrobacteria and to *Rhizobium ex planta*. J. Gen. Microbiol. 98:477–484.

HOOYKAAS, P.J.J., ROOBOL, C. and SCHILPEROORT, R.A. (1979) Regulation of the transfer of Ti plasmids of *Agrobacterium tumefaciens*. J. Gen. Microbiol. 110:99–109.

HOOYKAAS, P.J.J., DEN DULK-RAS, H., OOMS, G., and SCHILPEROORT, R.A. (1980) Interactions between octopine and nopaline plasmids in *Agrobacterium tumefaciens*. J. Bacteriol. 143:1295–1306.

HOOYKAAS, P.J.J., VAN BRUSSEL, A.A.N., DEN DULK-RAS, H., VAN SLOGTEREN, G.M.S. and SCHILPEROORT, R.A. (1981) Sym plasmid of *Rhizobium trifolii* expressed in different rhizobial species and *Agrobacterium tumefaciens*. Nature 291:351–353.

HOOYKAAS, P.J.J., SNIJDEWINT, F.G.M. and SCHILPEROORT, R.A. (1982) Identification of the Sym plasmid of *Rhizobium leguminosarum* strain 1001 and its transfer to and expression in other rhizobia and *Agrobacterium tumefaciens*. Plasmid 8:73–82.

HOOYKAAS, P.J.J., HOFKER, M., DEN DULK-RAS, H. and SCHILPEROORT, R.A. (1984) A comparison of virulence determinants in an octopine Ti plasmid, a nopaline Ti plasmid, and an Ri plasmid by complementation analysis of *Agrobacterium tumefaciens* mutants. Plasmid 11:195–205.

HOOYKAAS, P.J.J. and SCHILPEROORT, R.A. (1984) The molecular genetics of crown gall tumorigenesis. In: "Molecular Genetics of Plants" (Advances in Genetics, Vol. 22) Publ. Academic Press, Orlando, USA. Ed: J.G.Scandalios, pp. 209–283.

HOOYKAAS, P.J.J., DEN DULK-RAS, H., REGENSBURG-TUINK, A.J.G., VAN BRUSSEL, A.A.N. and SCHILPEROORT, R.A. (1985) Expression of a *Rhizobium phaseoli* Sym plasmid in *R.trifolii* and *Agrobacterium tumefaciens*: incompatibility with a *R.trifolii* Sym plasmid. Plasmid 14:47–52.

HUFFMAN, G.A., WHITE, F.F., GORDON, M.P. and NESTER, E.W. (1984) Hairy-root inducing plasmid: physical map and homology to tumour-inducing plasmids. J. Bacteriol. 157:269–276.

IYER, V.N., KLEE, H.J. and NESTER, E.W. (1982) Units of genetic expression in the virulence region of a plant tumour-inducing plasmid of *Agrobacterium tumefaciens*. Mol. Gen. Genet. 188:418-424.

JANSSENS, A., ENGLER, G., ZAMBRYSKI, P. and VAN MONTAGU, M. (1984) The nopaline C58 T-DNA region is transcribed in *Agrobacterium tumefaciens*. Mol. Gen. Genet. 195:341-350.

JAYNES, J.M. and STROBEL, G.A. (1981) The position of *Agrobacterium rhizogenes*. Int. Rev. Cytol. 13:105-125.

JOOS, H., INZÉ, D., CAPLAN, A., SORMANN, M., VAN MONTAGU, M. and SCHELL, J. (1983) Genetic analysis of T-DNA transcripts in nopaline crown galls. Cell 32:1057-1067.

JORGENSEN, R., SNYDER, C. and JONES, J.D.G. (1987) T-DNA is organized predominantly in inverted repeat structures in plants transformed with *Agrobacterium tumefaciens* C58 derivatives. Mol. Gen. Genet. 207:471-477.

JOUANIN, L. (1984) Restriction map of an agropine-type Ri plasmid and its homologies with Ti plasmids. Plasmid 12:91-102.

JOUANIN, L., VILAINE, F., D'ENFERT, C. and CASSE-DELBERT, F. (1985) Localization and restriction maps of the replication origin regions of the plasmids of *Agrobacterium rhizogenes* strain A4. Mol. Gen. Genet 201:370-374.

JOUANIN, L., GUERCHE, P., PAMBOUKJIAN, N., TOURNEUR, C., CASSE-DELBART, F. and TOURNEUR, J. (1987) Structure of T-DNA in plants regenerated from roots transformed by *Agrobacterium rhizogenes* strain A4. Mol. Gen. Genet. 206:387-392.

KARCHER, S.J., DIRITA, V.J. and GELVIN, S.B. (1984) Transcript analysis of T_R-DNA in octopine-type crown gall tumours. Mol. Gen. Genet. 194:159-165.

KEANE, P.J., KERR, A., NEW, P.B. (1970) Crown gall of stone fruit. II. Identification and nomenclature of *Agrobacterium* isolates. Aust. J. biol. Sci. 23:585-595.

KEMPER, E., WAFFENSCHMIDT, S., WEILER, E.W., RAUSCH, T. and SCHRÖDER, J. (1985) T-DNA encoded auxin formation in crown-gall cells. Planta 163:257-262.

KENNEDY, B.W. and ALCORN, S.M. (1980) Estimates of USA crop losses to prokaryotic plant pathogens. Plant. Dis. 64:674-676.

KERR, A. and HTAY, K. (1974) Biological control of crown gall through bacteriocin production. Physiol. Plant. Pathol. 4:37-44.

KERR, A. and ROBERTS, W.P. (1976) *Agrobacterium*: correlations between and transfer of pathogenicity, octopine and nopaline metabolism and bacteriocin 84 sensitivity. Physiol. Plant Pathol. 9:205-221.

KERR, A. and PANAGOPOULOS, C.G. (1977) Biotypes of *Agrobacterium radiobacter* var. *tumefaciens* and their biological control. Phytopath. Z. 90:172-179.

KERSTERS, K., DE LEY, J., SNEATH, P.H.A., SACKIN, M. (1973) Numerical taxonomic analysis of *Agrobacterium*. J. gen. Microbiol. 78:227-239.

KERSTERS, K. and DE LEY, J. (1984) Genus III. *Agrobacterium* Conn 1942. In: Bergey's Manual of Systematic Bacteriology, Vol. I, The Williams and Wilkins Co. Baltimore, pp 244-254.

KLAPWIJK, P.M., OUDSHOORN, M. and SCHILPEROORT, R.A. (1977) Inducible permease involved in the uptake of octopine, lysopine and octopine acid by *Agrobacterium tumefaciens* strains carrying virulence-associated plasmids. J. Gen. Microbiol. 102:1-11.

KLAPWIJK, P.M., SCHEULDERMAN, T. and SCHILPEROORT, R.A. (1978) Coordinated regulation of octopine degradation and conjugative transfer of Ti plasmids in *Agrobacterium tumefaciens*: evidence for a common regulatory gene and separate operons. J. Bacteriol. 136:775-785.

KLEE, H.J., GORDON, M.P. and NESTER, E.W. (1982) Complementation analysis of *Agrobacterium tumefaciens* Ti plasmid mutations affecting oncogenicity. J. Bacteriol. 150:327-331.

KLEE, H.J., WHITE, F.F., IYER, V.N., GORDON, M.P. and NESTER, E.W. (1983) Mutational analysis of the virulence region of an *Agrobacterium tumefaciens* Ti-plasmid. J. Bacteriol. 153:878-883.

KLEE, H., MONTOYA, A., HORODYSKI, F., LICHTENSTEIN, C., GARFINKEL, D., FULLER, S., FLORES, C., PESCHON, J., NESTER, E. and GORDON, M. (1984) Nucleotide sequence of the *tms* genes of the pTiA6NC octopine Ti plasmid: two gene products involved in plant tumorigenesis. Proc. Natl. Acad. Sci. USA. 81:1728-1732.

KNAUF, V.C. and NESTER, E.W. (1982) Wide host range cloning vectors: a cosmid clone bank of an *Agrobacterium* Ti plasmid. Plasmid 8:45-54.

KNAUF, V., YANOFSKY, M., MONTOYA, A. and NESTER, E. (1984) Physical and functional map of an *Agrobacterium tumefaciens* tumor-inducing plasmid that confers a narrow host-range. J. Bacteriol. 160:564-568.

KOEKMAN, B.P., OOMS, G., KLAPWIJK, P.M. and SCHILPEROORT, R.A. (1979) Genetic map of an octopine Ti-plasmid. Plasmid 2:347-357.

KOEKMAN, B.P., HOOYKAAS, P.J.J. and SCHILPEROORT, R.A. (1980) Localization of the replication control region on the physical map of the octopine Ti plasmid. Plasmid 4:184-195.

KOEKMAN, B.P., HOOYKAAS, P.J.J. and SCHILPEROORT, R.A. (1982) A functional map of the replicator region of the octopine Ti plasmid. Plasmid 7:119-132.

KOMARI, T., HALPERIN, W. and NESTER, E.W. (1986) Physical and functional map of supervirulent *Agrobacterium tumefaciens* tumour-inducing plasmid pTiB0542. J. Bacteriol. 166:88-94.

KOMRO, C.T., DIRITA, V.J., GELVIN, S.B. and KEMP, J.D. (1985) Site-specific mutagenesis in the T_R-DNA region of octopine-type Ti plasmid. Plant. Mol. Biol. 4:253-263.

KONCZ, C. and SCHELL, J. (1986) The promoter of T_L-DNA gene 5 controls the tissue-specific expression of chimaeric genes carried by a novel type of *Agrobacterium* binary vector. Mol. Gen. Genet. 204:383-396.

KOPLOW, J., BYRNE, M.C., JEN, G., TEMPÉ, J. and CHILTON, M.-D. (1984) Physical map of the *Agrobacterium rhizogenes* strain 8196 virulence plasmid. Plasmid 11:17-27.

KOUKOLIKOVA-NICOLA, Z., SHILLITO, R.D., HOHN, B., WANG, K., VAN MONTAGU, M. and ZAMBRYSKI, P. (1985) Involvement of circular intermediates in the transfer of T-DNA from *Agrobacterium tumefaciens* to plant cells. Nature 313:191-196.

KRIKOS, A., MUTOH, N., BOYD, A. and SIMON, M.I. (1983) Sensory transducers of *E. coli* are composed of discrete structural and functional domains. Cell 33:615-622.

KWOK, W.W., NESTER, E.W. and GORDON, M.P. (1985) Unusual plasmid DNA organization in an octopine crown gall tumour. Nucleic Acids Res. 13:459-471.

LAHNERS, K., BYRNE, M.C. and CHILTON, M.-D. (1984) T-DNA fragments of hairy root plasmid pRi8196 are distantly related to octopine and nopaline Ti plasmid T-DNA. Plasmid 11:130-140.

LEEMANS, J., DEBLAERE, R., WILLMITZER, L., DE GREVE, H., HERNALSTEENS, J.P., VAN MONTAGU, M. and SCHELL, J. (1982) Genetic identification of functions of T_L-DNA transcripts in octopine crown galls. EMBO J. 1:147–152.

LEMMERS, M., DE BEUCKELEER, M., HOLSTERS, M., ZAMBRYSKI, P., DEPICKER, A., HERNALSTEENS, J.P., VAN MONTAGU, M. and SCHELL, J. (1980) Internal organization, boundaries and integration of Ti-plasmid DNA in nopaline crown gall tumors. J. Mol. Biol. 144:353–376.

LEROUX, B., YANOFSKY, M.F., WINANS, S.C., WARD, J.E., ZIEGLER, S.F. and NESTER, E.W. (1987) Characterization of the *virA* locus of *Agrobacterium tumefaciens*: a transcriptional regulator and host range determinant. EMBO J. 6:849–856.

LICHTENSTEIN, C., KLEE, H., MONTOYA, A., GARFINKEL, D., FULLER, S., FLORES, C., NESTER, E. and GORDON, M. (1984) Nucleotide sequence and transcript mapping of the *tmr* gene of the pTiA6NC octopine Ti-plasmid: a bacterial gene involved in plant tumorigenesis. J. Mol. Appl. Genet. 2:354–362.

LILJESTRÖM (1986) The EnvZ protein of *Salmonella typhimurium* LT-2 and *Escherichia coli* K-12 is located in the cytoplasmic membrane. FEMS Microbiol. Letters 36:145–150.

LIPPINCOTT, B.B. and LIPPINCOTT, J.A. (1969) Bacterial attachment to a specific wound site as an essential stage in tumour initiation by *Agrobacterium tumefaciens*. J. Bacteriol. 97:620–628.

LIU, S.-T. and KADO, C.I. (1979) Indoleacetic acid production: a plasmid function of *Agrobacterium tumefaciens*. Biochem. Biophys. Res. Commun. 90:171–178.

LIU, S.T., PERRY, K.L., SCHARDL, C.L. and KADO, C.I. (1982) *Agrobacterium* Ti-plasmid indoleacetic acid gene is required for crown gall oncogenesis. Proc. Natl. Acad. Sci. USA 79:2812–2816.

LOPER, J.E. and KADO, C.I. (1979) Host range conferred by the virulence-specifying plasmid of *Agrobacterium tumefaciens*. J. Bacteriol. 139:591–596.

LUNDQUIST, R.C., CLOSE, T.J. and KADO, C.I. (1984) Genetic complementation of *Agrobacterium tumefaciens* Ti plasmid mutants in the virulence region. Mol. Gen. Genet. 193:1–7.

MACHIDA, Y., SAKURAI, M., KIYOKAWA, S., UBASAWA, A., SUZUKI, Y. and IKEDA, J.-E. (1984) Nucleotide sequence of the insertion sequence found in the T-DNA region of mutant Ti plasmid pTiA66 and distribution of its homologues in octopine Ti plasmid. Proc. Natl. Acad. Sci. USA 81:7495–7499.

MACHIDA, Y., USAMI, S., YAMAMOTO, A., NIWA, Y. and TAKEBE, I. (1986) Plant-inducible recombination between the 25 bp border sequences of T-DNA in *Agrobacterium tumefaciens*. Mol. Gen. Genet. 204:374–382.

MATTHYSSE, A.G., HOLMES, K.V. and GURLITZ, R.H.G. (1981) Elaboration of cellulose fibrils by *Agrobacterium tumefaciens* during attachment to carrot cells. J. Bacteriol. 145:583–595.

MATTHYSSE, A.G. (1983) Role of bacterial cellulose fibrils in *Agrobacterium tumefaciens* infection. J. Bacteriol. 154:906–915.

MATTHYSSE, A.G. (1987) Characterization of nonattaching mutants of *Agrobacterium tumefaciens*. J. Bacteriol. 169:313–323.

MELCHERS, L.S., THOMPSON, D.V., IDLER, K.B., SCHILPEROORT, R.A. and HOOYKAAS, P.J.J. (1986) Nucleotide sequence of the virulence gene *virG* of the *Agrobacterium tumefaciens* octopine Ti plasmid: significant homology between *virG* and the regulatory genes *ompR*, *phoB* and *dye* of *E. coli*. Nucleic Acids Res. 14:9933–9942.

MELCHERS, L.S., THOMPSON, D.V., IDLER, K.B., NEUTEBOOM, S.T.C., DE MAAGD, R.A., SCHILPEROORT, R.A. and HOOYKAAS, P.J.J. (1987) Molecular characterization of the virulence gene *vir*A of the *Agrobacterium tumefaciens* octopine Ti plasmid. Plant Mol. Biol.9:635–646.

MESSENS, E., LENAERTS, A., VAN MONTAGU, M. and HEDGES, R.W. (1985) Genetic basis for opine secretion from crown gall tumour cells. Mol. Gen. Genet. 199:344–348.

MESSENS, E., LENAERTS, A., VAN MONTAGU, M., DE BRUYN, A., JANS, A.W.H. and VAN BINST, G. (1986) ^{1}H and ^{31}P NMR spectroscopy of agrocinopine. J. Carbohydrate Chem. 5:683–699.

MURAI, N. and KEMP, J.D. (1982a) T-DNA of pTi15955 from *Agrobacterium tumefaciens* is transcribed into a minimum of seven polyadenylated RNAs in a sunflower crown gall tumor.   Nucleic Acids Res. 10:1679–1689.

MURAI, N. and KEMP, J.D. (1982b) Octopine synthase messenger RNA isolated from sunflower crown gall callus is homologous to the Ti plasmid of *Agrobacterium tumefaciens*. Proc. Natl. Acad. Sci. USA 79:86–90.

MURPHY, P.J. and ROBERTS, W.P. (1979) A basis for agrocin 84 sensitivity in *Agrobacterium radiobacter*. J. gen. Microbiol. 114:207–213.

MURPHY, P.J., TATE, M.E. and KERR, A. (1981) Substituents at N^6 and C–5' control selective uptake and toxicity of the adenine–nucleotide bacteriocin, agrocin 84, in agrobacteria. Eur.   J. Biochem. 115:539–543.

NEW, P.B. and KERR, A. (1972) Biological control of crown gall: field observations and glasshouse experiments. J. appl. Bact. 35:279–287.

NISHIGUCHI, R., TAKANAMI, M. and OKA, A. (1987) Characterization and sequence determination of the replicator region in the hairy-root inducing plasmid pRi4b. Mol. Gen. Genet. 206:1–8.

NIXON, B.T., RONSON, C.W. and AUSUBEL, F.M. (1986) Two-component regulatory systems responsive to environmental stimuli share strongly conserved domains with the nitrogen assimilation regulatory genes *ntr*B and *ntr*C. Proc. Natl. Acad. Sci. USA 83:7850–7854.

OFFRINGA, I.A., MELCHERS, L.S., REGENSBURG-TUINK, A.J.G., COSTANTINO, P., SCHILPEROORT, R.A. and HOOYKAAS, P.J.J. (1986) Complementation of *Agrobacterium tumefaciens* Ti *aux* mutants by genes from the T_R-region of the Ri plasmid of *Agrobacterium rhizogenes*. Proc.   Natl. Acld. Sci. USA 83:6935–6939.

OHTA, K. and NISHIYAMA, K. (1984) Studies on the crown gall diseases of flower crops. I Occurrence of the disease and the characterization of the causal bacterium. Ann. Phytopath. Soc.   Japan, 50:197–204.

OKKER, R.J.H., SPAINK, H., HILLE, J., VAN BRUSSEL, T.A.N., LUGTENBURG, B. and SCHILPEROORT, R.A. (1984) Plant-inducible virulence promoter of the *Agrobacterium tumefaciens* Ti plasmid. Nature 312:564–566.

OOMS, G., KLAPWIJK, P.M., POULIS, J.A. and SCHILPEROORT, R.A. (1980) Characterization of Tn904 insertions in octopine Ti plasmid mutants of *Agrobacterium tumefaciens*. J. Bacteriol. 144:82–91.

OOMS, G., HOOYKAAS, P.J.J., MOOLENAAR, G. and SCHILPEROORT, R.A. (1981) Crown gall plant tumours of abnormal morphology, induced by *Agrobacterium tumefaciens* carrying mutated octopine   Ti plasmids; analysis of T-DNA functions. Gene 14:33–50.

OOMS, G., BAKKER, A., MOLENDIJK, L., WULLEMS, G.J., GORDON, M.P., NESTER, E.W. and SCHILPEROORT, R.A. (1982a) T-DNA organization in homogenous and heterogenous octopine-type crown gall tissues of *Nicotiana tabacum*. Cell 30:589–597.

OOMS, G., HOOYKAAS, P.J.J., VAN VEEN, R.J.M., VAN BEELEN, P., REGENS-
BURG-TUINK, A.J.G., SCHILPEROORT, R.A. (1982b) Octopine Ti-plasmid
deletion mutants of *Agrobacterium tumefaciens* with emphasis on the
right side of the T-region. Plasmid 7:15-29.
OOMS, G., TWELL, D., BOSSON, M.E., HOGE, J.H.C. and BURRELL, M.M.
(1986) Developmental regulation of Ri T_L-DNA gene expression in
roots, shoots and tubers of transformed potato (*Solanum tuberosum* cv.
Desiree). Plant Mol. Biol. 6:321-330.
OTTEN, L.A.B.M. and SCHILPEROORT, R.A. (1978) A rapid micro scale
method for the detection of lysopine   and nopaline   dehydrogenase
activities. Biochim. Biophys. Acta. 527:497-500.
OTTEN, L., DEGREVE, H., LEEMANS, J., HAIN, R., HOOYKAAS, P., SCHELL,
J. (1984) Restoration of virulence of *vir* region mutants of
Agrobacterium tumefaciens strain B6S3 by coinfection with normal and
mutant *Agrobacterium* strains. Mol. Gen. Genet. 195:159-163.
OTTEN, L., PIOTROWIAK, G., HOOYKAAS, P., DUBOIS, M., SZEGEDI, E. and
SCHELL, J. (1985) Identification of an *Agrobacterium tumefaciens*
pTiB6S3 vir region fragment that enhances the virulence of pTiC58.
Mol. Gen. Genet. 199:189-193.
PABO, C.O. and SAUER, R.T. (1984) Protein-DNA recognition. Ann. Rev.
Biochem. 53:293-321.
PANAGOPOULOS, C.G. and PSALLIDAS, P.G. (1973) Characteristics of
Greek isolates of *Agrobacterium tumefaciens* (Smith and Townsend)
Conn. J. Appl. Bact. 36:233-240.
PEERBOLTE, R. (1986) The fate of T-DNA during vegetative and
generative propagation - crown gall and hairy root tissues of
Nicotiana spp. Thesis, Leiden University, The Netherlands, 218 pp.
PEERBOLTE, R., LEENHOUTS, K., HOOYKAAS-VAN SLOGTEREN, G.M.S., HOGE,
J.H.C., WULLEMS, G.J. and SCHILPEROORT, R.A. (1986) Clones from a
shooty tobacco crown gall tumour I: deletions, rearrangements and
amplifications resulting in irregular T-DNA structures and
organizations. Plant. Mol. Biol. 7:265-284.
PERALTA, E.G. and REAM, L.W. (1985) T-DNA border sequences required
for crown gall tumorigenesis. Proc. Natl. Acad. Sci. USA,
82:5112-5116.
PERALTA, E.G., HELLMISS, R. and REAM, W. (1986) *Overdrive*, a T-DNA
transmission enhancer on the *A. tumefaciens* tumor-inducing plasmid.
EMBO J. 5:1137-1142.
PERRY, K.L. and KADO, C.I. (1982) Characteristics of Ti plasmids from
broad host range and ecologically specific biotype 2 and 3 strains of
Agrobacterium tumefaciens. J. Bacteriol. 151:343-350.
PETIT, A. and TEMPÉ, J. (1978) Isolation of *Agrobacterium* Ti-plasmid
regulatory mutants. Mol. Gen. Genet. 167:147-155.
PETIT, A., DELHAYE, S., TEMPÉ, J. and MOREL, G. (1970) Recherches sur
les guanidines des tissus de crown-gall. Mise en evidence d'une
relation biochimique specifique entre les souches d'*Agrobacterium
tumefaciens* et les tumeurs qu'elles induisent. Physiol. Veget.
8:205-213.
PETIT, A., TEMPÉ, J., KERR, A., HOLSTERS, M., VAN MONTAGU, M. and
SCHELL, J. (1978) Substrate induction of conjugative activity of
Agrobacterium tumefaciens Ti plasmid. Nature 271:570-572.
PETIT, A., DAVID, C., DAHL, G.A., ELLIS, J.G., GUYON, P.,
CASSE-DELBART, F. and TEMPÉ, J. (1983) Further extension of the opine
concept: plasmids in *Agrobacterium rhizogenes* cooperate for opine
degradation. Mol. Gen. Genet. 190:204-214.

PETIT, A., BERKALOFF, A. and TEMPÉ, J. (1986) Multiple transformation of plant cells by *Agrobacterium* may be responsible for the complex organization of T-DNA in crown gall and hairy root. Mol. Gen. Genet. 202:388-393.
PLESSIS, A., ROBAGLIA, C., DIOLEZ, A., BEYOU, A., LEACH, F., CASSE-DELBART, F. and RICHAUD, F. (1985) *lacZ* gene fusions and insertion mutagenesis in the T_L region of *Agrobacterium rhizogenes* Ri plasmid. Plasmid 14:17-27.
PLOTKIN, G.R. (1980) *Agrobacterium radiobacter* prosthetic valve endocarditis. Ann. Int. Med. 93:839-840.
POMPONI, M., SPANO, L., SABBADINI, M.G. and COSTANTINO, P. (1983) Restriction endonuclease mapping of the root-inducing plasmid of *Agrobacterium rhizogenes* 1855. Plasmid 10:119-129.
PUVANESARAJAH, V., SCHELL, F.M., STACEY, G., DOUGLAS, C.J. and NESTER, E.W. (1985) Role for 2-linked-β-D-glucan in the virulence of *Agrobacterium tumefaciens*. J. Bacteriol 164:102-106.
REAM, L.W., GORDON, M.P. and NESTER, E.W. (1983) Multiple mutations in the T region of the *Agrobacterium tumefaciens* tumor-inducing plasmid. Proc. Natl. Acad. Sci. USA 80:1660-1664.
REGIER, D.A. and MORRIS, R.O. (1982) Secretion of *trans*-zeatin by *Agrobacterium tumefaciens*: a function determined by the nopline Ti plasmid. Biochem. Biophys. Res. Commun. 104:1560-1566.
RIKER, A.J. (1930) Studies on infectious hairy root of nursery apple trees. J. Agr. Res. 41:507-540.
RILEY, P.S. and WEAVER, R.E. (1977) Comparison of thirty-seven strains of Vd-3 bacteria with *Agrobacterium radiobacter*: morphological and physiological observations. J. clin. Microbiol. 5:172-177.
RISULEO, G., BATTISTONI, P., COSTANTINO, P. (1982) Regions of homology between tumorigenic plasmids from *Agrobacterium rhizogenes* and *Agrobacterium tumefaciens*. Plasmid 7:45-51.
ROGOWSKY, P., CLOSE, T.J. and KADO, C.I. (1987) Dual regulation of virulence genes of *Agrobacterium* plasmid pTiC58. In: Molecular Genetics of Plant-Microbe Interactions. Eds: D.P.S. Verma and N.Brisson, M.Nijhoff Publ., Dordrecht, pp 14-19.
RUBIN, R.A. (1986) Genetic studies on the role of octopine T-DNA border regions in crown gall tumour formation. Mol. Gen. Genet. 202:312-320.
RYDER, M.H., TATE, M.E. and JONES, G.P. (1984) Agrocinopine A, a tumour-inducing plasmid-coded enzyme product, is a phosphodiester of sucrose and l-arabinose. J. Biol. Chem. 259:9704-9710.
RYDER, M.H., TATE, M.E. and KERR, A. (1985) Virulence properties of strains of *Agrobacterium* on the apical and basal surfaces of carrot root discs. Plant Physiol. 77:215-221.
SALOMON, F., DEBLAERE, R., LEEMANS, J., HERNALSTEENS, J.-P., VAN MONTAGU, M. and SCHELL, J. (1984) Genetic identification of functions of T_R-DNA transcripts in octopine crown galls. EMBO J. 3:141-146.
SCHELL, J., VAN MONTAGU, M., DE BEUCKELEER, M., DE BLOCK, M., DEPICKER, A., DE WILDE, M., ENGLER, G., GENETELLO, C., HERNALSTEENS, J.P., HOLSTERS, M., SEURINCK, J., SILVA, B., VAN VLIET, F. and VILLARROEL, R. (1979) Interactions and DNA transfer between *Agrobacterium tumefaciens*, the Ti plasmid and the plant host. Proc. R. Soc. London Ser B 204:251-266.
SCHILPEROORT, R.A. (1969) Investigations on plant tumours - Crown Gall. On the biochemistry of tumour induction by *Agrobacterium*

215

tumefaciens. Thesis, Leiden University, The Netherlands.

SCHRÖDER, J., SCHRÖDER, G., HUISMAN, H., SCHILPEROORT, R.A. and SCHELL, J. (1981) The mRNA for lysopine dehydrogenase in plant tumor cells is complementary to a Ti-plasmid fragment. FEBS Lett. 129:166-168.

SCHRÖDER, G., WAFFENSCHMIDT, S., WEILER, E.W. and SCHRÖDER, J. (1984) The T-region of Ti-plasmids codes for an enzyme synthesizing idole-3-acetic acid. Eur. J. Biochem. 138:387-391.

SCIAKY, D., MONTOYA, A.L. and CHILTON, M.-D. (1978) Fingerprints of *Agrobacterium* Ti plasmids. Plasmid 1:238-253.

SCIAKY, D. and THOMASHOW, M.F. (1984) The sequence of the *tms* transcript 2 locus of the *Agrobacterium tumefaciens* plasmid pTiA6 and characterization of the mutation in pTiA66 that is responsible for auxin attenuation. Nucleic Acids Res. 12:1447-1461.

SHARP, P.M. and LI, W.H. (1986) Codon usage in regulatory genes in *Escherichia coli* does not reflect selection for rare codons. Nucleic Acids Res. 14:7737-7749.

SHAW, C.H., CARTER, G.H., WATSON, M.D. and SHAW, C.H. (1984a) A functional map of the nopaline synthase promoter. Nucleic Acids Res. 12:7831-7846.

SHAW, C.H., WATSON, M.D., CARTER, G.H. and SHAW, C.H. (1984b) The right hand copy of the nopaline Ti-plasmid 25 bp repeat is required for tumour formation. Nucleic Acids Res. 12:6031-6041.

SHEIKHOLESLAM, S.N. and WEEKS, D.P. (1987) Acetosyringone promotes high efficiency transformation of *Arabidopsis thaliana* explants by *Agrobacterium tumefaciens.* Plant. Mol. Biol. 8:291-298.

SHINAGAWA, H., MAKINO, K. and NAKATA, A. (1983) Regulation of the pho regulon in *Escherichia coli* K-12. J. Mol. Biol. 168:477-488.

SIMOENS, C., ALLIOTTE, TH., MENDEL, R., MULLER, A., SCHIEMANN, J., VAN LIJSEBETTENS, M., SCHELL, J., VAN MONTAGU, M. and INZÉ, D. (1986) A binary vector for transferring genomic libraries to plants. Nucleic Acids Res., 14:8073-8090.

SIMPSON, R.B., O'HARA, P.J., KWOK, W., MONTOYA, A.M., LICHTENSTEIN, C., GORDON, M.P. and NESTER, E.W. (1982) DNA from the A6S/2 crown gall tumor contains scrambled Ti-plasmid sequences near its junctions with plant DNA. Cell 29:1005-1014.

SKINNER, F.A. (1977) An evaluation of the nile blue test for differentiating *Rhizobia* from *Agrobacterium.* J. Appl. Bact. 43:91-98.

SKOOG, F. and MILLER, C.O. (1957) Chemical regulation of growth and organ formation in plant tissues cultured in vitro. Symp. Soc. Exp. Biol. 11:118-131.

SLIGHTOM, J.L., JOUANIN, L., LEACH, F., DRONG, R.F. and TEPFER, D. (1985) Isolation and identification of T$_L$-DNA/plant junctions in *Convolvulus arvensis* transformed by *Agrobacterium rhizogenes* strain A4. EMBO J. 4:3069-3077.

SLIGHTOM, J.L., DURAND-TARDIF, M., JOUANIN, L. and TEPFER, D. (1986) Nucleotide sequence analysis of T$_L$-DNA of *Agrobacterium rhizogenes* agropine type plasmid: identification of open-reading frames. J. Biol. Chem. 261:108-121.

SMITH, E.F. and TOWNSEND, C.O. (1907) A plant tumour of bacterial origin. Science 25:671-673.

SPANO, L., POMPONI, M., COSTANTINO, P., VAN SLOGTEREN, G.M.S. and TEMPÉ, J. (1982) Identification of T-DNA in the root-inducing plasmid of the agropine type *Agrobacterium rhizogenes* 1855. Plant Mol. Biol. 1:291-300.

216

SPIELMANN, A. and SIMPSON, R.B. (1986) T-DNA structure in transgenic tobacco plants with multiple independent integration sites. Mol. Gen. Genet. 205:34–41.

STACHEL, S.E. and NESTER, E.W. (1986) The genetic and transcriptional organization of the vir region of the A6 Ti plasmid of *Agrobacterium tumefaciens.* EMBO J. 5:1445–1454.

STACHEL, S.E. and ZAMBRYSKI, P.C. (1986) *virA* and *virG* control the plant-induced activation of the T-DNA transfer process of *Agrobacterium tumefaciens.* Cell 46:325–333.

STACHEL, S.E., AN, G., FLORES, C. and NESTER, E.W. (1985a) A Tn3 *lacZ* transposon for the random generation of β–galactosidase gene fusions: application to the analysis of gene expression in *Agrobacterium.* EMBO J. 4:891–898.

STACHEL, S.E., MESSENS, E., VAN MONTAGU, M. and ZAMBRYSKI, P. (1985b) Identification of the signal molecules produced by wounded plant cells that activate T-DNA transfer in *Agrobacterium tumefaciens.* Nature 318:624–629.

STACHEL, S.E., NESTER, E.W. and ZAMBRYSKI, P.C. (1986a) A plant cell factor induced *Agrobacterium tumefaciens vir* gene expression. Proc. Natl. Acad. Sci. USA 83:379–383.

STACHEL, S.E., TIMMERMAN, B. and ZAMBRYSKI, P. (1986b) Generation of single-stranded T-DNA molecules during the initial stages of T-DNA transfer from *Agrobacterium tumefaciens* to plant cells. Nature 322:706–712.

STACHEL, S.E., TIMMERMAN, B. and ZAMBRYSKI, P. (1987) Activation of *Agrobacterium tumefaciens vir* gene expression generates multiple single-stranded T-strand molecules from the pTiA6 T-region: requirement for 5' *virD* gene products. EMBO J. 6:857–863.

SÜLE, S. (1978) Biotypes of *Agrobacterium tumefaciens* in Hungary. J. appl. Bact. 44:207–213.

SWANN, R.A., FOULKES, S.J., HOLMES, B., YOUNG, J.B., MITCHELL, R.G. and REEDERS, S.T. (1985) "*Agrobacterium* yellow group" and *Pseudomonas paucimobilis* causing peritonitis in patients receiving continuous ambulatory peritoneal dialysis. J. Clin. Pathol. 38:1293–1299.

TATE, M.E., MURPHY, P.J., ROBERTS, W.P. and KERR, A. (1979) Adenine N[6]-substituent of agrocin 84 determines its bacteriocin-like specificity. Nature 280:697–699.

TAYLOR, B.H., WHITE, F.F., NESTER, E.W. and GORDON, M.P. (1985) Transcription of *Agrobacterium rhizogenes* A4 T-DNA. Mol. Gen. Genet. 201:546–553.

TEMPÉ, J.,, GOLDMANN, A. (1982) Occurrence and biosynthesis of opines. In: Kahl, G., SCHELL, J. (eds). Molecular biology of plant tumors. Acad. Press, New York, pp. 427–449.

TEPFER, D.A. and TEMPÉ, J. (1981) Production d'agropine par des racines formées sous l'action d'*Agrobacterium rhizogenes*, souche A4. C.R. Acad. Sci. 292D:153–156.

TEPFER, D. (1984) Transformation of several species of higher plants by *Agrobacterium rhizogenes*: sexual transmission of the transformed genotype and phenotype. Cell 37:959–967.

THOMASHOW, M.F., PANAGOPOLOUS, C.G., GORDON, M.P. and NESTER, E.W. (1980a) Host range of *Agrobacterium tumefaciens* is determined by the Ti plasmid. Nature 283:794–796.

THOMASHOW, M.F., NUTTER, R., MONTOYA, A.L., GORDON, M.P. and NESTER, E.W. (1980b) Integration and organization of Ti plasmid sequences in crown gall tumors. Cell 19:729–739.

THOMASHOW, L.S., REEVES, S. and THOMASHOW, M.F. (1984) Crown gall oncogenesis: evidence that a T-DNA gene from the *Agrobacterium* Ti plasmid pTiA6 encodes an enzyme that catalyzes synthesis of indoleacetic acid. Proc. Natl. Acad. Sci. USA 81:5071–5075.

THOMASHOW, M.F., HUGLY, S., BUCHHOLZ, W.G. and THOMASHOW, L.S. (1986) Molecular basis for the auxin-independent phenotype of crown gall tumour tissues. Science 231:616–618.

THOMASHOW, M.F., KARLINSEY, J.E., MARKS, J.R. and HURLBERT, R.E. (1987) Identification of a new virulence locus in *Agrobacterium tumefaciens* that affects polysaccharide composition and plant cell attachment. J. Bacteriol. 169:3209–3216. UNGER, L., ZIEGLER, S.F., HUFFMAN, G.A., KNAUF, V.C., PEET, R., MOORE, L.W., GORDON, M.P. and NESTER, E.W. (1985) New class of limited-host range *Agrobacterium* mega-tumour-inducing plasmids lacking homology to the transferred DNA of a wide-host-range tumour-inducing plasmid. J. Bacteriol. 164:723–730.

URSIC, D., SLIGHTOM, J.L. and KEMP, J.D. (1983) *Agrobacterium tumefaciens* T-DNA integrates into multiple sites of the sunflower crown gall genome. Mol. Gen. Genet. 190:494–503.

USAMI, S., MORIKAWA, S., TAKEBE, I. and MACHIDA, Y. (1987) Absence in monocotyledonous plants of the plant factors inducing T-DNA circularization and *vir* gene expression in *Agrobacterium*. Mol. Gen. Genet. 209:221–226. VAN BRUSSEL, A.A.N., TAK, T., WETSELAAR, A., PEES, E. and WIJFFELMAN, C.A. (1982) Small *leguminosae* as test plants for nodulation of *Rhizobium leguminosarum* and other rhizobia and agrobacteria harbouring a leguminosarum Sym-plasmid. Plant Sc. Lett. 27:317–325.

VAN HAAREN, M.J.J., PRONK, J.T., SCHILPEROORT, R.A. and HOOYKAAS, P.J.J. (1986) T-region transfer from *Agrobacterium tumefaciens* to plant cells: functional characterization of border repeats. Recognition in Microbe – Plant Symbiotic and Pathogenic Interactions, Ed.: B.Lugtenberg, NATO ISI Series H, Cell Biology, Vol. 4:203–214.

VAN HAAREN, M.J.J., PRONK, J.T., SCHILPEROORT, R.A. and HOOYKAAS, P.J.J. (1987) Functional analysis of the *Agrobacterium tumefaciens* octopine Ti-plasmid left and right T-region border fragments. Plant. Mol. Biol. 8:95–104.

VAN LAREBEKE, N., GENETELLO, C., HERNALSTEENS, J.P., DEPICKER, A., ZAENEN, I., MESSENS, E., VAN MONTAGU, M. and SCHELL, J. (1977) Transfer of Ti plasmids between *Agrobacterium* strains by mobilisation with the conjugative plasmid RP4. Mol. Gen. Genet. 152:119–124.

VAN LIJSEBETTENS, M., INZÉ, D., SCHELL, J. and VAN MONTAGU, M. (1986) Transformed cell clones as a tool to study T-DNA integration mediated by *Agrobacterium tumefaciens*. J. Mol. Biol. 188:129–145.

VAN SLOGTEREN, G.M.S., HOOYKAAS, P.J.J. and SCHILPEROORT, R.A. (1984) Silent T-DNA genes in plant lines transformed by *Agrobacterium tumefaciens* are activated by grafting and by 5-azacytidine treatment. Plant Mol. Biol. 3:333–336.

VAN VEEN, R.J.M., DEN DULK-RAS, H., SCHILPEROORT, R.A. and HOOYKAAS, P.J.J. (1987) Chromosomal nodulation genes: Sym-plasmid containing *Agrobacterium* strains need chromosomal virulence genes (*chv*A and *chv*B) for nodulation. Plant Mol. Biol. 8:105–108.

VELUTHAMBI, K., JAYASWAL, R.K. and GELVIN, S.B. (1987) Virulence genes, A,G and D mediate the double-stranded border cleavage of T-DNA from the *Agrobacterium* Ti plasmid. Proc. Natl. Acad. Sci. USA

218

84:1881–1885.

VILAINE, F. and CASSE-DELBART, F. (1987) Independent induction of transformed roots by the T_L and T_R regions of the Ri plasmid of agropine type *Agrobacterium rhizogenes*. Mol. Gen. Genet. 206:17–23.

WALDRON, C. and HEPBURN, A.G. (1983) Extra DNA in the T–region of crown gall Ti plasmid pTiA66. Plasmid 10:199–203.

WALLROTH, M., GERATS, A.G.M., ROGERS, S.G., FRALEY, R.T. and HORSCH, R.B. (1986) Chromosomal localization of foreign genes in *Petunia hybrida*. Mol. Gen. Genet. 202:6–15.

WANG, K., HERRERA-ESTRELLA, L., VAN MONTAGU, M. and ZAMBRYSKI, P. (1984) Right 25 bp terminus sequence of the nopaline T–DNA is essential for and determines direction of DNA transfer from *Agrobacterium* to the plant genome. Cell 38:455–462.

WANG, K., STACHEL, S.E., TIMMERMAN, B., VAN MONTAGU, M. and ZAMBRYSKI, P.C. (1987) Site–specific nick in the T–DNA border sequence as a result of *Agrobacterium vir* gene expression. Science 235:587–591.

WEISBURG, W.G., WOESE, C.R., DOBSON, M.E. and WEISS, E. (1985) A common origin of rickettsiae and certain plant pathogens. Science 230:556–558.

WHITE, L.O. (1972) The taxonomy of the crown–gall organism *Agrobacterium tumefaciens* and its relationship to *Rhizobia* and other *Agrobacteria*. J. gen. Microbiol. 72:565–574.

WHITE, F.F. and NESTER, E.W. (1980) Relationship of plasmids responsible for hairy root and crown gall tumorigenicity. J. Bacteriol. 144:710–720.

WHITE, F.F., GHIDOSSI, G., GORDON, M.P. and NESTER, E.W. (1982) Tumour induction by *Agrobacterium rhizogenes* involves the transfer of plasmid DNA to the plant genome. Proc. Natl. Acad. Sci. USA 79:3193–3197.

WHITE, F.F., GARFINKEL, D.J., HUFFMAN, G.A., GORDON, M.P. and NESTER, E.W. (1983) Sequences homologous to *Agrobacterium rhizogenes* T–DNA in the genomes of uninfected plants. Nature 301:348–350.

WHITE, F.F., TAYLOR, B.H., HUFFMAN, G.A., GORDON, M.P. and NESTER, E.W. (1985) Molecular and genetic analysis of the transferred DNA regions of the root–inducing plasmid of *Agrobacterium rhizogenes*. J. Bacteriol. 164:33–44.

WILLMITZER, L., SIMONS, G. and SCHELL, J. (1982a) The T_L–DNA in octopine crown–gall tumours for seven well–defined polyadenylated transcripts. EMBO J. 1:139–146.

WILLMITZER, L., SANCHEZ-SERRANO, J., BUSCHFELD, E. and SCHELL, J. (1982b) DNA from *Agrobacterium rhizogenes* is transferred to and expressed in axenic hairy root plant tissues. Mol. Gen. Genet. 186:16–22.

WILLMITZER, L., DHAESE, P., SCHREIER, P.H., SCHMALENBACH, W., VAN MONTAGU, M. and SCHELL, J. (1983) Size, location and polarity of T–DNA encoded transcripts in nopaline crown gall tumours; common transcripts in octopine and nopaline tumours. Cell 32:1045–1056.

WINANS, S.C., EBERT, P.R., STACHEL, S.E., GORDON, M.P. and NESTER, E.W. (1986) A gene essential for *Agrobacterium* virulence is homologous to a family of positive regulatory loci. Proc. Natl. Acad. Sci. USA 83:8278–8282.

WINANS, S.C., ALLENZA, P., STACHEL, S.E., McBRIDE, K.E. and NESTER, E.W. (1987) Characterization of the *virE* operon of the *Agrobacterium* Ti plasmid pTiA6. Nucleic Acids Res. 15:825–837.

WINTER, J.A., WRIGHT, R.L. and GURLEY, W.B. (1984) Map locations of five transcripts homologous to T$_R$-DNA in tobacco and sunflower crown gall tumours. Nucleic Acids Res. 12:2391-2406.
YADAV, N.S., VANDERLEYDEN, J., BENNETT, D.R., BARNES, W.M. and CHILTON, M.-D. (1982) Short direct repeats flank the T-DNA on a nopaline Ti plasmid. Proc. Natl. Acad. Sci. USA 79:6322-6326.
YAMAMOTO, A., IWAHASHI, M., YANOFSKY, M.F., NESTER, E.W., TAKEBE, I. and MACHIDA, Y. (1987) The promoter proximal region in the *virD* locus of *Agrobacterium tumefaciens* is necessary for the plant-inducible circularization of T-DNA. Mol. Gen. Genet. 206:174-177. YANOFSKY, M., LOWE, B., MONTOYA, A., RUBIN, R., KRUL, W., GORDON, M. and NESTER, E. (1985a) Molecular and genetic analysis of factors controlling host range in *Agrobacterium tumefaciens*. Mol. Gen. Genet. 201:237-246.
YANOFSKY, M., MONTOYA, A., KNAUF, V., LOWE, B., GORDON, M. and NESTER, E. (1985b) Limited-host-range plasmid of *Agrobacterium tumefaciens*: molecular and genetic analysis of transferred DNA. J. Bacteriol. 163:341-348.
YANOFSKY, M.F., PORTER, S.G., YOUNG, C., ALBRIGHT, L.M., GORDON, M.P. and NESTER, E.W. (1986a) The virD operon of *Agrobacterium tumefaciens* encodes a site-specific endonuclease. Cell 47:471-477.
YANOFSKY, M.F. and NESTER, E.W. (1986b) Molecular characterization of a host-range-determining locus from *Agrobacterium tumefaciens*. J. Bacteriol. 168:244-250.
ZAENEN, I., VAN LAREBEKE, N., TEUCHY, H., VAN MONTAGU, M. and SCHELL, J. (1974) Supercoiled circular DNA in crown-gall inducing *Agrobacterium* strains. J. Mol. Biol. 86:109-127.
ZAMBRYSKI, P., DEPICKER, A., KRUGER, K. and GOODMAN, H.M. (1982) Tumour induction by *Agrobacterium tumefaciens*: analysis of the boundaries of T-DNA. J. Mol. Appl. Genet. 1:361-370.
ZAMBRYSKI, P., JOOS, H., GENETELLO, C., LEEMANS, J., VAN MONTAGU, M. and SCHELL, J. (1983) Ti plasmid vector for the introduction of DNA into plant cells without alteration of their normal regeneration capacity. EMBO J. 2:2143-2150.

Oxford Surveys of Plant Molecular and Cell Biology Vol.4 (1987) 221-272

LONG–TERM PRESERVATION OF PLANT CELLS, TISSUES AND ORGANS

Lyndsey A. Withers

Department of Agriculture and Horticulture, University of Nottingham School of Agriculture, Sutton Bonington, Loughborough, LE12 5RD, UK

INTRODUCTION

The provision of experimental material of known phenotypic and genotypic characters is a fundamental requirement in many aspects of plant biology. In laboratories concerned with cell and molecular biological investigations using *in vitro* systems, the maintenance of cultures has, for years, been a major consumer of time, manpower, space and material resources. The importance of the identity of experimental material has increased in recent times as technological developments have moved from the examination of model systems to the examination of subjects chosen for their applied value. This evolution can be seen in the history of, for example, mass propagation, protoplast technology, transformation and anther culture. As a result, a demand for the development of storage techniques capable of conserving exact genotypes has been generated.

The investigation of storage techniques themselves has followed a similar pattern with well worked out model systems first being used to develop a framework of basic rules. Only recently has the threshold been crossed from model systems to the examination of subjects less amenable to, but with a greater need for, storage. Inevitably, this process results in some of the basic rules being revealed as less than sacrosanct. However, this should not give cause for concern as the considerable heterogeneity of the systems requiring storage is likely to be reflected in a diversity of requirements. Model systems help in the early stages of technological development but by their very amenability often fail to offer the challenges that lead to technological refinement. Only through discovering differences between subjects do we gain a feeling for, and understanding of, the nature of storage phenomena and learn to control them.

The development of *in vitro* storage of plants is taking place in parallel with that of other materials including cells, tissues, organs and embryos of humans and other mammals. Here, problems are encountered that are of a similar type and consequence to those found in plant work. A beneficial exchange of ideas is clearly possible.

221

In vitro storage has been reviewed extensively by several authors in recent years (Bajaj 1983; Bajaj and Reinert, 1977; Bhojwani and Radzan, 1983; Dereuddre, 1985; Kartha, 1985; Nitzsche, 1983; Sakai, 1985; Withers, 1980a; 1982a; 1985a; 1985b; 1986a; 1986b; 1987a; 1987b; Withers and Street, 1977a). This review attempts to evaluate the importance and contribution of different approaches to storage, examine impediments to progress and suggest some avenues of investigation that may help to resolve present difficulties.

GERMPLASM STORAGE NEEDS

Genetic Instability *In Vitro*

One of the attractive features of *in vitro* plant systems to the experimental biologist and the agricultural scientist is the ability to produce and maintain clones, thus avoiding problems caused by heterozygosity and other sources of heterogeneity in working populations. Whilst *in vitro* cloning at both the cell and plantlet levels is undoubtedly a powerful means of enhancing uniformity, it has to be accepted that absolute clonal homogeneity is a myth (Scowcroft, 1985). Genetic flux, the fuel of evolution, is a more widespread phenomenon than generally accepted and is an important fact of life in plant systems including *in vitro* cultures (Cullis, 1986; Scowcroft, 1985). To a certain extent, this flux is welcomed as it offers the possibility of developing new genotypes (Scowcroft, 1986). However, until the nature and control of the processes that fall within the very broad embrace of somaclonal variation are fully understood, some means of avoiding genetic drift in cultured material must be found. Even then, storage will have a role in stabilizing material subject to known levels of instability.

Several clear areas of need for *in vitro* storage can be identified. These needs have developed for various practical reasons. In some cases genetic instability is the primary reason for seeking a storage method. In others it is a complicating factor, but in none can it be ignored. It should also be realised that the manipulations involved in storage must themselves be considered to carry some degree of risk of instability. Therefore, storage does not remove the need for vigilance in the evaluation of the experimental material by whatever means are at the disposal of the scientist (see below: Stability in cryopreserved material).

Before describing the various needs, a further caution is given that whilst *in vitro* storage can offer great benefits, it is still at a very early stage of development. Whenever possible, valuable genetic material should be stored by as many diverse methods as is practicable. Untried methods should not be adopted without adequate testing and, most important of all, the method chosen should be logical. It is not logical, for example to conserve a potato clone as

222

frozen protoplasts, or a fully developed cereal variety as any form of *in vitro* culture. Technical complexity is not a virtue in itself, quite the opposite.

Experimental Research Material

Any laboratory using *in vitro* cultures for experimental work will need to maintain cultures under conditions that minimise risks of change or loss. It may seem impractical to undertake a new approach to culture maintenance, but when the costs of continuous growth and the risks of contamination, equipment failure and human error are taken into consideration, the need for stable, secure storage is very evident. Continuous, more or less rapid growth will lead to progressive change, particularly in unorganised cultures (Bayliss, 1980; d'Amato, 1978; Scowcroft, 1984). Some changes, such as loss of totipotency, will be obvious; others are more subtle but no less consequential (Scowcroft et al., 1985). Some material is intrinsically ephemeral leading to an enhanced need to suspend time-related changes for reasons of convenience and to provide timed controls. Examples here include isolated protoplasts at particular stages of development and material undergoing treatments that form part of a sequence that has practical bottlenecks such as transformation followed by selection.

Secondary Product Synthesis

Recent years have seen dramatic progress in the use of *in vitro* systems to synthesize products of pharmaceutical and nutritional importance (Fowler, 1986). *In vitro* systems can avoid problems of erratic supply, climatic influences on cropping, market forces and other factors governing agricultural production. However, it is an extremely expensive option that must have the provision of consistent, high yielding culture stocks. These may take the form of cells, callus, somatic embryos or organs such as roots. All will be special in terms of their synthetic capacity; some will be selections from populations of variable quality, others will be genetically engineered.

A tendency to instability has been noted in many secondary product synthesizing cultures, necessitating periodic reselection (Deus-Neumann and Zenk, 1984). This will not always be successful and valuable lines will be lost. As well as the practical need to store stock cultures, there is also the necessity to place stocks in recognised depositories to facilitate patent registration.

Crop Production

In vitro cloning already makes a significant contribution to the mass propagation of a number of horticultural crops and is becoming important in other areas including forestry (Bonga and Durzan, 1982; Conger, 1981; George and Sherrington, 1984; Schäfer-Menuhr, 1985; Walkey, 1987). The culture systems in use here are intrinsically more stable than many of those used for secondary product synthesis, for example. Nonetheless, the ability to store material for indefinite periods of time and then recover and multiply it at will is extremely attractive. Storage applications are not limited to material that is conventionally multiplied vegetatively. For example, it would be convenient to be able to store hybrid brassicas that have been cloned in vitro, bulked-up parental lines for F_1 production material, clones of mature, long-lived forest trees that have proven superior, and stocks of micropropagated plants that are subject to seasonal demand (Withers, 1987b; Zeldin and McCown, 1985)

Fluid drilling and the production of artificial seed are opening up new possibilities for the handling of in vitro cloned material (Kitto and Janick, 1985a; 1985b; Lawrence, 1981; Redenbaugh et al., 1986). One of the problems encountered in these technologies is the phasing of the various stages of production, preparation and delivery. The ability to hold material for periods of time without deterioration would be highly beneficial.

The presence of pathogens, particularly viruses, is a serious problem in the production of some crops, notably those that are vegetatively propagated as they lack the 'filter' of seed production. In vitro techniques such as meristem-tip culture, sometimes combined with chemotherapy or thermotherapy, are used to clean up clonal material (Kartha, 1986). Keeping this material pathogen-free is difficult. In vitro storage offers a means of both isolating clean material from potential sources of reinfection and avoiding the propagation events that risk the reintroduction of the pathogen.

Plant Breeding

As each year passes, the definition of plant breeding becomes broader. Culture storage has a part to play in several areas (Doré, 1987; Withers, 1987b). These include mass propagation of parental lines for the production of F_1 hybrids, anther culture (Dunwell, 1986), somaclonal variation (Scowcroft, 1986; Semal, 1986), and, perhaps most significant of all, transformation technology (Draper et al., 1986; Peacock, 1987). Many exciting possibilities are opened up by the ability to transfer genes from, for example, wild relatives with high levels of disease or stress resistance, to otherwise fully developed cultivars. Some of the most dramatic applications of transformation may be in the improvement of clonal crops that are not amenable to conventional breeding processes. Here, the storage of somatic material will be very important.

224

Genetic Conservation

This area overlaps with plant breeding in that conservation is not just preservation for its own sake but an important facilitator for efficient crop improvement (Peeters and Williams, 1984). For many crops, including most of the major world food sources, genetic conservation can be carried out very efficiently in seed genebanks (Ellis et al., 1985a; 1985b; Williams, 1984). However, there are some notable exceptions (De Langhe, 1984; Withers, 1980a; 1984a; Withers and Williams, 1985). Firstly, there are crops that produce so-called 'recalcitrant' seeds that are unresponsive to standard seed storage techniques involving desiccation and cold storage (Roberts et al., 1982; 1984). Examples include mango, rubber, cocoa, coconut and oak. Secondly, some clonal crops are either sterile or fail to breed true due to gross heterozygosity. Many root and tuber crops of importance in developing countries (e.g. yam, potato, cassava, sweet potato) and fruits such as banana and plantain come into this category. In examples where clones could theoretically be reconstituted from seed, such as some of the root and tuber crops, and apple, pear and other temperate fruits, success would not be guaranteed and there would be a considerable wastage of breeding effort.

At present, problem crops are conserved either in field genebanks (plantations, orchards etc.), as vegetative propagules, or in nature reserves ('*in situ* conservation'). Arguments of cost, safety, practicality and adequacy can be levelled against these approaches to conservation. *In vitro* storage has been proposed as a means of overcoming problems here (De Langhe, 1984; IBPGR, 1983; Withers, 1980a; Withers and Williams, 1985). This application of storage technology has, in fact, been a stronger motivator of planning, research and development than the other applications listed above.

As the working systems involved in this area are not naturally *in vitro*, unlike in secondary product synthesis for example, this new approach to genetic conservation has led to the development of a system of handling that parallels all of the conventional processes from germplasm acquisition to storage and distribution. Such aspects of *in vitro* genetic conservation are discussed further below (see: Strategies of *in vitro* conservation).

In principle, all culture systems from the disorganised cell suspension to the *in vitro* plantlet are available for use in conservation. However, the question of appropriateness must be addressed. As both the starting point and the final product are whole plants, the maintenance of a high level of organisation during storage is logical. Furthermore, as the objective is to conserve genetic diversity in a representative, stable condition, none of the storage techniques should put stability at risk; nor should there be a risk of narrowing of genetic diversity by selection. Thus, it is more appropriate to use culture systems that minimise risks of somaclonal variation, such as shoots, embryos and plantlets (Scowcroft, 1984; 1985). Having said this, it should be acknowledged that practical limitations are imposed by difficulties in maintaining high levels of viability and organisation particularly in some cryopreserved systems. Therefore, it is still pragmatic to look at other systems. In the future, as protoplast and cell technologies for crop improvement become more routine, it may be appropriate in some specialised cases to conserve unorganised systems, or even DNA sequences themselves (Peacock, 1987). The latter possibility is discussed further in a later section (see: Strategies of *in vitro* conservation).

TECHNICAL APPROACHES TO STORAGE

Choice of Approach

Cautions were given earlier regarding the dangers of technical over-complexity for its own sake and the need to chose storage methods appropriate to the system and situation in question. This is best done by defining the objectives in using *in vitro* storage and the likely consequences of so doing. These relate to time scale (short, medium or long-term) facilities available (basic *in vitro* propagation facilities, a range of environmental variations or dedicated storage equipment) the culture system in use and, finally, whether the genotype has already been stored *in vitro* with an adequate level of success.

This and the following section attempt to provide the information needed to chose an appropriate storage method on the basis of some of the criteria given above. The reader is referred to other reviews (Bajaj and Reinert, 1977; Kartha, 1985; Nitzsche, 1983; Sakai, 1985; Withers, 1980a; 1982a; 1985a; 1986a) and the International Board for Plant Genetic Resources database on *in vitro* conservation (Wheelans and Withers, 1984; 1987) for crop by crop detail.

The techniques available can be divided, broadly, into those that reduce the rate of growth and those that suspend growth. As will become apparent, not all of the techniques are fully developed; only slow growth and cryopreservation can be considered seriously for adoption. However, by presenting the full spectrum of approaches, it is hoped that some momentum can be provided towards the exploration and improvement of a comprehensive range of storage techniques to suit all potential users.

Storage in the Growing State

For material that originates *ex vitro* and that has a low viability potential in its natural state, e.g. recalcitrant seed embryos, the very act of inoculation into culture provides a storage mode with advantages over conventional alternatives. However, in general, cultures cannot be considered to be truly 'in storage' unless some modification be made to the rate of growth. Slow growth can be induced by a reduction in temperature or modification of the culture medium, usually by the addition of osmotic or hormonal retardants.

The aim of growth retarding treatments is simple, namely to increase the time interval between transfer events. However, behind that simplicity lies a more complex situation. For a culture to remain healthy, it should be in continuous growth. Therefore, the increased transfer interval should be achieved by changing the gradient of the growth curve rather than extending the lag or stationary phases. Adverse selection for genotypes tolerant of stress could occur during the latter phases, and the possibility should not be ignored of competition between genotypes within a heterogeneous population leading to the more successful competitors becoming numerically better-represented during the growth phase (Bayliss, 1980). These phenomena have implications for the subculturing routines and management procedures for *in vitro* collections. Extension of the period of growth can be achieved by alternation of retarding and normal conditions on a frequent (diurnal) or longer-phased (e.g. fortnightly) basis.

As an alternative to modifying growth rates via the culture medium or through temperature reduction, some sporadic attempts have been made to use oxygen limitation. Taking success with fungal cultures as a model, mineral oil overlay has been tried with callus cultures (Agereau et al., 1986; Caplin, 1959). Although intrinsically uncomplicated in concept, this approach is not without problems of practicality and convenience.

Restricting the oxygen supply by applying a low overall atmospheric pressure or low partial pressure of oxygen in a carrier gas has been examined with some limited success for callus and shoot cultures (Bridgen and Staby, 1981; 1983). This approach derives from the use of similar techniques to extend the storage life of various foodstuffs (meat, fish and vegetables) and horticultural materials (potted plants, cuttings). A requirement for specialised equipment, frequent monitoring and the maintenance of very precise storage conditions must be recognised as drawbacks.

Overall, slow growth by medium modification or temperature reduction is recommended as the most readily adopted method of maintaining cultures in the short to medium term. If successive transfers can be carried out without loss of material, relatively long-term storage can be achieved. Slow growth utilizes comparatively unsophisticated equipment and, in the case of growth retardation via medium additives, can involve stored cultures being maintained alongside experimental material, thereby imposing no additional equipment requirements. Some examples of the successful use of slow growth, its limitations in terms of culture systems served and the assessment of stability in slowly grown material are given in a later section (see: Slow growth).

227

Suspension of Growth

All approaches to the suspension of growth involve modifying the status of water, the mediator of cellular reactions. Three strategies are feasible, namely substitution, removal and inactivation. Substitution of water by organic solvents has been used for pollen storage (Iwanami, 1972) but has not been explored for other higher plant subjects. Partial substitution of water by cryoprotectant compounds to preserve macromolecular structures and avoid denaturation is a feature of cryopreservation procedures, as discussed below.

Removal of water by desiccation has precedents in nature, for example, in seed drying and in the survival of xerophytic plants and their propagules exposed to environmental extremes (Bewley, 1979; Schwab and Gaff, 1986). The highly specialised nature of these structures must be recognised with consequent limitations upon the extent to which drying might be successful with other subjects. Success has been recorded in the use of freeze–drying for pollen storage (Akihama and Omura, 1986). This is undoubtedly due to the extreme simplicity of structure of pollen and its adaptation to survival in a hostile environment in the course of pollination. Larger, more complex structures would probably fall victim to desiccation gradients, physical tensions and risks of localised damage if subjected to freeze–drying, even if able to withstand the initial freezing stage. Freeze–drying is used relatively widely in microbiology but the prospects for its adoption for higher plants cultures is virtually nil in the forseeable future, particularly since serious experimentation is absent in this area. Whilst there would be considerable practical advantages to a storage method that involves a relatively simple preparative stage followed by storage under undemanding and not necessarily physically stable conditions, it should be noted that genetic instability has been reported in freeze–dried material (Ashwood–Smith and Grant, 1976). The drying component of the process is thought to be the source of damage.

Desiccation at elevated temperatures, perhaps mimicing the processes that occur in seed maturation has been reported for somatic embryos of carrot. They were allowed to desiccate by natural drying of the culture and supporting semi–solid medium and were then revived after periods of up to 2 years by the addition of a sucrose solution (Jones, 1974). Callus cultures of carrot have also been stored at a temperature of $-80°C$ after preculture on medium containing a high level of sucrose and abscisic acid (ABA) and desiccation in air (Nitzsche, 1978). If there were interest in pursuing the possibility of desiccation further, useful information might be found in the comparative physiologies of orthodox and recalcitrant seeds which are respectively tolerant and intolerant of drying. Recalcitrant seeds appear to lack the programmed closing down of their metabolism that precedes and accompanies drying and, rather, adopt a strategy of resisting desiccation and providing metabolite reserves for rapid germination under advantageous conditions (Nkang and Chandler, 1986). In some of these terms, higher plant cultures are 'recalcitrant'.

228

One further point to be made concerning desiccation is that it is a progressive process that must be terminated when certain optimal conditions have been reached. The specimen must then be held in circumstances that maintain the level of hydration and suspend whatever minimal metabolic activity remains. Neither objective is easily achieved without using freezing, given which conclusion, the use of cryopreservation proper would appear to be a more logical approach to storage.

Inactivation of water is achieved by severe temperature reduction below the level at which metabolic activity can be sustained. Undercooling and cryopreservation are storage methods that involve such a suspension of metabolism. Undercooling of cells dispersed in an emulsion, first described 12 years ago (Rasmussen et al., 1975) has been re-examined and refined more recently (Mathias et al., 1983; 1985). The operating principle of undercooling is that biological material may be cooled to a temperature below that at which ice formation would normally be expected to occur, but maintained in an ice-free condition by the dispersal of the specimen in an aqueous phase emulsified with mineral oil. At the storage temperature (ca. -20°C) the mineral oil solidifies, providing a stable matrix that impedes the propagation of ice from one droplet to another. Storage times reported to date are relatively modest (in the region of 2 days) but the principle has been established and a foundation for improvement provided. Some of the treatments used to improve tolerance of cryopreservation conditions might be adopted; pregrowth in the presence of osmotically active medium additives has already been used (Mathias et al., 1985).

Heterogeneity in response to undercooling has been observed. Yeast cells are more amenable to storage in this way than higher plant cells, and shoot-tips less so, reflecting the pattern in cryo-preservation (Mathias et al., 1985). The advantages of undercooling include the comparative simplicity of equipment requirements and the avoidance of potentially damaging extracellular or intracellular solute concentrations. However, long-term viability and stability in physical, physiological and genetic terms have yet to be proven; nor has a wide taxonomic range of material been examined. It cannot be assumed that all metabolism will be suspended at the storage temperatures used (Douzou, 1983) or that free radical activity will have ceased. Furthermore, damage repair mechanisms will be suspended in the stored material. Undercooling cannot yet be recommended even as a back-up storage method although it is acknowledged as a useful analytical tool in cryobiology.

The latter point concerning repair processes is one of the few arguments of principle that can be levelled against cryopreservation as a storage method. The objective of cryopreservation procedures is to take biological material from normal physiological temperatures to ultralow temperatures and back again without incurring damage during the temperature excursions. Very few cellular systems can survive such transitions unaided by chemical additives or strict control of the cooling and warming conditions. Those that require minimal support, such as dormant buds (see: Cryopreservation procedures, Shoot cultures) are interesting exceptions to a rule that dominates the planning and execution of cryopreservation procedures.

The phenomena accompanying cryopreservation are described in detail elsewhere, as are the underlying biochemical and biophysical principles of chemical and physical protection (Finkle et al., 1985a, Meryman, 1966; Meryman and Williams, 1982; Farrant, 1980; James, 1983). A practical approach to the development of working procedures is taken here by outlining the stages of cryopreservation, and the prophylactic measures taken and problems encountered at each stage. As the nature of the specimens subjected to cryopreservation have a strong influence upon the details of the procedures adopted in each case, a more detailed examination system by system is given later (see: Cryopreservation procedures).

The Stages of Cryopreservation

Cryopreservation procedures can be broken down into the following stages: pregrowth, cryoprotection, cooling, storage, warming and recovery. Pregrowth covers the period of time in culture during which measures may be taken to enhance freeze-tolerance through influence upon the growth and metabolism of the culture, and from which the most freeze-tolerant stages of growth may be selected. Generally, material in active growth is more resistant to freezing stress, although dormant buds are highly resistant and some pregrowth treatments actually depress the rate of cell division.

Cryoprotection involves the application of diverse compounds during the period immediately preceding cooling. The most frequently used is dimethyl sulphoxide (DMSO) on its own or in combination with various sugars, sugar alcohols and other compounds. Some high molecular weight polymers are also used infrequently (Finkle et al., 1985a). It is beneficial to prepare cryoprotectants in culture medium rather than water (Withers, 1980b). As a result, cryoprotectant applications are, in fact, highly complex mixtures including low to moderate levels of sucrose, mineral salts, amino acids and possibly hormones. Where a pregrowth additive with cryoprotectant activity has been used, it should be remembered that this may be carried over to the cryopreservation stage. The highest grade available of cryoprotectant chemicals should be used. In the case of DMSO, where toxicity by impure stocks has been reported, it is advisable to use spectrophotometry grade or locally purified stocks. Cryoprotectant mixtures are prone to caramelisation if autoclaved. They should, therefore, be sterilized by filtration. The modes of action of cryoprotectants are highly complex and, in some cases, far from well understood (Finkle et al., 1985a; McGann, 1978; Nash, 1966).

Cooling can be carried out at a range of relatively slow or rapid rates (Farrant, 1980; Meryman and Williams, 1982). In simplest terms, slow cooling induces extracellular freezing that then causes cellular dehydration. The relative benefits and dangers of this, that lead to the existence of an optimal rate, relate to the extent of cellular shrinkage and the concentration of intracellular solutes. Rapid cooling causes intracellular freezing. Whether this is damaging or not is determined by the amount, crystal size and location of the ice. Optimal rates are best determined empirically. Slow cooling need not be carried out at a constant, linear rate and stepped cooling is often more convenient and controllable than continuous cooling (Withers, 1986a). Slow cooling will normally be terminated at a temperature well above that of liquid nitrogen. The choice of cooling profile will depend, to some extent, upon the available apparatus. Simple units (Withers, 1984b) can achieve highly reproducible slow cooling, usually, at a decelerating rate or stepped cooling. Programmable freezers can cover all of the options, have a high throughput capacity, rapid turn-round time and only present disadvantages in their capital cost.

Storage must be carried out at an adequately low temperature to prevent progressive deterioration resulting from ice recrystal- lization. Below -100°C is recommended and this is achieved most easily by the use of a liquid nitrogen cooled refrigerator in which the liquid phase is stable at -196°C and the vapour phase at about -150°C. Storage in the vapour phase has advantages in that penetration of storage ampoules by liquid nitrogen and the consequent risk of explosion upon warming is avoided. Metabolism is suspended at ultralow temperatures but molecular changes due to the ionizing effect of radiation, and consequent free radical activity, may lead to cumulative damage in the very long term. Experiments with bacteria and mammalian cells and embryos indicates that the time scale is unlikely to be of concern (Ashwood-Smith, 1985; Ashwood-Smith and Friedman, 1979; Whittingham et al., 1977) but precautions in the form of glass and metal shielding and the use of radioprotectant chemicals (free radical scavengers) are means of combating the risk.

Warming is as critical as cooling in the influence that it can exert upon survival. Indeed, there is a strong interaction between the two temperature excursions. It is essential that warming be carried out at a rate that prevents recrystallization of any ice that is present intracellularly. A sensible precaution is to always warm specimens rapidly by the use of a warm water bath (+40°C) or warm medium (+20°C or higher) for naked specimens. Exceptions that benefit from slow warming are given later.

The recovery phase commences with the first treatments to which the thawed specimen is exposed. Washing to remove potentially toxic cryoprotectants has been recommended but there is ample evidence that washing can be deleterious and little to suggest its benefits. The first culture passage after thawing is normally carried out using semi-solid medium, the composition of which is often modified to bring about a desired pattern of recovery. This is most important in the case of organised or unorganised but morphogenic cultures.

No single stage of the cryopreservation process is more important than any other. However, it must be said that progress in the evolution of cryopreservation protocols in recent years has gained most from our increased understanding of requirements at the pregrowth and recovery stages. Whether this is likely to continue in the future is in question as there are obviously limitations at each

231

stage. The author is of the opinion that potential remains for enhancing survival and the quality of recovery by attention to pregrowth and post-thaw conditions but that for organised cultures in particular, serious attention must be given to cryoprotection and the cooling and warming stages.

Having outlined the options available in storage techniques, the following section will now examine attempts that have been made to develop and apply storage protocols within the two most successful options of slow growth and cryopreservation. Major differences are found in the response of different culture systems to storage procedures leading to a conclusion that slow growth is only really suitable for shoot cultures and that, whilst cryopreservation is much more generally applicable, only for cell suspension cultures can a wide range of genotypes be cryopreserved with confidence at present.

SLOW GROWTH

Storage by slow growth has been applied with most promising results to shoot cultures, and less successfully to callus cultures. Other than one report of the storage of isolated mesophyll cells of asparagus at 4°C in static liquid culture for up to 45 days (Jullien, 1983), no other types of culture have been found to respond favourably.

Shoot Cultures

The range of shoot cultures maintained in slow growth includes root and tuber crops, temperate and tropical fruits, ornamentals and other horticultural, forestry and economic species. Slightly different approaches are taken in the various cases, undoubtedly due in part to the experimental nature of the work and the facilities available to the scientists in question. This makes extrapolation to untried species less than easy. However, as can be seen in the most closely examined system, namely potato (Table 1a), diverse growth retarding conditions can be applied with success in any one genus or species. Some caution is needed in the use of slow growth. Intergenotypic differences have been observed in potato, indicating the importance of carrying out trials when storing new material. In a study of the response of a range of potato genotypes subjected to the same storage conditions (Henshaw et al., 1980), loss of viability at 6 months ranged from 9 – 69%, and at 18 months from 29 – 100%. A number of other root and tuber crops have been stored in slow growth (Table 1b); these include some important world germplasm collections of the crops. The principles of using low temperature, osmotic and hormonal additives and intermittent warming are illustrated in the examples in Table 1a and 1b.

Most examples of slow growth have used culture on semi-solid medium and this is generally recommended for its convenience, physical support, resistance to desiccation and control of contamination (although ideally all contaminated cultures should be discarded). However, one of the earliest studies of slow growth storage in strawberry (Mullin and Schlegel, 1976) illustrates the success that can be achieved using liquid medium. Cultures were stored for periods of up to 6 years with the media being replenished drop by drop as necessary. The benefits of avoiding wholesale transfer of the culture at intervals might be considered to be outweighed by the risks of introducing contaminants when adding fresh medium. Woody temperate fruits are also amenable to slow growth storage and some are listed in Table 1c, along with other fruit subjects. Further miscellaneous examples of successful slow growth storage are given in Table 1d.

A combination of stressing treatment such as low temperature with osmotic stress, has been used with success in some species (e.g. Staritsky et al., 1986; Table, 1b). However, in the storage of pear (Wanas et al., 1986) and cassava the application of mannitol at low temperatures has proved toxic, in the latter case, only in the absence of sucrose.

It is suggested here that it would be helpful for the future development and wider application of slow growth if a study could be carried out to evaluate the effect of a range of storage conditions on a given species in shoot culture or, better still, several representative species. Data would be collected at intervals on viability, biomass accumulation, height, shoot production, node production, degree of necrosis, chlorosis and morphological modification. This would build up a pattern of responses linking early symptoms to eventual viability and uniformity or divergence in morphology. It might then be possible to construct nomographs for use with data gathered by short-term testing to expedite the determination of suitable storage conditions for the long term.

Efforts to develop slow growth storage for shoot cultures have concentrated upon the development of treatments that maximise transfer intervals. Little consideration been given yet to stability except in the gross sense of viability. Because growth restriction of necessity induces morphological change, it is difficult to evaluate stability on the basis of the appearance of cultures. Some attempts have been made to carry out cytological and isozyme analysis of material in storage (CIAT, 1985; Wanas, 1987) but this must be amplified if confidence is to be placed in the method. Regular stability monitoring will form part of the management routines of *in vitro* genebanks.

Table 1: Slow growth procedures for shoot cultures. (In some cases, selected data are presented).

Species	Storage conditions	Storage period	Survival	Reference
(a) Potato				
Solanum spp. (13 cultivars)	Reduced temperature (from 22 to 6°C)	18 months	0-71% mean: 29%	Westcott, 1981a
"	Cycling temperature regime (12°C day/6°C night)	12 months	60-100% mean: 83%	"
"	ABA (5-10 mg/1); increased medium volume (from 20 to 60 ml)	12 months	mean: 73%	Westcott, 1981b
S. tuberosum	Reduced temperature (from 20 to 10°C); N-dimethyl-succinamic acid (50 mg/1); reduced light level (from 4000 to 2000 lux)	24 months	not specified	Mix, 1982; 1985
(b) Other root and tuber crops				
Ipomoea batatas (Sweet potato)	Reduced temperature (down to 16-20°C); mannitol (6%)	8-24 months	not specified	IITA, 1981; Ng and Hahn, 1985
Dioscorea sp. (Yam)	"	"	"	"
Manihot esculenta (Cassava)	Reduced temperature (from 30 to 20°C); 4% sucrose	15-18 months	90%	Roca et al., 1982

Table 1:

Species	Storage conditions	Storage period	Survival	Reference
Colocasia esculenta (Taro)	Mannitol (6%)	14 months	98%	Staritsky et al., 1986
"	Reduced temperature (from 28 to 9°C)	2 years	100%	"
"	Reduced temperature (from 28 to 3°C), with intermittent warming (48 h every 2 weeks)	14 months	79%	"
(c) Fruits				
Fragaria vesca (Strawberry)	Reduced temperature (down to 1 or 4°C); dark incubation	up to 6 years	100%	Mullin and Schlegel, 1976
Actinidia chinensis (Kiwi)	Reduced temperature (from 23 to 8°C); dark incubation; low salts and vitamins; modified hormones	12 months	100%	Monette, 1986
Malus sp. (Apple)	Reduced temperature (from 26 to 1 or 4°C)	12 months	70-100%	Lundergan and Janick, 1979
Pyrus communis (Pear)	Reduced temperature (from 28 to 4°C)	18 months	94%	Wanas et al., 1986
Prunus spp. (5 species)	Reduced temperature (from 18 to 2°C; dark incubation; shoots defoliated	28 months	46-100% mean: 78%	Druart, 1985

Table 1:

Species	Storage conditions	Storage period	Survival	Reference
Musa spp. (Banana, plantain)	Reduced temperature (from 30 to 15°C; reduced light level (from 3000 to 1000 lux)	13 months	25–100%	Bannerjee and De Langhe, 1985
(d) Miscellaneous species				
Chrysanthemum morifolium	Reduced temperature (from 24 to 2°C); reduced light level (down to 10–50 lux)	at least 2 years	not specified	Preil and Hoffmann, 1985
Trifolium repens	Reduced temperature (from 25 to 5°C); dark incubation	10 months	100%	Bhojwani, 1981
Pinus radiata	Reduced temperature (from 23–27 to 2°C)	8 months	not specified	Smith et al., 1982

Callus Cultures

Slow growth storage of callus cultures by maintenance at a reduced temperature has received only a limited amount of attention but some important observations have been made in relation to stability. It is reassuring that slow growth can help retain morphogenic potential in callus cultures (Hiraoka and Kodama, 1984) but data relating to secondary product synthesis are less encouraging. In a comparison of the performance of 5 species in terms of their ability to undergo recovery growth and resume secondary product synthesis when returned to normal growth conditions after increasing periods of time in slow growth at 4°C, all species showed symptoms of progressive impairment, but variations in its severity were observed (Hiraoka and Kodama, 1982; 1984). *Bupleurum falcatum* gradually lost both growth and synthetic potential. In *Phytolacca americana* and *Nicotiana tabacum*, a progressive loss of growth potential occurred over 10 months, being most steep in the latter species; synthetic capacity in both was severely impaired from the first sampling interval (2 months). *Mallotus japonicus* was severely impaired in both respects from the outset and, finally, *Lithospermim erythrorhizon* performed erratically throughout.

No general predictions or rules for the further application of slow growth to callus cultures can be made from the data. It can only be concluded that the routine storage of callus cultures by slow growth is some way off and that until improvements are made in procedures, cryopreservation is recommended for those species in which it is successful. For those unresponsive to cryopreservation or untried in that respect, only further research can help. However, the author feels that slow growth will never be a satisfactory approach to the storage of a system such as callus which is prone to soma-clonal variation and which presents an ideal situation for the generation of genetic instability and selection under stress (Cullis, 1986).

There is a dilemma relating to approaches that might be taken to avoiding progressive deterioration and selection. In seeds, it is known that storage in the imbibed state can be very successful as it allows repair to take place (Villiers, 1975). Also, observations with taro (Staritsky et al., 1986; Table 1b) suggest that periodic warming can alleviate the lethal effects of slow growth at a low temperature, indicating that repair is beneficial. However, within a heterogeneous population, the circumstances that promote repair might also favour the progressive increase in prominence of good competitors within that population leading to a change in the proportions of different genotypes. (Bayliss, 1980).

Figure 1. (opposite) Cryopreservation procedure for cell suspension cultures indicating the variables available at each stage. Compatible treatments from one report (Withers and King, 1980) are marked *. The original culture (1) is transferred to (2) a pregrowth passage in standard or modified medium [plus 0.02M asparagine; 0.05M alanine; 0.01M serine; 0.02M proline; 5-10% proline*; 0.15M mannitol; 3-6% mannitol; 6-15% sorbitol; 1M sorbitol; 7.5x10⁻⁵M ABA; 5% DMSO; 5-10% trehalose]. After a period of exposure [6h - 7 days*] at the normal* culture temperature or at a reduced temperature, cryoprotectants (3) are added [5-10% DMSO; 5-10% glycerol; 2.5-10% DMSO + 5-10% glycerol; 20% sucrose; 2.5-10% DMSO + 10-20% glycerol + 20% sucrose; 0.5M DMSO + 0.5M glycerol + 1M sucrose*; 10-12% DMSO + 5-8% glucose; 10-23% glycerol + 10-17% sucrose; 2M DMSO + 0.5M sucrose; 10% proline; 0.5M DMSO + 0.5M glycerol + 1M proline; 5-12% DMSO + 8-10% glucose; 5% DMSO + 1M sorbitol; 1M sorbitol; 20-40% trehalose; 5-10% polyethylene glycol ('PEG 6000'); 10% DMSO + 8% glucose + 10% polyethylene glycol]. The addition of cryoprotectants can be carried out at room temperature or at 4°C* and slowly* or in one aliquot. After a period of exposure of 1h* or, exceptionally, longer [up to 16h], the cells are (4) transferred to sterile polypropylene ampoules, sealed, labelled and transferred to the freezing unit (5). Slow cooling is carried out [0.1-2°C/min; 1 C/min*] until a suitable transfer temperature is reached [-23 to -100 C; -35°C*]. The specimens are then either held at that temperature [for 30-60 mins; 40 mins*] or transferred directly to liquid nitrogen. Warming(6) is carried out rapidly [in water at +40 C*] or, exceptionally, slowly [air at room temperature]. The cells are then transferred, usually without washing*, to semi-solid medium (7) [of normal formulation*; normal formulation but with cells supported on a filter; with a feeder cell layer]. Exceptionally, they can survive direct transfer to liquid medium. After one* or more passages on semi-solid medium, the cells can be reinoculated into liquid medium (8). (Sources in addition to those quoted in text: Sugawara and Sakai,1974; Weber et al., 1983. All % values are w/v except those for DMSO and glycerol, which are normally v/v.)

CRYOPRESERVATION PROCEDURES

Unlike slow growth, cryopreservation is applicable to all types of culture. Differences are seen between systems, with unorganised materials generally performing better than organised. The contrast between this pattern and that seen in slow growth derives from the very different stresses that are incurred in the two approaches to storage and the different risks of time-related deterioration. Cell suspension and shoot cultures have received most attention and are examined in detail below. Summaries of findings with callus, protoplast, somatic and zygotic embryo, anther and pollen embryo cultures are also included.

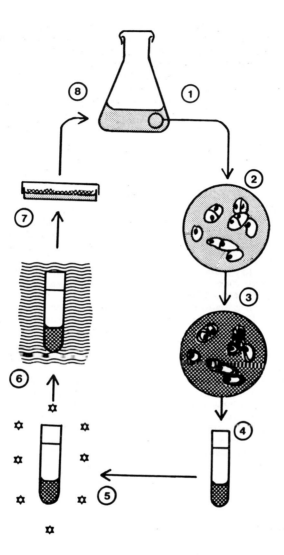

Cell suspension cultures

Cryopreservation techniques for cell suspension cultures have progressed since the first report in 1973 (Nag and Street, 1973) that described a procedure that owed its simplicity to the very amenable nature of the species under investigation, carrot. Since then, much has been learned about the role of the culture phases before and after freezing, the need for more complex cryoprotectant applications and the factors contributing to cellular injury. Figure 1 illustrates the framework of a cryopreservation procedure for cell suspension cultures and summarises the range of applications that can be used at each stage. The background to these practical details is given below.

Pregrowth

The importance of selecting the most suitable pregrowth stage is paramount. Cells in late lag phase or early exponential phase are superior to the larger, more highly vacuolated cells present elsewhere in the growth cycle (Withers and Street, 1977b). Several passages of frequent subculturing can be beneficial, although cell synchronization should be avoided (Withers, 1978; Withers and Street, 1977b). The influence of the duration of the penultimate passage upon the achievement of peak freeze-tolerance in the pregrowth passage is seen in rice, illustrating the importance of a good knowledge of the growth kinetics of the culture (Sala et al., 1979).

In some cases, the pregrowth passage can be carried out in unmodified medium but there is ample evidence that supplementation of the medium with additives as diverse as DMSO, mannitol or sorbitol, trehalose, sucrose and proline is beneficial (Bhandal et al., 1979; Butenko et al., 1984; Chen et al., 1984b; Kartha et al., 1982a; Withers and King, 1979; 1980). That such additives are effective is beyond question but their modes of action remain unresolved. Low levels (3-6%) of sugar alcohols are effective over relatively long periods of application (Withers and Street, 1977b). Progressive reduction in the mean cell size is observed in some but not all species (Diettrich et al., 1986; Pritchard at al., 1986a; Withers and Street, 1977b). Depression of the intracellular freezing point is similarly inconsistent in occurrence (Chen et al., 1984a; Mathias et al., 1985; Pritchard et al., 1986c). Other effects noted in some species include osmotic adjustment and changes in cell cycle duration, cell wall thickness and flexibility (the latter leading to cellular shrinkage rather than plasmolysis under further osmotic stress) and changes in total cell lipid, proline and soluble protein content (Pritchard, 1986b; 1986c). The efficacy of the short-term (up to 20h) application of a stronger concentration of sorbitol (Chen et al., 1984b) cannot be attributed to growth related phenomena and may either be strictly osmotic in action (reducing intracellular water levels) or conditioning in that it induces endogenous solutes with protective activity. A significant reduction in viability is noted after about 6 hours of exposure to high levels of sorbitol (Chen et al., 1984b). If survival is then calculated on the basis of the number of cells recorded as alive after pregrowth, it should be noted that some of the apparent enhancement in survival will be cosmetic as osmotically sensitive cells will have been eliminated before freezing.

Amino acids have been found to be effective pregrowth additives, again at two levels of application. At a level of 10%, proline was certainly acting osmotically if only in part (Withers and King, 1979) but more recent reports of the improvement of freeze tolerance by the application of much lower levels (0.01-0.05%) of proline and other amino acids suggest a more subtle protective effect, perhaps on membrane structure (Butenko et al., 1984; Heber et al., 1971). DMSO can be effective in pregrowth over 24h or longer, but cultures should be monitored for toxic effects (Kartha, 1985; Kartha et al., 1982a). The mechanism of enhancement of freeze-tolerance is unknown; cell loading and consequent enhancement of DMSO's cryoprotective effect are unlikely in view of its extreme permeability. Hardening by growth at a reduced temperature with or without the addition of ABA (Butenko et al., 1984; Chen et al., 1985; Keith and McKenzie, 1986; Orr et al., 1985) are less widely explored as possible pregrowth treatments and warrant further examination.

Cryoprotection

Mixtures of cryoprotectants, often of a very high total osmolarity are most effective. A common combination consists of two or three compounds including one or both of the classical cryoprotectants (DMSO or glycerol) with one or more sugars or sugar alcohols (Bhandal et al., 1985; Butenko et al., 1984; Diettrich et al., 1986; Finkle et al., 1985a; Hauptmann and Widholm 1982; Nag and Street, 1975a; Seitz et al., 1983; Withers, 1985a; Withers and King, 1980). Some question as to the need for DMSO and its possible toxic effects has been raised (Diettrich et al., 1986). In the author's experience, as long as an adequately pure grade is used (>99.5%) and stocks are maintained in the liquid state and replenished frequently, benefits outweigh disadvantages. Nonetheless, consideration should always be given to possible toxic effects that may contribute to the total damage experienced. The use of DMSO in pregrowth and cryoprotection is discussed in relation to its toxicity in a later section (see: Future developments in cryopreservation). Cryoprotectants have been applied successfully at room temperature or on ice, and either slowly or in one addition and even immediately before commencing cooling (Hauptmann and Widholm 1982; Withers, 1984b; Withers and King, 1980). Considerable flexibility is possible here but levels of uptake, plasmolysis and toxic effects will undoubtedly vary with mode of application. Sometimes very high levels of toxicity are experienced (e.g. 63%; Kartha et al., 1982a).

Cryoprotected cells are transferred to ampoules for freezing. Polypropylene ones are preferable to glass for reasons of safety. Glass has a higher thermal conductivity but this is relatively unimportant in slow cooling. Several reports quote cell densities (e.g. 50% settled cell volume; Watanabe et al., 1983) but no systematic study has been carried out to determine the optimal density of cells in the freezing ampoule. This has implications for behaviour during cooling and recovery growth. It can be beneficial to filter out large cell aggregates before freezing (Diettrich et al., 1986).

Cooling

Cooling has been carried out at a slow continuous rate or step-wise, but a net rate of 0.5–1°C/min is generally suitable. Seeding can be carried out at ca. –5°C but it has not been a routine feature of reported procedures. Slow cooling should be terminated at an intermediate temperature and transfer to liquid nitrogen carried out immediately or after a holding period (Withers and King, 1980). Viability may plateau or even fall if slow cooling is carried out to too low a temperature (Chen et al., 1984b; Diettrich et al., 1982; Withers, 1980b). A fall is attributed to either intracellular ice damage or excessive dehydration.

Warming

Cells are usually thawed by warming rapidly in warm water at +40°C A rapid transition through temperatures between –100°C and the thawed state is particularly important (Diettrich et al., 1985). Interactions have been observed between storage temperature and thawing rate (Nag and Street, 1975b), underlining the importance of storing at a suitably low temperature. Exceptionally, thawing can be carried out slowly and may even be beneficial (Withers, 1980b). However, it is not recommended unless there be reason to suspect that the cells have been excessively dehydrated during cryoprotection and cooling, rendering them susceptible to deplasmolysis injury.

Recovery

In early studies, cultures were invariably washed to remove potentially damaging cryoprotectants. The apparent logic of this is accepted. Unfortunately though, two factors work against this being of overall benefit. Firstly, the use of washing assumes that cryoprotectants have no role to play in the post-thaw situation. On the contrary, DMSO in particular may continue to be protective in its function as a free radical scavenger (Benson and Withers, 1987). Carbohydrates and amino acids may continue to stabilise membranes (Schwab and Gaff, 1986; Womersley et al., 1985). Secondly, washing is likely to trigger deplasmolysis injury, a widespread phenomenon in biological systems subjected to freezing and thawing. Evidence from animal tissue culture suggests that glycerol in particular may predispose cells to this lesion (Armitage, 1986; Penninckx et al., 1984). It appears that slow dilution may be beneficial (Armitage, 1986) and that some alleviation of injury may be achieved by controlling Na^+/K^+ balances and using an additional osmoticum such as sorbitol during washing (de Loecker et al., 1987; Penninckx et al., 1984). Slow dilution has been used with some success for plant cells and work with callus would suggest that warm washing may be less damaging (Finkle and Ulrich, 1982).

Despite there being possible alternative treatments, it would seem sensible to avoid washing and simply layer the cell suspension over semi-solid medium, allowing the cells to equilibrate osmotically, resorbing water and solutes without external interference. Data from studies of cells of rice and other species indicate that lesions including depletion of metabolite pools, impairment of respiration and weakness of the plasmalemma may require a recovery period of at

242

least two days (Cella et al., 1982; Diettrich et al., 1985). There is certainly a time lag of such duration or longer in the resumption of cell division. If a reduction in the concentration of cryoprotectants is felt to be essential, cells can be layered over a filter paper and moved periodically to plates of fresh medium (Chen et al., 1984b).

In view of the inevitability of a reduction in the viable cell density, compounded by deleterious influences of neighbouring dead cells, care should be taken to ensure an adequate inoculum density. Nursing with feeder plates or the use of conditioned medium may be beneficial (Hauptman and Widholm, 1982; Maddox et al., 1983). Once cells have recovered on semi-solid medium, they can be transferred to liquid culture. In one study, cells were inoculated directly into a liquid medium designed to support growth from a very low density (Caboche, 1980; Madox et al., 1983), as recovery on semi-solid medium resulted in the formation of cell aggregates that were difficult to break up when reinoculated into liquid medium. However, in the experience of the author, this is a very infrequent occurrence.

Freshly thawed and recovering cell cultures are often assessed by viability tests (Withers, 1986a) but growth is the most satisfactory criterion of recovery. Although viability tests can give a rapid indication of the number of surviving cells, there can be difficulties in clearly distinguishing between living and slowly dying cells and, in the case of the TTC test, changes in respiratory physiology can give the illusion of changes in the viable cell population (Cella et al., 1982; Diettrich et al., 1985). The assessment of qualitative aspects of recovery, i.e. stability, is discussed below (see: Stability in cryopreserved material).

Callus Cultures

Far less attention has been given to the cryopreservation of callus cultures. However, the general principles established for cell cultures apply; indeed, it is possible to fractionate a callus and subject a filtered suspension to a procedure designed for a cell culture. This has been done with great success by at least two research groups (Watanabe et al., 1983; Ziebolz and Forche, 1985). As well as avoiding the problems of poor cryoprotectant penetration, inadequate dehydration and physical damage that may be incurred by large masses, this approach has advantages in that it allows the concentration of superficial cells from aggregates, discarding central cells that are likely to be older, possibly senescent and less amenable to freezing.

In studies dedicated to callus, there have been some notable successes and the work is characterised by a concentration upon the use of a cryoprotectant mixture containing the polymer polyethylene glycol combined with DMSO and glucose (Finkle et al., 1985a; 1985b; Ulrich et al., 1979). It is not felt, however, that this is a special requirements of callus, rather that it reflects the efforts of one research group.

243

Protoplasts

A similar degree of neglect has been shown towards protoplast cultures and, again, the principles found with cell cultures apply (Bajaj, 1983; Hauptman and Widholm, 1982; Takeuchi et al., 1982; Withers, 1985c). The osmotic stabiliser present in the protoplast culture medium will add to the cryoprotective effect of any other additives used. The process of protoplast isolation may provide a stress filter, rather like some pregrowth treatments that presents a population with a relatively high freeze tolerance. However, other than in special cases of, for example, a leaf blade that would be virtually impossible to cryopreserve intact, or a protoplast preparation that for reasons of convenience is to be placed in suspended animation, or the preservation of evacuolate protoplasts from cells too large to freeze intact, little purpose can be seen in using protoplasts in this way. The problems encountered in relation to fragility of the plasmalemma should not be underestimated.

One of the most important use of protoplasts in the context of cryopreservation may turn out to be their potential as a model system to aid understanding of freezing injury and other cellular phenomena (Bartolo et al., 1987; Steponkus, 1984).

Shoot Cultures

Despite receiving a considerable amount of attention comparable to that given to cell suspension cultures, shoots have proven far more difficult to cryopreserve. No single procedure has emerged as superior and only the loosest of guidelines can be offered for the extrapolation of reported methods to untried specimens. It is likely that two or even three approaches may need to be tried and refinements then made empirically to develop a suitable procedure.

The magnitude of the task undertaken in attempting to cryopreserve a shoot culture should not be underestimated. It is clear on theoretical grounds alone that cryopreservation of an entire shoot culture or *in vivo* shoot would be a formidable task. Furthermore, as evidence accumulates, it appears that the cryopreservation even of a shoot-tip comprising the apical dome and several leaf primordia may not be possible without severe localised damage. Figure 2 attempts to summarise the procedures used with some success for the cryopreservation of shoot-tips. Most of the stages will be recognised from the cell suspension method but resemblances of detail are few.

Pregrowth

Dissection forms the first stage in all procedures. For shoot cultures this involves standard shoot-tip dissection. It is recommended that the specimen should not be reduced to a meristem-tip as this is likely to be prone to callusing, and would, in any case,

244

be very demanding to prepare in reasonable numbers. Where the starting material is budwood taken from a tree or a herbaceous *in vivo* shoot, dissection must be allied to surface sterilization and inoculation into culture. As shown in Figure 2, it may be logical to multiply material originating *in vivo* by several passages of shoot culture to provide adequate explants and ensure the absence of contaminants. Much of the work on the cryopreservation of shoot-tips has involved material of *in vivo* origin. This should be borne in mind when adopting reported methods as such material is likely to be more amenable to cryopreservation than *in vitro* counterparts.

One report cites the cold pretreatment of an entire culture (Kuo and Lineberger, 1985); otherwise such hardening treatments have been confined to source plants (Katano et al., 1983; Sakai and Uemura, 1982; Seibert and Wetherbee, 1977). Pregrowth for a period of 1–4 days on standard or modified medium is a common but not universal treatment (e.g. Grout and Henshaw, 1977; c.f. Kartha et al., 1982b). It may involve incubation at a reduced temperature for part of the time (Henshaw et al., 1985). In potato, it has been found that the duration of the pre growth phase can be critical, an optimum being reached at 24 - 96h depending upon the genotype. For dormant buds and similar material, a pregrowth passage might be detrimental by 'de-hardening' the explant. Suitable pregrowth additives include DMSO, mannitol and sorbitol.

Cryoprotection

Dormant buds appear to be able to survive cryopreservation without any cryoprotection; otherwise, it is essential (Katano et al., 1983; Sakai, 1985). DMSO is effective when used alone far more often than is the case for cell suspensions. Some cryoprotectant mixtures have been used successfully, as given in the legend to Figure 2.

Cooling

Major divergences in treatment are introduced at the post-cryoprotection stage with respect to enclosure of the specimen and the rate of cooling employed. Enclosure in an ampoule clearly is beneficial in terms of convenience and security but there is evidence that direct exposure of the specimen to the coolant can enhance survival, presumably by enhancing the rate of heat extraction. This can involve carrying the shoot-tip on a hypodermic needle (Grout and Henshaw, 1977; Withers, 1982b), cover glass (Uemura and Sakai, 1980), on a piece of cellulose tissue or fabric (Henshaw et al., 1985), or in droplets on a metal foil sheet (Kartha et al., 1982b). Nonetheless, relatively rapid freezing procedures have involved the use of specimens in ampoules. The author has found that a thin-walled, polypropylene ampoule with a tapered base, containing a very small amount of liquid can cool adequately rapidly.

Suitable slow cooling rates vary considerably from species to species and individual requirements can be very precise with viability dropping very severely at too rapid or slow a rate.

*Figure 2.(opposite) Cryopreservation procedure for shoot–cultures
indicating the variables available at each stage. Compatible
treatments from two reports are indicated: *Kartha et al., 1979;
**Sakai and Nishiyama, 1978. Terminal or axillary buds are
dissected from the source material (1). These are used to
establish a shoot culture (2), from which shoot–tips are dissected
and inoculated into the pregrowth phase (3) on standard or
modified medium [plus 4–5%* DMSO; 1M sorbitol] at room
temperature* or at a reduced temperature. Alternatively,
shoot–tips can be dissected directly from source plants* (4) which
may be naturally** or artificially cold hardened. After pregrowth
[1, 2*, 4 days] the explants are (5) treated with cryoprotectants
[4–16% DMSO; 5% DMSO*; 5–10% glycerol; 5% DMSO + 5% glycerol; 5%
DMSO + 5% glycerol + 5% sucrose; 10% glycerol + 10% glucose + 10 %
polyethylene glycol; 10% DMSO + 5% glucose; 10% DMSO + 5% sucrose;
10% DMSO + 1M sorbitol] on ice* or at room temperature for
approximately 1 h. Exceptionally, explants can be cryoprotected
without pregrowth (6) or frozen without pregrowth or cryoprotection**(7). Explants are transferred to ampoules (8)* ** or
placed on a carrier (9) [hypodermic needle; cover glass; cellulose
fabric or paper; foil]. Cooling is carried out (10) slowly
[0.2–1°C/min; 0.6°C/min*; unspecified rate**], or (11) rapidly
[ampoule or carrier directly immersed in liquid nitrogen; rates
between 50 and >1000°C/min]. After storage in liquid nitrogen,
warming is carried out (12) rapidly [in water at +40°C* for
ampoules; in medium at +20 to +40°C for specimens on carriers], or
slowly [in air at room temperature**]. Explants are then
transferred [with* or without** washing] to (13) semi–solid medium
of normal* composition or with hormone supplements. When
appropriate, explanted buds can be grafted onto rootstocks (14)**.
Sources in addition to those quoted in the text: Bajaj, 1977;
1979; Grout et al., 1978; Kuo and Lineberger, 1985; Moriguchi et
al., 1985; Reed and Lagerstedt, personal communication; Sakai et
al., 1978; Withers and Benson, unpublished observations. All %
values are w/v except for DMSO and glycerol which are normally
v/v.*

A rate of 0.84°C/min is optimal for strawberry (Kartha et al., 1980),
0.6°C /min for pea (Kartha et al., 1979) and 0.27°C/min for pear 1987
(Wanas, 1987; Wanas, Withers and Callow, unpublished observations).
Optimal transfer temperatures vary similarly. The upper limit of
suitable transfer temperatures is likely to be raised by hardening
and other beneficial pregrowth treatments, indicating that hardened
material has a lesser requirement for protective dehydration (Sakai,
1986).

Rapid cooling in the region of 50°C/min and ultra rapid cooling at
>1000°C/min have been used less frequently than slow cooling but they
can be dramatically effective in terms of net survival and the
preservation of organisation (Grout et al., 1978; Seibert and
Wetherbee, 1977; Withers, 1982b; unpublished observations).

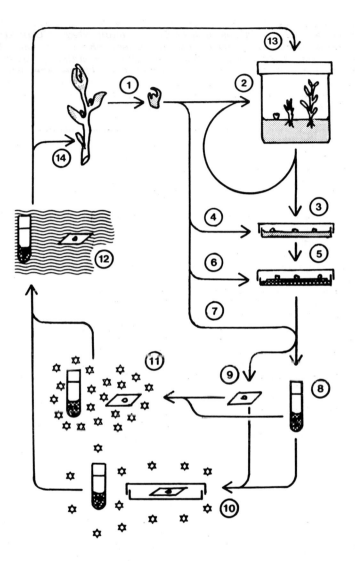

Considerable flexibility in cooling requirements is observed in studies of potato with, somewhat paradoxically, strong variations in response between varieties. Ultrarapid, rapid and slow cooling have been successful (Grout and Henshaw, 1977; Henshaw et al., 1985; Towill, 1981; 1983; 1984; Withers, unpublished observations). The necessity for seeding in this material is unclear. It has been reported that seeding is essential to avoid the damage incurred when a supercooled specimen seeds spontaneously (Keefe and Henshaw, 1984). However, the author has found that rapid cooling from 0°C or as low as -40°C without seeding (induced or spontaneous) is superior to cooling from within this range but after seeding (Withers, unpublished observations). A tendency for all of the specimens in one ampoule to behave similarly, yet differently from other ampoules has been noted (Henshaw et al., 1985; Withers, unpublished observations). This may relate to a phenomenon such as seeding.

Warming

After storage, rapid thawing is most commonly used, although slow thawing can be successful in material that has been dehydrated adequately during slow cooling (Katano et al., 1983; Moriguchi et al., 1985; Uemura and Sakai, 1980).

Recovery

The question of benefits and disadvantages of post-thaw washing is less clear than for cell suspension cultures; unenclosed specimens will inevitably be exposed to liquid medium as a part of the thawing process. Unless there are clear indications to the contrary, though, it is felt that washing should be avoided or at least minimised.

The composition of the recovery medium is of the greatest importance for shoot-tips as the object is to both maintain viability and induce the desired pattern of growth. Attempts to determine the extent of damage in recovering shoots indicate that the base of the shoot-tip is massively damaged and that recovery ensues from a relatively small apical region, often from just the leaf primordia (Grout and Henshaw, 1980; Haskins and Kartha, 1980; Withers, 1985a). As a result, standard medium formulations are unlikely to be adequate. This is borne out by many observations of the need to supplement the medium with hormones. In many cases, recovery commences with the formation of a callus mass, followed by adventitious shoot production. This is a far from satisfactory situation due to the risk of somaclonal variation. Direct shoot formation is a rare occurrence and seems more frequent in woody than herbaceous material (Katano et al., 1983; Wanas, 1987; Wanas, Withers and Callow, unpublished observations). The example of potato illustrates the problem that can be encountered in achieving adequate quantity and quality of regeneration. A high level of recovery of up to 100% can be achieved by supplementing the culture medium with an auxin (NAA), but extensive callusing is caused. When auxin is omitted from the medium, the level of recovery is dramatically lower at ca. 50% but callus was entirely absent. The best recovery in terms of quantity combined with quality (40-60% plant regeneration) occurred in the presence of gibberellic acid (Withers and Benson, unpublished observations). A further significant observation with potato, as well as other species, is the link between organised recovery and a rapid rate of cooling either from the unfrozen state or from an inter-mediate transfer temperature (Henshaw et al., 1985; Withers, 1982b; Withers unpublished observations).

248

A strong interaction between rate of cooling and frequency and pattern of recovery has also been noted in shoot-tips of oil seed rape (Withers, unpublished observations). Ultra-rapidly frozen specimens recover very well on both plain and hormone-supplemented medium, although organised growth resumes only on the former. After slow cooling, no survival was observed on plain medium but a level of survival insignificantly different from the ultra-rapidly frozen material was found on the hormone-supplemented medium. Clearly, there are two types of survival that differ in qualitative terms. They are revealed by their differing responses to a less supportive recovery medium. A hormone-supplemented medium will pull through surviving cells despite a degree of tissue disorganisation that would render the shoot-tip unable to recover on plain medium. Furthermore, the location of the cells in the apex is less likely to be critical; virtually any region of the shoot-tip will probably respond to hormonal support. In contrast, an unsupplemented medium will filter out specimens with more than a certain level of organisational disruption.

Linkage between the pattern of recovery and cooling rate may relate to the differing mechanisms of survival. For example, a rapidly frozen shoot-tip is unlikely to have suffered the plasmolysis damage and consequent disruption of the symplasm that would be incurred by a slowly frozen shoot-tip with a similar level of cell survival.

The question must be asked as to what extent it is wise to support recovery growth from a low level of survival. Such a situation almost certainly involves recovery from very few cells that have been subjected to severe stress and that will be exposed to high levels of hormones. There are serious implications for genetic stability.

In woody temperate fruits, organised recovery has, until recently, been found only in material dissected directly from plants rather than cultures. Recovery has been achieved both *in vitro* and by budding onto rootstocks (Katano et al., 1983; Sakai and Nishiyama, 1978). In one report, material originating *in vitro* would only produce callus after cryopreservation (Kuo and Lineberger , 1985). However, recent studies using pear have shown that organised recovery can ensue from cultured material when the base level of survival is high (Wanas, 1987; Wanas, Withers and Callow, unpublished observations).

Somatic Embryos

Two isolated studies of the cryopreservation of somatic embryos illustrate important practical points that might be translated into use with other systems. Somatic embryos of carrot are relatively unresponsive to methods found successful for cell suspensions of the species. As the embryos mature they become increasingly prone to viability loss and incur damage in an increasingly large proportion of the embryo structure (Withers, 1979; Withers and Street, 1977a).

249

Eventually, a point is reached when recovery only occurs from secondary embryogenesis in superficial cells. However, if the embryos are drained free of the suspending cryoprotectant (DMSO) solution, blotted dry on filter paper, enclosed in a foil envelope, cooled slowly and then thawed slowly, very high levels of recovery can be achieved. Strong interactions are seen between cooling rate, transfer temperature and warming rate, showing that survival is achieved by ensuring the correct degree of protective dehydration. Drying in a desiccator before freezing can aid the dehydration process. Rapid thawing is in fact deleterious, possibly through causing gross physical damage. The frozen material is very brittle and is vulnerable due to the size of the unit specimen and the lack of supporting frozen medium. Overly rapid rehydration resulting in expansion-related injury may also be occurring in rapidly thawed material. Recovery in the presence of activated charcoal enhances organised growth, presumably by adsorbing compounds with hormone-like activity produced by lethally damage tissue regions. Beneficial cryoprotectant adsorption may also be occurring.

Somatic embryos and embryogenic clusters of callus of oilpalm have been cryopreserved very successfully by another method that diverges from conventional procedures. The material is pregrown for two months in the presence of a relatively high level of sucrose (0.3M), rising to 0.75M in the 7 days immediately before freezing. Cooling is carried out either rapidly or slowly without further cryoprotection (Engelmann, 1986; Engelmann et al; 1985). Recovery on semi-solid medium with a reducing level of sucrose (0.3M for 1 week, then 0.1M) proceeds by continued development and embryogenesis. Embryogenic clusters recover without hormonal support but embryos require the presence of 2,4-dichlorophenoxyacetic acid (2,4-D).

Zygotic Embryos

Cryopreservation of zygotic embryos has two aspects in relation to whether the original seed is orthodox or recalcitrant. Embryos of many orthodox seeds with a relatively low water content can be cryopreserved without serious difficulty (Stanwood, 1985; Styles et al. 1982; Withers, 1987a). Even immature orthodox embryos and some imbibed mature or even germinating embryos respond to cryo-preservation given some support by cryoprotection or re-drying (Grout, 1979; Lockett and Luyett, 1951; Sun, 1958). In cereal embryos, the pattern of regeneration after thawing can be manipulated to some extent by the cooling procedure used. Rapid cooling tends to favour survival of the embryonic axis, whereas slow cooling favours recovery via callusing from the scutellum (Withers, 1982b). These differences, enhanced of course by the use of appropriate media, reflect those seen in shoot-tips exposed to different cooling regimes.

A case can be made for committing seed stocks in the form of entire seeds to storage in liquid nitrogen, for security, stability and a reduced need for regeneration (Stanwood, 1985). However, other

250

than for the provision of inoculum material for *in vitro* culture, little purpose can be seen in using cryopreservation for orthodox seed embryos. In contrast, though, cryopreservation of embryonic axes is likely to make an important contribution to the genetic conservation of species that produce recalcitrant seeds (Grout, 1986). Studies of truly recalcitrant seeds have been few. Citrus and oilpalm have in the past been classified as recalcitrant, but this classification has now been revised. Nonetheless, they still present handling problems due to unusual drying and or germination characteristics and a study of their cryopreservation has been instructive (Grout et al., 1983; Mumford and Grout, 1979).

Embryonic axes of rubber, a truly recalcitrant species, can be cryopreserved if desiccated to a water content of 14–20%. Survival levels of 20–69% have been recorded (Normah et al., 1986). Another recalcitrant species, *Araucaria hunsteinii*, has proved more difficult to handle. A relationship between moisture content and freezing tolerance has been observed but, as yet, recovery growth has only ensued in the root meristem region (Pritchard and Prendergast, 1986).

Anthers and Pollen Embryos

Some attention has been given to the cryopreservation of anthers and pollen embryos but few generalisations can be made (Bajaj, 1980; 1981; 1982; 1984; Withers 1985b; 1987a). It appears that the induction of androgenesis is beneficial, although pollen embryos show declining survival potential with increasing development, as in the case of somatic embryos. Bisected pollen sacs yield more surviving embryos than do intact ones, suggesting problems relating to cryoprotectant penetration or water flux. If attempting to cryopreserve an untried species, a sensible approach would be to devise a slow cooling/rapid warming procedure using cryoprotectants found to be suitable for other types of culture of the species. Recovery on a standard medium, possibly with the addition of activated charcoal is recommended.

Stability in cryopreserved material

The main concern in cryopreservation is damage incurred during cooling and warming. Against this, cumulative damage during storage is of very little consequence as long as suitable conditions are maintained. Therefore, whilst it can give confidence to quote data on the duration of storage, this should be of far less interest than those relating to the viability and quality of recovered material. Until relatively recently, qualitative aspects of recovery were neglected. This remains largely the case for organised cultures. Isolated examples to the contrary are the field testing of regenerant plants from cryopreserved cultures of oilpalm (Engelmann, personal communication; Grout et al., 1983), strawberry (Kartha, personal communication) and potato (Withers, unpublished observations). However, as storage procedures have been refined, making assessments of qualitative factors feasible, and as progress has been made in the cryopreservation of cell cultures of importance in biotechnology, more emphasis has been placed on stability. Two aspects warrant consideration here, genotypic differences in response to storage and genetic stability in storage. The two are not unrelated in that if the different components of a heterogeneous population respond differently to cryopreservation, the composition of the recovering culture may be genetically different from the original.

Vast differences in levels of survival between species and genera have been reported but it is difficult to draw firm conclusions about the genotypic contribution to survival potential in broad terms. Until a greater body of data accumulates relating the responses of a range of material to a given set of cryopreservation conditions, the question will remain largely unresolved. A few observations do imply differences in response between related genotypes but it is suggested here that they may just be amplifications of differences that might be expected in general tissue requirements between genotypes. In a comparison of the performance of a range of 6 species and 10 cultivars of potato, differences were found in overall survival levels (0-96%) and in the tendency to form shoots from the callus that regenerated initially (0-70%) under identical recovery conditions (Towill, 1984). A cryopreservation method found to be successful for a low secondary product synthesizing line of *Catharanthus roseus* was not effective for a high yielding culture; a modified pregrowth treatment was required (Kartha et al., 1982a; Chen et al., 1984b). Differences in cell size and degree of vacuolation were probably sufficient to explain the differing requirements. Different line of *Digitalis lanata* have been found to have differing cryopreservation requirements (Diettrich et al., 1986).

Despite one report of a slight, progressive increase in freezing tolerance in a repeatedly cryopreserved culture (Watanabe et al., 1985), data on stability in unorganised cultures stored by cryopreservation are very encouraging, in contrast to the observations of slowly grown callus (see above). Nuclear DNA contents and chromosome numbers have been found to be unchanged when examined (Diettrich et al., 1982; 1985; Nag and Street, 1973). In the case of cell and callus cultures with a high level of cell survival, normal growth characteristics can be resumed within one or two passages after thawing (e.g. Sala et al., 1979; Ulrich et al., 1982). The retention of embryogenic or shoot regenerating capacity has been confirmed in relevant cases (e.g. Diettrich et al., 1985; Nag and Street, 1973; Ulrich et al., 1982). Antimetabolite resistance and secondary product synthesis have been examined in several species. In the case of the former, retention of resistance and, where tested, dose responses have been confirmed in cryopreserved material (Hauptmann and Widholm, 1982; Strauss et al., 1985). In the case of the latter, all examples studied indicate that synthetic capacity is retained by cryopreservation. The compounds and species examined include anthocyanin in carrot (Dougall and Whitten, 1980), biotin in lavender (Watanabe et al., 1983), alkaloids in *C. roseus* and transformed cardenolides in *D. lanata* (Butenko et al., 1984; Seitz et al., 1983). The biotransforming properties of a cell line of *D. lanata* have been tested in small and large scale culture (Seitz et al., 1983). All aspects of culture growth kinetics, nutrient and substrate consumption, and product accumulation appear to be stable.

Some slight question might be raised concerning the quality of secondary products. One reassuring report (Butenko et al., 1984), is balanced by another suggesting that the composition of a mixture of secondary products differed from controls, probably as a result of changes in the cell population (Chen et al., 1984b). An aspect of stability that is similar to the latter point in that it relates to the contribution of cell types to a specimen is that of chimeral stability in shoots. *In vitro* culture can lead to the breakdown of chimeras (e.g. Cassells, 1985; Rosati et al., 1986). Preliminary findings with the potato variety Golden Wonder indicate that cryopreservation may enhance the rate of chimeral breakdown (Withers, unpublished observations).

Accepting the caution above, findings of stability studies conducted to date generally bear out reports from microbial and mammalian systems (see above) and give cause for optimism. However, it must be said that the answers found in studies of stability will only be as meaningful as the questions posed. Thus, the retention of the synthesis of a given compound suggests that the region of the genome governing the relevant processes has not been changed, but it does not carry the assurance that all other regions of the genome are equally stable; it only indicates the probability of that being so. Data on stability of isozymes (Ulrich et al., 1982), are similarly encouraging but limited. To gain a comprehensive picture, studies of the total genome using appropriate techniques such as analysis of restriction fragment length polymorphism and the use of nucleic acid probes will be needed.

To evaluate cryopreservation properly, any level of instability detected must be viewed against the background of, firstly, the level of inherent instability in the type of culture in use and, secondly, the level of cumulative variation that might be expected in alternative storage modes such as normal or slow growth. The security and saving in resources achieved by cryopreservation are also part of the equation, as cultures stored otherwise might have to be sacrificed due to demands on resources or even lost accidentally whilst in routine maintenance.

FUTURE DEVELOPMENTS IN CRYOPRESERVATION

The Development of Routine Methodology

Although cryopreservation has seen great progress over the past decade, there is still considerable room for improvement. In the case

253

of cell suspension cultures, we are still not at the point where an untried species could be cryopreserved without some experimentation, although it is possible to produce a relatively brief list of procedural variations to test within a standard protocol. The situation for shoot cultures is more complex and is likely to remain so. The best that can be envisaged is the devising of a method-ological key through which to work by testing a limited number of variables at each stage.

Drawing upon developments that have taken place in plant cryopreservation and other aspects of cryobiology, some thoughts are now offered upon the ways in which procedures might be improved by attention to the various stages of cryopreservation procedures.

Choice of Culture System

In general, the scientist will have little choice as to which culture system to use, particularly with respect to its being organised or unorganised. However, within the constraints operating, some choices can be made. A cell suspension culture is likely to be much more amenable to cryopreservation than a callus culture. A well established shoot culture is likely to be more freeze-tolerant in the case of a herbaceous species and less freeze-tolerant in woody material. In both cases, explants dissected from germinated seeds, seedlings or established plants are likely to be more freeze-tolerant than cultures. Where the initial choice is for shoot cultures, the option of using a highly morphogenic (shoot-producing or, preferably, embryogenic) callus or cell culture might be taken. This would be with the view to taking advantage of the amenability of the culture to cryopreservation and balancing that against the risk of somaclonal variation, which would relate in magnitude to the length of time that the material has been in unorganised form.

Pregrowth Treatments

It is clear that modification of pregrowth conditions can enhance freeze-tolerance greatly in both unorganised and organised material. However, the underlying mechanisms are far from understood. This is particularly the case for shoot-tips where the relative contribution of healing from dissection damage and physiological change akin to cold hardening is unknown. In comparing the performance and procedural demands of temperate fruit buds from hardened trees with that of unhardened material or *in vitro* cultures, it is clear that cold hardening in the conventional sense has a profound effect upon freeze-tolerance (Sakai, 1986). Efforts might, therefore, be directed more towards trying to induce a state of cold hardiness or even dormancy, drawing upon a knowledge of the characteristics of the species in question and general knowledge of plant physiology. The relatively recent use of ABA in pregrowth treatments is a promising

254

development in this respect. The compound cartolin, a cytokinin, has been used to promote induced dormancy in several woody species (Kuzina et al., 1984); it might be tried with *in vitro* material. The phenomenon of cross-protection between stresses is probably involved in the beneficial effect of osmotic stress during pregrowth. Heat stress can confer cold tolerance (Wu and Walner, 1985) and could be investigated in the context of cryopreservation.

Differences in freeze-tolerance in relation to cell cycle stage have been observed in animal and plant cells (McGann and Kruuv, 1977; Withers, 1978). Perturbations in progress through the cell cycle may be implicated in the generation of somaclonal variation (Gould, 1984). It might be helpful if a reproducible and readily applied method were available to arrest cells at a particular stage of the cell cycle to minimise damage. Indeed, this may be one of the mechanisms whereby some pregrowth treatments operate.

The use of the developing technology of artificial seed production is an attractive possibility. As yet, the specimens encapsulated during artificial seed preparation have a relatively low tolerance of drying (Kitto and Janick, 1985a; 1985b), thus limiting the duration of storage before 'germination'. However, these studies have shown the material to be amenable to treatments such as the application of ABA. Loading the specimen in advance with protective solutes such as proline or glycinebetaine also might enhance its tolerance of drying at this stage and during cooling (Lone et al., 1987; Withers and King, 1979). Beneficial effects of cinnamic acid, chlormequat (CCC) and sodium dikegulak upon artificially aged seeds may have relevance here too (Bhattacharjee and Choudhuri, 1086). A collaboration between scientists working in this field and in cryopreservation would be mutually beneficial, either through the development of better drying procedures, the introduction of a cryopreservation stage in the artificial seed production process, or both.

Cryoprotection

Improvement in other aspects of the cryopreservation procedure, notably pregrowth, may lead to the situation where the cryoprotection stage can be omitted for certain amenable specimens, as is now the case for oilpalm (Engelmann et al., 1985). However, for the majority of subjects it is likely that cryoprotection will remain a necessity. Concern has been expressed over the possible dangers of using DMSO as a cryoprotectant. Its use as a pregrowth additive amplifies concern here as much longer periods of exposure are involved. The undoubtedly beneficial properties of DMSO cannot be set aside; indeed, it is unlikely that any progress would have been made with the cryopreservation of cultured shoot-tips without this compound. However, it is felt that attention should be given to minimising the duration of periods of exposure and concentrations used. Rapid addition and dilution of DMSO at relatively elevated temperatures has recently been shown to be effective for rabbit embryos (Kojima et al., 1987).

255

There is some evidence that problems experienced in relation to viability relate to purity (Matthes and Hackensellner, 1981; Withers, unpublished observations). That matter is relatively easily resolved; however, effects of DMSO itself rather than accompanying contaminants still have to be considered. Reports of the induction of division in protoplasts (Hahn and Hoffmann, 1984, multinucleation (Fukui and Katsumaru, 1980), structural disturbance of microfibrils and microtub- ules (Hahn and Hoffmann, 1984; Vannini et al., 1985), conformational changes in macromolecules (Nilson, 1980), disturbance of protein synthesis (Sibly et al., 1977), breakdown of cell compartmentation (Delmer, 1979) cannot be ignored. Some of these phenomena will be particularly consequential in the shoot-tip, where organised development is a highly complex process dependent upon tight regulation of the sequence and orientation of cell division, itself dependent upon aspects of the cellular architecture (Green, 1986; Lyndon and Cunninghame, 1986). Examples of gene activation and repression, virus activation and structural alterations in DNA occurring under DMSO concentrations and conditions of exposure similar to those used in cryopreservation give further cause for concern (Ashwood-Smith, 1985).

The option of combining several cryoprotectants in low concentrations to give a high net cryoprotective effect could be explored further, and new cryoprotectant compounds, antioxidants (Matthes et al., 1981) and other cryoprotectant toxicity neutralisers (Fahy, 1983) might also be enlisted. Trehalose has recently been shown to have promising cryoprotectant effects (Bhandal et al., 1985). This and other membrane stabilising carbohydrates could be tested.

Cryoprotectants are clearly more effective when prepared in culture medium rather than water (Withers, 1980b). However, beyond that, little has been done to investigate the effects of organic and mineral components of the culture medium upon survival. Evidence from animal systems would suggest that, as mentioned earlier, Na^+/K^+ balances might be investigated. Ca^{2+} ions have been shown to help stabilise plant membranes under conditions of cryopreservation (Sugawara and Sakai, 1975). A further prophylactic measure might be to support energy metabolism through monitoring and, if necessary, boosting ATP levels (Southard et al., 1985).

Preparation for Freezing

Packing effects have been observed in cryopreserved animal cells (Pegg et al., 1984). This has not been examined for plant cells where similar phenomena might well occur in addition to the effects of cell density upon reinoculation onto recovery medium. A related phenomenon is that of the beneficial effect of dry freezing in air rather than cryoprotectant solution. Interestingly, the very different system of the mammalian cornea responds well to dry freezing (Taylor and Hunt, 1985). This approach and droplet freezing (Kartha et al., 1982b) could be explored further for use with all types of plant culture.

The encapsulation of specimens in wafers of a water soluble resin as used to enclose somatic embryos of carrot (Kitto and Janick, 1985a) might be used to overcome some of the practical problems of ultra-rapid freezing. The wafer could provide a stable but highly thermally conductive support. Microcarriers have been used successfully in the cryopresevation of animal cells (Duda, 1982). Similar structures might be made from sodium alginate or a similar material as used to immobilise plant cells.

Cooling and Warming

There seems to be little scope for modifying cooling rates for cell suspension cultures outside the limits of current experimentation. However, shoot-tips remain problematic as they appear to be able to adopt various survival strategies. There is no clear concensus as to the general, let alone specific rate that gives superior survival. The potato shoot-tip system may be a suitable model for comparing the cellular and tissue responses to different rates at the electron microscopical level, thereby shedding some light upon differences in survival mechanisms and their consequences for organised recovery.

One approach that has been virtually neglected for plant material is that of using vitrification. This has been pursued for mammalian embryos to try and provide a cryopreservation method that preserves intercellular architecture (Rall and Fahy, 1985). Conventional cryopreservation techniques lead to the loss of important regions of complex structures. Vitrification involves the application of very high concentrations of cryoprotectants that can supercool to very low temperatures and then solidify without ice formation.

It has already been pointed out that slow warming can be beneficial in systems that have achieved an adequate degree of protective dehydration. Some scope for modifying warming rates lies in the use of containers of different material and different dimension and in the use of water baths of different temperatures. Other than one reported suggestion of overheating damage (Finkle et al., 1985b), little is known of the effects of varying the terminal thawing temperature. This could be investigated further, especially in the light of the finding that warm washing can be less damaging than cold (Finkle and Ulrich, 1982). Microwave warming has been explored for animal organs where satisfactory thawing of deep tissues can be difficult to achieve. Little benefit can be envisaged for plant material except, perhaps, in the case of relatively large recalcitrant seed embryos. However, for the latter, it may be more logical to cryopreserve the embryo shoot apex rather than the entire structure.

Post-thaw Treatments and Recovery

The advantages and disadvantages of post-thaw washing and interactions between washing effects and cryoprotection, cooling and warming rates require further exploration. The freshly thawed specimen is likely to be suffering from the interruption and uncoupling of metabolic processes, the presence of potentially damaging cell breakdown products and other molecular species with free radical activity. The effects of neighbouring dead and dying cells upon potential survivors could be investigated by the use of artificial mixtures of cells. One report suggests that in such mixtures a viable cell frequency of at least 30% is necessary for recovery (Diettrich et al., 1982).

The level of free radical activity in cryopreserved material has recently been investigated using gas chromatographic analysis of volatile hydrocarbons. Stress-induced ethylene, various membrane breakdown products and methane produced from DMSO by free radical scavenging have been monitored (Benson and Withers, 1987). Variable and sometimes very high levels of evolution are found at all stages of the cryopreservation procedure, and even in normal culture growth. This novel analytical approach should facilitate the comparison of cryopreservation procedures and help identify beneficial treatments.

The composition of the post-thaw recovery medium has been discussed at length earlier and little can be added here beyond saying that attempts should be made to minimise the generation of somaclonal variation. The post-thaw environment in terms of lighting and temperature has not been investigated in depth, although there has been a suggestion that low light levels are beneficial in the recovery period (Henshaw et al., 1985).

Having offered various possible means of improving cryopreservation procedures, it must be said that only by a serious increase in research effort will significant progress continue to be made. Contributions to date have been from a very small number of scientists, for many of whom cryopreservation has emerged as an essential adjunct to other biotechnological applications of *in vitro* culture. The proportion of research dedicated to cryopreservation *per se* has been minimal.

STRATEGIES OF *IN VITRO* CONSERVATION

The International Board for Plant Genetic Resources (IBPGR), one of the CGIAR International Centres has given attention to the development of an *in vitro*-based set of procedures that parallels the conventional way of handling plant germplasm (Withers and Williams, 1986). Tentative guidelines for the design, planning and operation of *in vitro* genebanks have been proposed (IBPGR, 1986). Those relating to material in slow growth are now under test in a joint project between IBPGR and the Centro International de Agricultura Tropical (CIAT) in Colombia using cultures of cassava. This project should provide useful information relating to maintenance procedures, stability monitoring and the development of culture management databases. Information relevant to *in vitro* genebank maintenance can also be found in the literature on microbial collections (Hunter et al., 1986; Kirsop, 1986; Wolff et al., 1986).

As indicated earlier, storage is just one aspect of conservation. The adoption of *in vitro* storage as a potential means of resolving problems with vegetatively propagated and recalcitrant seed producing material has generated interest that has led to the consideration of *in vitro* techniques in other contexts. Difficulties are experienced in collecting certain types of germplasm in the field. By the use of simplified *in vitro*-based methods, explants such as shoot cuttings and recalcitrant seed embryos can be put into a 'holding culture' until they can be returned to a genebank (Withers, 1987c).

Once collected, germplasm needs to be disease indexed and, if necessary, treated to eliminate pathogens. The use of *in vitro* techniques to clean up material is well know (Kartha, 1986) but indexing *in vitro* is a relatively under–developed concept. However, examples of the use of cultures as indicators of disease show how quarantine and the safe distribution of germplasm could be expedited by such means (Withers, 1987b). The use of *in vitro* culture to produce new genotypes has led to the development of screening procedures for stress, herbicide and disease resistance. These same tests could be used to screen germplasm collections for useful traits, thereby making the collections more useful to the breeder (Withers, 1987b). Finally, mention should be made of the now relatively well developed and tested methods of germplasm distribution in the form of *in vitro* cultures. In both the commercial and public sectors, material is moved internationally in the form of *in vitro* shoots, plantlets or mini–tubers (Withers, 1987b; Withers and Williams, 1985).

All of the strategies of *in vitro* conservation considered in this chapter have centred upon the use of cultures consisting of entire cells, tissues or organs. However, the consideration of germplasm storage would not be complete without giving attention to the possibility of developing 'gene libraries' (Peacock, 1987). Extracted, purified and partially enzyme digested DNA could be stored by freeze drying and then made available for use by gene transfer methods that are now under development. We are a long way from the situation when an entire genome could be conserved in this way but it should be possible for the majority if not all of the genes to be recoverable. If other approaches fail, DNA storage may eventually be the only feasible way of conserving certain problem materials. However, in the author's opinion, good *in vitro* conservation, particularly by cryopreservation, depends upon good general competence in culture methodology for the genotype in question. The utilisation of stored DNA sequences is likely to require a similar degree of *in vitro* competence. Thus, we may find that the conservation problem disappears once the material becomes amenable to culture. This is not to dismiss DNA storage as an important conservation option. Its potential convenience, security, compactness and relevance to developing techniques of *in vitro* breeding indicate that it clearly has a place in plant biotechnology. What is essential is that this approach to conservation, like all others, should be adopted when its use is logical and beneficial.

REFERENCES

AGEREAU, J.M., COURTOIS, D., PETIARD, V (1986). Long-term storage of callus cultures at low temperatures or under mineral oil layer. Plant Cell Rep. 5, 372–376.

AKIHAMA, T. and OMURA, M. (1985) Preservation of fruit tree pollen. In Biotechnology in Agriculture and Forestry. Ed. by Y.P.S.Bajaj, Springer Berlin, pp. 101–112.

ARMITAGE, W.J. (1986) Osmotic stress as a factor in the detrimental effect of glycerol on human platelets. Cryobiol. 23, 116–125.

ASHWOOD–SMITH, M.J. (1985) Genetic damage is not produced by normal cryopreservation involving either glycerol or dimethyl sulphoxide: a cautionary note, however, on possible effects of dimethyl sulphoxide. Cryobiol. 22, 427–433.

ASHWOOD–SMITH, M.J. and FRIEDMAN, G.B. (1979) Lethal and chromosomal effects of freezing, thawing, storage time, and X-irradiation on mammalian cells preserved at -196 °C in dimethyl sulphoxide. Cryobiol. 16. 132–140.

ASHWOOD–SMITH, M.J. and GRANT, E. (1976) Mutation induction in bacteria by freeze drying. Cryobiology 13, 206–213.

BAJAJ, Y.P.Ş. (1977) Clonal multiplication and cryopreservation of cassava through tissue culture. Crop Improv. 4, 198–204.

BAJAJ, Y.P.S. (1979) Freeze preservation of meristems of *Arachis hypogaea* and *Cicer arietinum*. Ind. J. Exp. Biol. 17, 1405–1407.

BAJAJ, Y.P.S. (1980) Induction of androgenesis in rice anthers frozen at -196°C. Cereal Res. Comm. 8, 365–369.

BAJAJ, Y.P.S. (1981) Regeneration of plants from ultra-low frozen anthers of *Primula obconica*. Sci. Hort. 14, 93–95.

BAJAJ, Y.P.S. (1982) Survival of anther and ovule derived cotton callus frozen in liquid nitrogen. Current Sci. 51, 139–140.

BAJAJ, Y.P.S. (1983a) Cryopreservation and international exchange of germplasm. In Plant Cell Culture in Crop Improvement. Ed. by S.K.Sen and K.L.Giles, Plenum Press, New York, pp. 19–41.

BAJAJ, Y.P.S. (1983b) Survival of somatic hybrid protoplasts of wheat x pea and rice x pea subjected to -196°C. Ind. J. Exp. Biol. 21, 120–122.

BAJAJ, Y.P.S. (1984) The regeneration of plants from frozen pollen embryos and zygotic embryos of wheat and rice. Theor. Appl. Gen. 67, 525–528.

BAJAJ, Y.P.S. and REINERT , J. (1977) Cryobiology of plant tissue cultures and establishment of gene banks. In Applied and Fundamental Aspects of Plant Cell, Tissue and Organ Culture. Ed. by J.Reinert and Y.P.S.Bajaj, Springer, Berlin, pp. 757-777.

BANERJEE, N. and DE LANGHE, E. (1985) A tissue culture technique for rapid clonal propagation and storage under minimal growth conditions of *Musa* (banana and plantain). Plant Cell Rep. 4, 351-354.

BARTOLO, M.E., WALLNER, S.J. and KETCHUM, R.E. (1987) Comparison of freeze tolerance in cultured plant cells and their respective protoplasts. Cryobiol. 24, 53-57.

BAYLISS, M.W. (1980) Chromosome variations in plant tissues in culture. Int. Rev. Cytol. Suppl. 11A, 13-144.

BENSON, E.E. and WITHERS, L.A. (1987) Gas chromatographic analysis of volatile hydrocarbon production by cryopreserved plant tissue cultures: a non-destructive method for assessing stability. Cryolett. 8, 35-46.

BEWLEY, J.D. (1979) Physiological aspects of desiccation tolerance. Ann. Rev. Plant Physiol. 30, 195-238.

BHANDAL, I.S., HAUPTMANN, R.M. and WIDHOLM, J.M. (1985) Trehalose as a cryoprotectant for the freeze preservation of carrot and tobacco cells. Plant Physiol. 78, 430-432.

BHATTACHARJEE, A. and CHOUDHURI, M.A. (1986) Chemical manipulation of seed longevity and stress tolerance capacity of seedlings of *Corchorus capsularis* and *C. olitorius*. J. Plant Physiol. 125, 391-400.

BHOJWANI, S.S. (1981) A tissue culture method for propagation and low temperature storage of *Trifolium repens* genotypes. Phys. Plant. 52, 187-190.

BHOJWANI, S.S. and RADZAN, M.K. (1983) Germplasm storage. In Plant Tissue Culture: Theory and Practice. Ed. by S.S.Bhojwani and M.K.Radzan, Elsevier, Amsterdam, pp. 373-385.

BONGA, J.M. and DURZAN, D.J. (Eds.) (1982) Tissue culture in forestry. Nijhoff/Junk, The Hague.

BRIDGEN, M.P. and STABY, G.L. (1981) Low pressure and low oxygen storage of *Nicotiana tabacum* and *Chrysanthemum x morifolium* tissue cultures. Plant Sci. Lett. 22, 177-186.

BRIDGEN, M.P. and STABY, G.L. (1983) Protocols of low pressure storage. In Handbook of Plant Cell Culture, Volume 1, Techniques for Propagation and Breeding. Ed. by D.A.Evans, W.R.Sharp, P.V.Ammirato and Y.Yamada, Macmillan, New York, pp. 816-827.

BUTENKO, R.G., POPOV, A.S., VOLKOVA, L.A., CHERNYAK, N.D. and NOSOV, A.M. (1984) Recovery of cell cultures and their biosynthetic capacity after storage of *Dioscorea deltoidea* and *Panax ginseng* in liquid nitrogen. Plant Sci. Lett. 33, 285-292.

CABOCHE, M. (1980) Nutritional requirements of protoplast derived, haploid tobacco cells grown at low cell densities in liquid medium. Planta 149, 7-18.

CAPLIN, S.M. (1959) Mineral-oil overlay for conservation of plant tissue cultures. Am. J. Bot. 46, 324-329.

CASSELLS, A.C. (1985) Genetic, epigenetic and non-genetic variation in tissue culture derived plants. In *In Vitro* Techniques - Propagation and Long Term Storage. Ed. B. A Schääfer-Menuhr, Nijhoff/Junk for CEC, Dordrecht, pp., 111-120.

CELLA, R., R . COLOMBO, R., GALLI, M.G., NIELSON, E., ROLLO, F. and SALA, F. (1982) Freeze-preservation of rice cells: a physiological study of freeze-thawed cells. Physiol. Plant. 55, 279-284.

CENTRO INTERNACIONAL DE AGRICULTURA TROPICAL (1985) Tissue culture. In Annual Report for 1984: Cassava Program, CIAT, Cali, Colombia, pp. 197-217.

CHEN, T.H.H., KARTHA, K.K., CONSTABEL, F. and GUSTA, L.V. (1984a) Freezing characteristics of cultured *Catharanthus roseus* (L).G. Don. cells treated with dimethylsulphoxide and sorbitol in relation to cryopreservation. Plant Physiol. 75, 720-725.

CHEN, T.H.H., KARTHA, K.K., LEUNG, N.L., KURZ, W.G.W., CHATSON, K.B. and CONSTABEL, F. (1984b) Cryopreservation of alkaloid producing cells of periwinkle (*Catharanthus roseus*). Plant Physiol. 75, 726-731.

CHEN, T.H.H., KARTHA, K.K. and GUSTA, L.V. (1985) Cryopreservation of wheat suspension culture and regenerable callus. Plant Cell, Tissue and Organ Culture 4, 101-109.

CONGER, B.V. (Ed.) (1981) Cloning agricultural plants via *in vitro* techniques. CRC Press, Boca Raton.

CULLIS, C.A. (1986) Unstable genes in plants. In Plasticity in Plants. SEB Symposium XXXX. Ed. by D.H. Jennings and A.J.Trewavas, Company of Biologists, Cambridge, pp. 211-232.

D'AMATO, F. (1978) Chromosome number variation in cultured cells and regenerated plants. In Frontiers of Plant Tissue Culture, 1978. Ed. by T.A.Thorpe, International Association for Plant Tissue Culture/Calgary University Press, Calgary, pp. 287-291.

DE LANGHE, E.A.L. (1984) The role of *in vitro* techniques in germplasm conservation. In Crop Genetic Resources: Conservation and Evaluation. Ed. by J.H.W.Holden and J.T.Williams, Allen and Unwin, London, pp. 131-137.

DE LOECKER, R., PENNINCKX, F. and KERREMANS, R. (1987) Osmotic effects of rapid dilution of cryoprotectants. 1. Effects on human erythrocyte swelling. Cryoletters, 8, 130-139.

DELMER, D.D. (1979) Dimethylsulphoxide as a potential tool for analysis of compartmentation in living cells. Plant Physiol. 64, 623-629.

DEREUDDRE, J. (1985) Problemes posés par la conservation des souches cellulaires et des organes cultivés *in vitro*. Bull. Soc. Bot. Fr. 132, 123-140.

DEUS-NEUMANN, B. and ZENK, M.H. (1984) Instability of indole alkaloid production in *Catharanthus roseus* cell suspension cultures. Planta Medica 50, 427-431.

DIETTRICH, B., POPOV, A.S., PFEIFFER, B., NEUMANN, D., BUTENKO, R. and LUCKNER, M., (1982) Cryopreservation of *Digitalis lanata* cell cultures. Planta Medica 46, 82-87.

DIETTRICH, B., HAACK, U., POPOV, A.S., BUTENKO, R.G., and LUCKNER, M. (1985) Long term storage in liquid nitrogen of an embryogenic strain of *Digitalis lanata*. Biochem. Physiol. Pfl. 180, 33-43.

DIETTRICH, B. HAACK, U, and LUCKNER, M. (1986) Cryopreservation of *Digitalis lanata* cells grown *in vitro*. Precultivation and recultivation. J. Plant Physiol. 126, 63-73.

DORÉ, C. (1987) Application of tissue culture to vegetable crop improvement. In Plant Tissue and Cell Culture. Ed. by C.E. Green,

D.A. Somers, W.P. Hackett and D.D. Biesboer, Alan R. Liss, New York, pp. 419–432. R. Liss, New York.

DOUGALL, D.H. and WHITTEN, G.H. (1980) The ability of wild carrot cell cultures to retain their capacity for anthocyanin synthesis after storage at –140°C. Planta Medica (Suppl.), pp. 129–135.

DOUZOU, P. (1983) Cryoenzymology. Cryobiol. 20, 625–635.

DRAPER, J., DAVEY, M.R. and FREEMAN, J.P. (1986) Delivery systems for the transformation of plant protoplasts. In Plant Cell Culture Technology. Ed. by M.M. Yeoman, Blackwell, Oxford, pp. 271–301.

DRUART, P.H. (1985) In vitro germplasm techniques for fruit trees. In In Vitro Techniques – Propagation and Long Term Storage. Ed. by A. Schäfer-Menuhr, Nijhoff/Junk for CEC, Dordrecht, pp. 167–171.

DUDA, E. (1982) Microcarrier improves viability of cryopreserved cells. Cryolett. 3, 67–70.

DUNWELL, J.M. (1986) Pollen, ovule and embryo culture as tools in plant breeding. In Plant Tissue Culture and its Agricultural Applications. Ed. by L.A.WITHERS and P.G.Alderson, Butterworths, London, pp. 375–404.

ELLIS, R.H., HONG, T.D. and ROBERTS, E.H. (1985a) Handbook of seed technology for genebanks. Vol. 1. Principles and methodology. IBPGR, Rome.

ELLIS R.H., HONG, T.D. and ROBERTS, E.H. (1985b) Handbook of seed technology for genebanks. Vol. 2. Compendium of specific germination information and test recommendations. IBPGR, Rome.

ENGELMANN, F. (1986) Cryoconservation des embryons somatiques de Palmier à huile (Elaeis guineensis Jacq.) mise au point des conditions de survie et de reprise. Doctoral Thesis, University of Paris VI.

ENGELMANN, F., DUVAL, Y. and DERUEDDRE, J. (1985) Survie et prolifération d'embryons somatiques de Palmier à huile (Elaeis guineensis Jacq.) après congélation dans l'azote liquide. C. R. Acad. Sci. 301, Sér. III, 111–116.

FAHY, G.M. (1983) Cryoprotectant toxicity neutralizers reduce freezing damage. Cryolett. 4, 300–314.

FARRANT, J. (1980) General observations on cell preservation. In Low Temperature Preservation in Medicine and Biology. Ed. by M.J.Ashwood-Smith and J.Farrant. Pitman Medical, Tunbridge Wells, pp. 1–18

FINKLE, B.J. and ULRICH, J.M. (1979) Effects of cryoprotectants in combination on the survival of frozen sugarcane cells. Plant Physiol. 63, 598–604.

FINKLE, B.J. and ULRICH, J.M. (1982) Cryoprotectant removal temperature as a factor in the survival of frozen rice and sugar cane cells. Cryobiol. 19, 329–335.

FINKLE, B.J., ULRICH, J.M., RAINS, W.R. and STAVAREK, S.S. (1985b) Growth and regeneration of alfalfa callus lines after freezing in liquid nitrogen. Plant Sci. 42, 133–140.

FINKLE, B.J., ZAVALA, M.E. and ULRICH, J.M. (1985a) Cryoprotectant compounds in the viable freezing of plant tissues. In Cryopreservation of Plant Cells and Organs. Ed. by K.K.Kartha, CRC Press, Boca Raton, pp. 75–113.

FOWLER, M.W. (1986) Industrial applications of plant cell culture. In Plant Cell Culture Technology. Ed. by M.M.Yeoman, Blackwell, Oxford, pp. 202–227.

FRANKS, F., MATHIAS, S.F., GALFRE, P., WEBSTER, S.D. and BROWN, D. (1983) Ice nucleation and freezing in undercooled cells. Cryobiol. 20, 298–309.

FUKUI, K. and KATSUMARU, H. (1980) Formation of mutinuclear cells induced by dimethyl sulphoxide: inhibition of cytokinesis and occurrence of novel nuclear division in *Dictyostelium* cells. J. Cell Biol. 86, 181–189.

GEORGE, E.F. and SHERRINGTON, P.D. (1984) Plant propagation by tissue culture, Handbook and directory of commercial laboratories. Exegetics Ltd., Basingstoke.

GOULD, A.R. (1984) Control of the cell cycle in cultured plant cells. CRC Critical Reviews in Plant Science 1, 315–344.

GREEN, P.B. (1986) Plasticity in shoot development: a biophysical view. In Plasticity in Plants. SEB Symposium XXXX. Ed. by D.H. Jennings and A.J.Trewavas, Company of Biologists, Cambridge, pp. 211–232.

GROUT, B.W.W. (1979) Low temperature storage of imbibed tomato seeds: a model for recalcitrant seed storage. Cryolett. 1, 71–76.

GROUT, B.W.W. (1986) Embryo culture and cryopreservation for the conservation of genetic resources of species with recalcitrant seed. In Plant Tissue Culture and its Agricultural Applications. Ed. by L.A.WITHERS and P.G.Alderson, Butterworths, London, pp. 303–309.

GROUT, B.W.W and HENSHAW, G.G. (1977) Freeze-preservation of potato shoot-tip cultures. Ann. Bot. 42, 1227–1229.

GROUT, B.W.W. and HENSHAW, G.G. (1980) Structural observations on the growth of potato shoot-tip cultures after thawing from liquid nitrogen. Ann. Bot. 46, 243–2 48.

GROUT, B.W.W., SHELTON, K. and PRITCHARD, H.W. (1983) Orthodox behaviour of oil palm seed and cryopreservation of the excised embryo for genetic conservation. Ann. Bot. 52, 381–384.

HAHN, G. and HOFFMANN, F. (1984) Dimethyl sulphoxide can initiate cell divisions of arrested callus protoplasts by promoting cortical microtubule assembly. Proc. Nat. Acad. Sci. 81, 5449–5453.

HASKINS, R.H. and KARTHA, K.K. (1980) Freeze preservation of pea meristems: cell survival. Can. J. Bot. 58, 833–84 0.

HAUPTMANN, R.M. and WIDHOLM, J.M. (1982) Cryopreservation of cloned amino acid analog-resistant carrot and tobacco suspension cultures. Plant Physiol. 70, 30–34.

HEBER, U., TYANKOVA, U.L. and SANTARIUS, K.A. (1971) Stabilisation and inactivation of biological membranes during freezing in the presence of amino acids. Biochem. Biophys. Acta. 241, 578–592.

HENSHAW, G.G., STAMP, J.A. and WESTCOTT, R.J. (1980) Tissue culture and germplasm storage. In Plant Cell Culture: Results and Perspectives. Ed. by F.Sala, B.Parisi, R.Cella and O.Ciferri, Elsevier, Amsterdam, pp. 277–282.

HENSHAW, G.G., KEEFE, D.P. and O'HARA, J.F. (1985a) Cryopreservation of potato meristems. In In Vitro Techniques – Propagation and Long Term Storage. Ed. by A. Schäfer-Menuhr, Nijhoff/Junk for CEC, Dordrecht, pp. 155–160.

HENSHAW, G.G., O'HARA, J.F. and STAMP, J.A. (1985b) Cryopreservation of potato meristems. In Cryopreservation of Plant Cells and Organs. Ed. by K.K.Kartha, CRC Press, Boca Raton, pp. 159–170.

HIRAOKA, N. and KODAMA, T. (1982) Effects of non-frozen cold storage on the growth, organogenesis and secondary metabolite production of callus cultures. In Plant Tissue Culture 1982. Ed. by A.Fujiwara, Maruzen Co., Tokyo, pp. 359-360.

HIRAOKA, N. and KODAMA, T. (1984) Effect of non-frozen cold storage on the growth, organogenesis and secondary metabolism of callus cultures. Plant Cell, Tiss. and Organ Cult. 3, 349-357.

HUNTER, J.C., BELT, A. and HALLUIN, A.P. (1986) Guidelines for establishing a culture collection within a biotechnology company. Trends in Biotechnol. 4. 5-11.

INTERNATIONAL BOARD FOR PLANT GENETIC RESOURCES (1983) Advisory Committee on In Vitro Storage - Report of the First Meeting. IBPGR, Rome.

INTERNATIONAL BOARD FOR PLANT GENETIC RESOURCES (1986) Design, Planning and Operation of In Vitro Genebanks. IBPGR, Rome.

INTERNATIONAL INSTITUTE OF TROPICAL AGRICULTURE (1981) Annual Report for 1980, pp. 64-82.

IWANAMI, Y. (1972) Retaining the viability of Camellia japonica pollen in various organic solvents. Plant and Cell Physiol. 13, 1139-1141.

JAMES, E. (1983) Low-temperature preservation of living cells. In Plant Biotechnology. SEB Seminar Series 18. Ed. by S.H.Mantell and H.Smith, Press Syndicate of the University of Cambridge, Cambridge, pp. 163-186.

JONES, L.H. (1974) Long term survival of embryoids of carrot. Plant Sci. Lett. 2, 221-224.

JULLIEN, M. (1983) Medium-term preservation of mesophyll cells isolated from Asparagus officinalis L.: Development of a simple method by storage at reduced temperature. Plant Cell, Tiss. and Organ Cult. 2, 305-316.

KARTHA, K. (1985) Meristem culture and germplasm preservation, In Cryopreservation of Plant Cells and Organs. Ed. by K.K.Kartha, CRC Press, Boca Raton, pp. 115-134.

KARTHA, K.K. (1986) Production and indexing of disease-free plants. In Plant Tissue Culture and its Agricultural Applications. Ed. by L.A.Withers and P.G.Alderson, Butterworths, London, pp. 219-238.

KARTHA, K.K., LEUNG, N.L. and GAMBORG, O.L. (1979) Freeze-preservation of pea meristems in liquid nitrogen and subsequent plant regeneration. Plant Sci. Lett. 15, 7-15.

KARTHA, K.K., LEUNG. N.L., GAUDET LAPRAIRIE, P. and CONSTABLE, F.(1982a) Cryopreservation of periwinkle, Catharanthus roseus cells cultured in vitro. Plant Cell Res. 1, 135-138.

KARTHA, K.K., LEUNG, N.L. and PAHL, K. (1980) Cryopreservation of strawberry meristems and mass propagation of plantlets. J. Am. Soc. Hort. Sci. 105, 481-484.

KARTHA, K.K., LEUNG, N.L. and MROGINSKI, L.A. (1982b) In vitro growth responses and plant regeneration from cryopreserved meristems of cassava (Manihot esculenta Crantz). Z. Pflanzenphysiol. 107, 133-140.

KATANO, M., ISHIHARA, A. and SAKAI, A. (1983) Survival of dormant apple shoot tips after immersion in liquid nitrogen. Hort.Sci.18, 707-708.

KEEFE, P.D. and HENSHAW, G.G. (1984) A note on the multiple role of artificial nucleation of the suspending medium during two-step cryopreservation processes. Cryolett. 5, 71-78.

265

KEITH, C.N. and McKENZIE, B. (1986) The effect of abscisic acid on the freeze tolerance of callus cultures of *Lotus corniculatus* L. Plant Physiol. 80, 766–770.
KIRSOP, B. (1986) Culture collections – their services to biotechnology. Trends in Biotechnol. 1, 4–8.
KITTO, S.L. and JANICK, J. (1985a) Production of synthetic seeds by encapsulating asexual embryos of carrot. J. Am. Soc. Hort. Sci. 110, 277–282.
KITTO, S.L. and JANICK, J. (1985b) Hardening treatments increase survival of synthetically-coated asexual embryos of carrot. J. Am. Soc. Hort. Sci. 110, 283–286.
KOJIMA, T., SOMA, T. and OGURI, N. (1987) Effect of rapid addition and dilution of dimethyl sulphoxide and 37°C equilibration on viability of rabbit morulae thawed rapidly. Cryobiol. 24, 247–255
KUO, C.-C. and LINEBERGER, R.D. (1985) Survival of *in vitro* cultured tissue of 'Jonathan' apples exposed to –196°C. HortSci. 20, 764–767.
KUZINA, G.V., KARNIKOVA, L.D. and KALININA, G.A. (1984) Effect of cartolin on dormancy and frost resistance of woody plants. Soviet Plant Physiol. 31, 451–461.
LAWRENCE, R.H. Jr (1981) *In vitro* cloning systems. Env. Exp. Bot. 21, 289–300.
LOCKETT, M.C. and LUYETT, B.J. (1951) Survival of frozen seeds of various water contents. Biodynamica 134, 67–76.
LONE, M.I., KEUH, J.S.H., WYN JONES, R.G. and BRIGHT, S.W.J. (1987) Influence of proline and glycine betaine on salt tolerance of cultured barley embryos. J. Ex. Bot. 38, 479–490.
LUNDERGAN, C. and JANICK, J. (1979) Low temperature storage of in vitro apple shoots (*Malus domestica* cv. Golden Delicious). HortSci. 14, 514.
LYNDON, R.F. and CUNNINGHAME, M.E. (1986) Control of shoot apical development via cell division. In Plasticity in Plants. SEB Symposium XXXX. Ed. by D.H.Jennings and A.J.Trewavas, Company of Biologists, Cambridge, pp 233–255.
MADDOX, A.D., GONSALVES, and SHIELDS, R. (1985) Successful preservation of suspension cultures of three *Nicotiana* species at the temperature of liquid nitrogen. Plant Sci. Lett. 28, 157–162.
MATHIAS, S.F., FRANKS, F, and HATLEY, R.H.M. (1985) Preservation of viable cells in the undercooled state. Cryobiol. 22, 537–546.
MATTHES, G. and HACKENSELLNER, K.D. (1981) Correlations between purity of dimethyl sulphoxide and survival after freezing and thawing. Cryolett. 2, 389–392.
MATTHES, G., HACKENSELLNER, H.A., JENTZSCH, K.D. and OEHME, P. (1981) Further studies on the cryoprotection supporting efficiency of selenium compounds. Cryolett. 2, 241– 245.
McGANN, L.E. (1978) Differing actions of penetrating and non–penetrating cryoprotective agents. Cryobiol. 15, 382–390.
McGANN, L.E. and KRUUV, J. (1977) Freeze-thaw damage in protected and unprotected synchronized mammalian cells. Cryobiol. 14, 503–505.
MERYMAN, H.T. (Ed.) (1966) Cryobiology. Academic Press, New York.
MERYMAN, H.T. and WILLIAMS, R.J. (1982) Mechanisms of freezing injury and natural tolerance and the process of artificial cryoprotection. In Crop Genetic Resources: the Conservation of Difficult Material. Ed. by L.A.Withers and J.T.Williams, IBPGR/IUBS/IGF, Paris, IUBS Series B42, pp. 5–37.

MIX, G. (1982) *In vitro* preservation of potato material. Plant Genetic Resources Newsletter 51, 6–8.
MIX, G. (1985) *In vitro* preservation of potato genotypes. In Efficiency in Plant Breeding. Ed. by W. Lange, A.C.Zeven and N.G.Hogenboom, Pudoc, Wageningen, pp. 194–195.
MONETTE, P.L. (1986) Cold storage of kiwifruit shoot tips *in vitro*. HortSci. 21, 1203–1205.
MORIGUCHI, T., AKIHAMA, T. and KOZAKI, I. (1985) Freeze–preservation of dormant pear shoot apices. Jap. J. Breeding 35, 194–199.
MULLIN, R.H. and SCHLEGEL, D.E. (1976) Cold storage maintenance of strawberry meristem plantlets. HortSci. 11, 100–101.
MUMFORD, P.M. and GROUT, B.W.W. (1979) Desiccation and low temperature (−196°C) tolerance of *Citrus limon* seeds. Seed Sci. and Technol. 7, 407–410.
NAG, K.K. and STREET, H.E. (1973) Carrot embryogenesis from frozen cultured cells. Nature 243, 270–272.
NAG, K.K. and STREET, H.E. (1975a) Freeze–preservation of cultured plant cells. I. The pretreatment phase. Physiol. Plant. 34, 254–260.
NAG, K.K. and STREET, H.E. (1975b) Freeze–preservation of cultured plant cells. II. The freezing and thawing phases. Physiol. Plant. 34, 261–265.
NASH, T. (1966) Chemical constitution and physical properties of compounds able to protect living cells against damage due to freezing and thawing. In Cryobiology. Ed. by H.T.Meryman, Academic Press, New York, pp. 197–211.
NG, S.Y. and HAHN, S.K. (1985) Applications of tissue culture to tuber crops at IITA. In Biotechnology in International Agricultural Research. International Rice Research Institute, Manila, pp. 29–40.
NILSSON, J.R. (1980) Effect of dimethyl sulphoxide on ATP content and protein synthesis in *Tetrahymena*. Protoplasma 103, 189–200.
NITZSCHE, W. (1978) Erhaltung der Lebensfaehigkeit in getrocknetem Kallus. Z. Pflanzenphysiol. 87, 469–472.
NITZSCHE, W. (1983) Germplasm preservation. In. Handbook of Plant Cell Culture. Volume 1: Techniques for propagation and breeding. Ed. by D.A.Evans, W.R.Sharp, P.V.Ammirato and Y.Yamada. Macmillan, New York, pp. 783–805.
NKANG, A. and CHANDLER, G. (1986) Changes during embryogenesis in rain forest seeds with orthodox and recalcitrant viability characteristics. J. Plant Physiol. 126, 243–256.
NORMAH, M.N., CHIN, H.F. and HOR, Y.L. (1986) desiccation and cryopreservation of embryonic axes of *Hevea brasiliensis* Muell. Arg. Pertanika 9, 299–303. ORR, W., SINGH, J. and BROWN, D.C.W. (1985) Induction of freeze tolerance in alfalfa cell suspension cultures. Plant Cell Rep. 4, 15–18.
PEACOCK, W.J. (1987) Molecular biology and genetic resources. In The Use of Crop Genetic Resources Collections. Ed. by A.D.H.Brown, O.H.Frankel, D.R.Marshall and J.T.Williams, Cambridge University Press, in press.
PEETERS, J.P., and WILLIAMS, J.T. (1984) Towards better use of genebanks with special reference to information. Plant Genetic Resources Newsletter 60, 22–32.
PEGG, D.E., DIAPER, M.P., SKAER, H. LE B. and HUNT, C.J. (1984) The effect of cooling rate and warming rate on the packing effect in

human erythrocytes frozen and thawed in the presence of 2M glycerol. Cryobiol. 21, 491-502.

PENNINCKX, F., POELMANS, S., KERREMANS, R. and DE LOEKKER, W. (1984) Erythrocyte swelling after rapid dilution of cryoprotectants and its prevention. Cryobiology 21, 25-32.

PREIL, W. and HOFFMANN, M. (1985) *In vitro* storage in chrysanthemum breeding and propagation. In *In vitro* Techniques - Propagation and Long Term Storage. Ed. by A. Schéfer-Menuhr, Nijhoff/Junk for CEC, Dordrecht, pp. 161-165.

PRITCHARD, H.W., GROUT, B.W.W. and SHORT, K.C. (1986a) Osmotic stress as a pregrowth procedure for cryopreservation. 1. Growth and ultrastructure of sycamore and soybean cell suspensions. Ann. Bot. 57, 41-48.

PRITCHARD, H.W., GROUT, B.W.W. and SHORT, K.C. (1986b) Osmotic stress as a pregrowth procedure for cryopreservation. 2. Water relations and metabolic state of sycamore and soybean cell suspensions. Ann. Bot. 57, 371-378.

PRITCHARD, H.W., GROUT, B.W.W. and SHORT, K.C. (1986c) Osmotic stress as a pregrowth procedure for cryopreservation. 3. Cryobiology of sycamore and soybean cell suspensions. An n. Bot. 57, 379-387.

PRITCHARD, H.W. and PRENDERGAST, F.G. (1986) Effect of desiccation and cryopreservation on the *in vitro* viability of embryos of *Araucaria hunsteinii* K. Schum. J. Ex. Bot. 37, 1388-1397.

RALL, W.F. and FAHY, G.M. (1985) Ice-free cryopreservation of mouse embryos at -196°C by vitrification. Nature 313, 573-575.

RASMUSSEN, D.H., MCAULAY, M.N. and MACKENZIE, A.P. (1975) Supercooling and nucleation of ice in single cells . Cryobiol. 12, 328-339.

REDENBAUGH, K., PAASCH, B.D. , NICHOL, J.W., KOSSLER, M.E., VISS, P.R. and WALKER, K.A. (1986) Somatic seeds: encapsulation of plant embryos. Biotechnol. 4, 797-810.

ROBERTS, E.H. and KING, M.W. (1982) Storage of recalcitrant seeds. In Crop Genetic Resources: The Conservation of Difficult Material. Ed. by L.A.Withers and J.T.Williams, IUBS/IBPGR/IGF, Paris, IUBS Series B42, pp. 39-48.

ROBERTS, E.H., KING, M.W. and ELLIS, R.J. (1984) Recalcitrant seeds: their recognition and storage. In Crop Genetic Resources: Conservation and Evaluation. Ed. by J.H.W.Holden and J.T.Williams, George Allen and Unwin, London, pp. 38-52.

ROCA, W.M., RODRIGUEZ, J., BELTRAN, J., ROA, J. and MAFLA, G. (1982) Tissue culture for the conservation and international exchange of germplasm. In Plant Tissue Culture 1982. Ed. by A. Fujiwara, Maruzen Co., Tokyo, pp. 771-772.

ROSATI, P., GAGGIOLI, D and GUINCHI, L. (1986) Genetic stability of micropropagated loganberry plants. J. Hort. Sci. 61, 33-41.

SAKAI, A. (1985) Cryopreservation of shoot tips of fruit trees and herbaceous plants. In Cryopreservation of Plant Cells and Organs. Ed. by K.K.Kartha, CRC Press, Boca Raton, pp. 135-158.

SAKAI, A. (1986) Cryopreservation of germplasm of woody plants. In Biotechnology in Agriculture and Forestry. Volume 1: Trees 1. Ed. by Y.P.S.Bajaj, Springer, Berlin, pp. 113-129.

SAKAI, A. and NISHIYAMA, Y. (1978) Cryopreservation of winter vegetative buds of hardy fruit trees in liquid nitrogen. HortSci. 13, 225-227.

SAKAI, A. and UEMURA, M. (1982) Recent advances of cryopreservation of apical meristems. In Plant Cold Hardiness and Freezing Stress. Ed. by P.H.Li and A. Sakai, Academic Press, London, pp. 635–64.

SAKAI, A., YAMAKAWA, M., SAKATA, D., HARADA, T. and YAKUWA, T. (1978) Development of a whole plant from an excised strawberry runner apex frozen to –196°C. Low Temp. Sci. Ser. B. 36, 31–38.

SALA, F., CELLA, R. and ROLLO, F. (1979) Freeze preservation of rice cells grown in suspension culture. Phys. Plant. 45, 170–176.

SCHÄFER-MENUHR, A. (Ed.) (1985) In vitro Culture – Preservation and Long Term Storage. Nijhoff/Junk for CEC, Dordrecht.

SCHWAB, K.B. and GAFF, D.F. (1986) Sugar and ion content in leaf tissues of several drought tolerant plants under water stress. J. Plant Physiol. 125, 257–265.

SCOWCROFT, W.R. (1984) Genetic variability in tissue culture: impact on germplasm conservation and utilization. IBPGR, Rome.

SCOWCROFT, W.R. (1985) Somaclonal variation: the myth of clonal uniformity. In Genetic Flux in Plants. Ed. By T.Hohn and E.S.Denis, Springer, Vienna, pp. 217–245.

SCOWCROFT, W.R. (1986) Tissue culture and plant breeding. In Plant Cell Culture Technology. Ed. by M.M.Yeoman, Blackwell, Oxford, pp. 67–95.

SCOWCROFT, W.R., DAVIES , P., RYAN, S.A., BRETTELL, R.I.S., PALLOTA, M.A. and LARKIN, P.J. (1985) The analysis of somaclonal mutants. In Plant Genetics: UCLA Series Vol. 35. Ed. by M.Freeling, Alan R. Liss, New York, pp. 799–815.

SEIBERT, M. and WETHERBEE, P.J. (1977) Increased survival and differentiation of frozen herbacaeous plant organ cultures through cold treatment. Plant Physiol. 59, 1043–1046.

SEITZ, U., ALFERMANN, A. W. and REINHART, E. (1983) Stability of biotransformation capacity in Digitalis lanata cell cultures after cryogenic storage. Plant Cell Rep. 2, 273–276.

SEMAL, J. (Ed.) (1986) Somaclonal variations and crop improvement. Nijhoff/Junk for CEC, Dordrecht.

SIBLEY, J.T., PAUL, M.D. and HANSON, E.D. (1977) Subcellular effect of cytochalasin B and dimethylsulphoxide on Paramecium aurelia. J. Protozool. 24, 595–604.

SMITH, D.R., HORGAN, K.J. and AITKEN-CHRISTIE, J. (1982) Micropropagation of Pinus radiata for afforestation. In Plant Tissue Culture 1982. Ed. by A.Fujiwara, Maruzen co., Tokyo. pp.723–724.

SOUTHARD, J.H., RICE, M.J. and BELZER, F.O. (1985) Preservation of renal function by adenosine-stimulated ATP synthesis in hypothermically perfused dog kidneys. Cryobiology 22, 237–242.

STANWOOD, P.C. (1985) Cryopreservation of seed germplasm for genetic conservation. In Cryopreservation of Plant Cells and Organs. Ed. by K.K.Kartha, CRC Press, Boca Raton, pp. 199–226.

STARITSKY, G., DEKKERS, A.J., LOUWAARS, N.P. and ZANDVOORT, E.A. (1986) In vitro conservation of aroid germplasm at reduced temperatures and under osmotic stress. In Plant Tissue Culture and its Agricultural Applications. Ed. by L.A.Withers and P.G.Alderson, Butterworths, London, pp. 277–2 83.

STEPONKUS, P.L. (1984) Role of the plasma membrane in freezing injury and cold acclimation. Ann. Rev. Plant Physiol. 35, 543–584.

STRAUSS, A., FRANKHAUSER, H. and KING, P.J. (1985) Isolation and cryopreservation of O-methylthreonine-resistant *Rosa* cell lines altered in the feedback sensitivity of L-threonine deaminase. Planta 163, 554–562.

STYLES, E.D., BURGESS, J.M., MASON, C. and HUBER, B.M. (1985) Storage of seed in liquid nitrogen. Cryobiol. 19, 195–199.

SUGAWARA, Y. and SAKAI, A. (1974) Survival of suspension-cultured sycamore cells cooled to the temperature of liquid nitrogen. Plant Physiol. 54, 722–724.

SUGAWARA, Y. and SAKAI, A. (1985) Effect of Ca^{++} on the survival of cultured plant cells after freezing and thawing. Low Temp. Sci. Ser. B. 33, 21–28.

SUN, C.N. (1958) The survival of excised pea seedlings after drying and freezing in liquid nitrogen. Bot. Gaz. 19, 234–236.

TAKEUCHI, M., MATSUSHIMA, H., and SUGAWARA, Y. (19 82) Totipotency and viability of protoplasts after long term freeze preservation. In Plant Tissue Culture 1982. Ed. by A.Fujiwara, Maruzen Co., Tokyo, pp. 797–798.

TAYLOR, M.J. and HUNT, C.J. (1985) A new preservation solution for storage of corneas at low temperatures. Curr. Eye Res. 4, 963–973.

TOWILL, L.E. (1981) *Solanum etuberosum*: a model for studying the cryobiology of shoot-tips in the tuber-bearing Solanum species. Plant Sci. Lett. 20, 315–324.

TOWILL, L.E. (1983) Improved survival after cryogenic exposure of shoot tips dissected from *in vitro* plantlet cultures of potato. Cryobiol. 20, 567–573.

TOWILL, L.E. (1984) Survival at ultra-low temperatures of shoot tips from *Solanum tuberosum* groups *andigena*, *phureja*, *stenotomum*, *tuberosum* and other tuber bearing *Solanum* species. Cryolett. 5, 319–326.

TOWILL, L.E. (1985) Low temperature and freeze-/vacuum-drying preservation of pollen. In Cryopreservation of Plant Cells and Organs. Ed. by K.K.Kartha, CRC Press, Boca Raton, pp. 171–198.

UEMURA, M. and SAKAI, A. (1980) Survival of carnation (*Dianthus caryophyllus* L.) shoot apices frozen to the temperature of liquid nitrogen, Plant and Cell Physiol. 21, 85–94.

ULRICH, J.M., FINKLE, B.J., MOORE, P.H. and GINOZA, H. (1979) Effect of a mixture of cryoprotectants in attaining liquid nitrogen survival of callus cultures of a tropical plant. Cryobiology 16, 550–556.

ULRICH, J.M., FINKLE, B.J. and TISSERAT, B.H. (1982) Effect of cryogenic treatment on plantlet production from frozen and unfrozen date palm callus. Plant Physiol. 69, 624–627.

VANNINI, G.L., POLI, F., PANCALDI, S. and DALL'OLIO, G. (1985) Alterations induced by dimethyl sulphoxide in *Euglena gracilis* with emphasis on the side effects. Protoplasma, 125, 199–204.

VILLIERS, T.A. (1975) Genetic maintenance of seeds in imbibed storage. In Crop Genetic Resources for Today and Tomorrow. Ed. by O.H.Frankel and J.G.Hawkes, Cambridge University Press, London, pp 297–316.

WALKEY, D.G.A. (1987) Micropropagation of vegetables. In Micropropagation in Horticulture: Practice and Commercial Problems. Ed. B.P.G.Alderson and W.M.Dullforce, Institute of Horticulture, University of Nottingham, pp. 97–112.

WANAS, W.H. (1987) Genetic conservation of *Pyrus* spp. using tissue culture techniques. PhD thesis, University of Birmingham.

WANAS, W.H., CALLOW, J .A. and WITHERS, L.A. (1986) Growth limitation for the conservation of pear genotypes. In Plant Tissue Culture and its Agricultural Applications. Ed. by L.A.Withers and P.G.Alderson, Butterworths, London, pp. 285-289.

WATANABE, K., MITSUDA, H. and YAMADA, Y. (1983) Retention of metabolic and differentiation potentials of green *Lavendula vera* callus after freeze-preservation. Plant and Cell Physiol. 24, 119-122.

WATANABE, K., YAMADA, Y., UENO, S. and MITSUDA, H. (1985) Change of freezing resistance and retention of metabolic and differentiation potentials in cultured green *Lavandula vera* cells which survived repeated freeze-thaw procedures. Agric. Biol. Chem. 49, 1727-1731.

WEBER, G., ROTH, E.J. and SCHWEIGER, H.G. (1983) Storage of cell suspensions and protoplasts of *Glycine max* (L.) Mer., *Brassica napus* (L.), *Datura innoxia* (Mill.) and *Daucus carota* (L.) by freezing. Z. Pflanzenphysiol. 10, 23-29.

WESTCOTT, R.J. (1981a) Tissue culture storage of potato germplasm. 1. Minimal growth storage. Potato Res. 24, 331-342.

WESTCOTT, R.J. (1981b) Tissue culture storage of potato germplasm. 2. Use of growth retardants. Potato Res. 24, 343-352.

WHEELANS, S.K. and WITHERS, L.A. (1984) The IBPGR International Database on In Vitro Conservation. Plant Genetic Resources Newsletter, 60, 33-38.

WHEELANS, S.K. and WITHERS, L.A. (1987) The IBPGR *In Vitro* Conservation Databases. In Proceedings of Advanced Study Institute on Plant Cell Biotechnology, sponsored by NATO, Scientific Affairs Division, Albufeira, Portugal, in press.

WHITTINGHAM, D.G., LYON, M.F. and GLENISTER, P.H. (1977) Long term storage of mouse embryos at -196°C: The effect of background radiation. Genet. Res. 29, 171-181.

WILLIAMS, J.T. (1984) A decade of crop genetic resources research. In Crop Genetic Resources: Conservation and Evaluation. Ed. by J.H.W.Holden and J.T.Williams, Allen and Unwin, London, pp. 1-17.

WITHERS, L.A. (1978) The freeze preservation of synchronously dividing cultured cells of *Acer pseudoplatanus* L. Cryobiol. 15, 87-92.

WITHERS, L.A. (1979) Freeze preservation of somatic embryos and clonal plantlets of carrot (*Daucus carota* L.). Plant Physiol. 63, 460-467.

WITHERS, L.A. (1980a) Tissue Culture Storage for Genetic Conservation. IBPGR, Rome.

WITHERS, L.A. (1980b) The cryopreservation of higher plant tissue and cell cultures - an overview with some current observations and future thoughts. Cryolett. 1 239-250.

WITHERS, L.A. (1982a) Storage of plant tissue cultures. In Crop Genetic Resources - The Conservation of Difficult Material. Ed. by L.A.Withers and J.T.Williams, IUBS/IBPGR/IGF, Paris, IUBS Series B42, pp. 49-82.

WITHERS, L.A. (1982b) The development of cryopreservation techniques for plant cell, tissue and organ cultures. In Plant Tissue Culture 1982. Ed. by A. Fujiwara, Maruzen Co., Tokyo, pp. 793-794.

WITHERS, L.A. (1984a) Germplasm conservation *in vitro*: present state of research and its application. In Crop Genetic Resources:

Conservation and Evaluation. Ed. by J.H.W.Holden and J.T.Williams, Allen and Unwin, London, pp. 138-157.
WITHERS, L.A. (1984b) Freeze preservation of cells. In Cell Culture and Somatic Cell Genetics of Plants. Vol. 1: Laboratory Procedures and Their Applications, Academic Press, Orlando, Florida, pp. 608-620.
WITHERS, L.A. (1985a) Cryopreservation of cultured cells and meristems. In Cell Culture and Somatic Cell Genetics of Plants. Vol. 2: Cell Growth, Nutrition, Cytodifferentiation and Cryopreservation. Ed. by I.K.Vasil, Academic Press, Orlando, Florida, pp. 253-316.
WITHERS, L.A. (1985b) Cryopreservation and storage of germplasm. In Plant Cell Culture: a Practical Approach. Ed. by R.A.Dixon, IRL Press, Oxford, pp. 169-191.
WITHERS, L.A. (1985c) Cryopreservation of cultured plant cells and protoplasts. In Cryopreservation of Plant Cells and Organs. Ed. by K.K.Kartha, CRC Press, Boca Raton, pp. 243-267.
WITHERS, L.A. (1986a) Cryopreservation and genebanks. In Plant Cell Culture Technology. Ed. by M.M.Yeoman, Blackwell, Oxford, pp. 96-140.
WITHERS, L.A. (1986b) In vitro approaches to the conservation of plant genetic resources. In Plant Tissue Culture and its Agricultural Applications. Ed. by L.A.Withers and P.G.Alderson, Butterworths, London, pp. 261-276.
WITHERS, L.A. (1987a) Low temperature preservation of plant cell, tissue and organ cultures for genetic conservation and improved agricultural practice. In The Effects of Low Temperatures on Biological Systems. Ed. by B.W.W.Grout and G.J.Morris, Edward Arnold, London, pp. 389-409.
WITHERS, L.A. (1987b) In vitro conservation and germplasm utilization. In The Use of Crop Genetic Resources Collections. Ed. by A.D.H.Brown, O.H.Frankel, D.R.Marshall and J.T.Williams, Cambridge University Press, in press.
WITHERS, L.A. (1987c) In vitro methods for collecting germplasm in the field. Plant Genetic Resources Newsletter 69, 2-6.
WITHERS, L.A. and KING, P.J. (1979) Proline: a novel cryoprotectant for the freeze preservation of cultured cells of Zea mays L. Plant Physiol. 64, 675-678.
WITHERS, L.A. and KING, P.J. (1980) A simple freezing unit and cryopreservation method for plant cell suspensions. Cryolett. 1, 213-220.
WITHERS, L.A. and STREET, H.E. (1977a) Freeze preservation of plant cell cultures. In Plant Tissue Culture and its Biotechnological Application. Ed. by W.Barz, E.Reinhard and M.H.Zenk, Springer, Berlin, pp. 226-244.
WITHERS, L.A. and STREET, H.E. (1977b) Freeze preservation of cultured plant cells. II I. The pregrowth phase. Physiol. Plant. 39, 171-178.
WITHERS, L.A. and WILLIAMS, J.T. (1985) Research on long-term storage and exchange of in vitro plant germplasm. In Biotechnology in International Agricultural Research. International Rice Research Institute, Manila, pp. 11-24.
WITHERS, L.A. and WILLIAMS, J.T. (1986) IBPGR Research Highlights 1964-1985: In Vitro Conservation. IBPGR, Rome.
WOLLF, B.R., GLICK, B.R. and PASTERNAK, J.J. (1986) A microcomputer

programme for establishing and maintaining databases for laboratory culture collections. Lett. in Appl. Microbiol. 3, 45–48.

WOMERSLEY, C., USTER, P., RUDOLPH, A. and CROWE, J. (1985) Inhibition of dehydration-induced fusion between freeze dried liposomal membranes by carbohydrates. Cryobiology, 22, 627.

WU, M.-T. and WALNER, S.J. (1985) Effect of temperature on freezing and heat stress tolerance of cultured plant cells. Cryobiol. 22, 191–195.

ZELDIN, E.L. and McCOWN, B.H. (1985) Use of lighted cold storage of plants derived from micro-culture to circumvent the problem of seasonal demand. Hort. Sci. 20, 585.

ZIEBOLZ, B. and FORCHE, E. (1985) Cryopreservation of plant cells to retain special attributes. In *In vitro* Culture - Propagation and Long Term Storage. Ed. by A. Schäfer-Menuhr, Nijhoff/Junk for CEC, Dordrecht, pp. 181–183.

Oxford Surveys of Plant Molecular & Cell Biology Vol.4 (1987) 275–334

PLANT PROTEINACEOUS INHIBITORS OF
PROTEINASES AND α-AMYLASES

*F. Garcia-Olmedo, G. Salcedo, R. Sanchez-Monge, L. Gomez,
J. Royo and P. Carbonero*

*Departamento de Bioquimica. E.T.S. Ingenieros Agronomos.
Universidad Politecnica de Madrid. 28040 Madrid. Spain.*

INTRODUCTION

Plant proteins which are inhibitory towards various types of enzymes from a wide range of organisms have been extensively studied for many years. Proteinase inhibitors have received particular attention and accordingly a number of reviews concerning their structure, activity, evolution, possible physiological roles and nutritional properties have appeared regularly in the literature (Ryan, 1973, 1981, 1984; Laskowski and Kato, 1980; Richardson, 1981; Boisen, 1983). Recent technical advances in molecular biology have accelerated the output of information about these inhibitors to the extent that entirely new types have been uncovered and previously unsuspected relationships have been established. These developments justify the present review that will emphasize the novel aspects, glossing over many important topics that have been adequately covered before. Among the most striking recent findings is the structural and evolutionary relationships of different α-amylase inhibitors with different types of proteinase inhibitors, which is the reason for their joint consideration in this survey.

The inhibitors have been grouped for our present purposes as listed in Table 1, following an eclectic criterion: the first nine groups are true protein families, based on sequence homology, and the last two are based on the mechanistic classes of the enzymes inhibited because not enough sequence information is available, and therefore may represent more than one protein family. A considerable number of reports deal with inhibitors that have not been sufficiently characterized to discern whether they belong to any of the listed groups or represent new types. These cases have not been comprehensibly included in this review.

Table 1. Groups of protein inhibitors of proteinases and α-amylases

Inhibitor group	Enzymes inhibited
1. Soybean trypsin inhibitor (Kunitz) family	Serine proteinases and endogenous α-amylases
2. Bowman-Birk inhibitor family	Serine proteinases
3. Cereal trypsin/α-amylase inhibitor family	Serine proteinases and heterologous α-amylases
4. Potato inhibitor I family	Serine proteinases
5. Potato inhibitor II family	Serine proteinases
6. Squash inhibitor family	Serine proteinases
7. Barley protein Z/α_1-antitrypsin family	Serine proteinases
8. Ragi I-2/maize bifunctional inhibitors family	Serine proteinases and heterologous α-amylases
9. Carboxypeptidase A, B inhibitor family	Metallo-carboxypeptidases
10. Thiol-proteinase inhibitors	Endogenous and heterologous thiol-proteinases
11. Cathepsin D and Pepsin inhibitors	Carboxyl-proteinases

All the serine proteinase inhibitors treated here seem to obey a standard mechanism, which has been thoroughly studied from different points of view (see Laskowski and Kato, 1980), including the detailed elucidation by X-ray crystallography of the three–dimensional structure of several enzyme-inhibitor complexes, as reviewed recently by Read and James (1986). In summary, the inhibitors are highly specific substrates for their target enzymes, which undergo a limited and extremely slow proteolysis, so that the system behaves as if the free enzyme and the inhibitor were in simple equilibrium with the enzyme/inhibitor complex. On the surface of each inhibitor there is at least one reactive bond (P1-P'1) which interacts with the active site of the enzyme. An inhibitor molecule which has undergone hydrolysis of its reactive bond is as active as the unhydrolised inhibitor and is able to form a stable complex with the enzyme. In most, but not all, of these inhibitors, the reactive site peptide bond is encompassed in at least one disulphide loop, which ensures that during conversion of the original to the modified inhibitor the two peptide chains cannot dissociate. The nature of the amino acid residue at the carboxyl side, P1 position, generally determines the proteinase inhibited: Lys or Arg for trypsin–like enzymes; Phe, Tyr, or Leu for chymotrypsin; and Ala for elastase. The amino side of the peptide bond, the P'1 position, does not seem to be involved in determining specificity (Laskowski and Kato, 1980).

More than one reactive site is sometimes present in a single polypeptide chain, in which case more than one enzyme molecule of the same or different specificity can be simultaneously inhibited by a single inhibitor molecule. In some of these cases, it is fairly obvious that the multiple reactive sites are associated with multiple protein domains that must have been generated from an ancestral domain by internal gene duplications, while in other cases a convergent evolutionary process cannot be excluded.

In subsequent pages, we will first deal with the different inhibitor groups, and then, their possible implication in plant metabolism, plant protection, and human and animal nutrition will be addressed.

SOYBEAN TRYPSIN INHIBITOR (KUNITZ) FAMILY

The crystallization of a trypsin inhibitor from soybean (STI) and of its complex with trypsin, which was carried out over forty years ago by M. Kunitz (1945, 1946, 1947a,b, 1949), was one of the major achievements in the early stages of research on protein inhibitors from plants. Further studies concerning its amino acid sequence (Koide and Ikenaka, 1973a,b; Koide et al., 1973), its three-dimensional structure (Sweet et al., 1974), and its mechanism of interaction with the enzyme (Sealock and Laskowski, 1969; Kowalski et al., 1974; Kowalski and Laskowski, 1976a,b; Hunkapiller et al., 1979; Baillargeon et al., 1980) made STI the first plant inhibitor to be well characterized. In spite of this early start, the identification of further members of the STI family in other species has been a rather slow process that has recently acquired momentum with the identification of a number of new inhibitors and, especially, with the realization that the inhibitors of subtilisin/endogenous α-amylase from cereals are indeed homologues of STI (Hejgaard et al., 1983; Svendsen et al., 1986; Maeda, 1986).

Distribution and Inhibitory Properties

Inhibitory proteins with M_r of about 20,000, two disulphide bridges and similar amino acid composition to that of STI have been identified in a wide range of species. Although STI appears as a single form for which two additional alleles have been found, mixtures of numerous isoforms with different inhibitory properties are present in many species (Table 2). A majority of the identified inhibitors are specific for either trypsin or chymotrypsin, with either weak or null activity for the second enzyme, while a few are about equally effective against both enzymes. Some members of this family are only weak inhibitors and it can be speculated that they might be active against other, unidentified enzymes. Only one of

Table 2.   Distribution and inhibitory properties of members of the STI (Kunitz) family

Species	Inhibitor	Specificity*	Inh/Enz	Selected references
Papilionideae				
Psophocarpus tetragonolobus	WTI1A,1B,2,3	T-s	1:1	Yamamoto et al.1983
(winged bean)	WTCI1	T-s, Ch-s	1:1	Shibata et al. 1986
	WCI1	Ch-s	1:1	
	WCI2,3	Ch-s	1:2	
	WCI4	Ch-w		
Erythrina latissima	DE1	Ch-s		Joubert et al. 1985
	DE3	T-s	1:1	
E. cristagalli	DE1,8	T-s, Ch-w	1:1	Joubert and Sharon,
	DE2,4	Ch-s	1:1	1985
	DE3	Ch-s	1:2	
E. corolladendron	DE1,5	Ch-s	1:1	
	DE6,7	T-s, Ch-w	1:1	
	DE8	T-s	1:1	
E. acanthocarpa	DE1	T-s, Ch-w		
	DE2	Ch-s		
E. caffra	DE1	T-s, Ch-s		
	DE2	Ch-s		
	DE3	T-s, Ch-w		
	DE4	T-s		
E. humeana	DE3	T-s, Ch-w		
E. lysistemon	DE1	Ch-s		
	DE2-4	T-s, Ch-w		
E. seyheri	DE1,3,5	T-s, Ch-w		
	DE2,4	Ch-s		
Glycine maxima	STIa,b,c	T-s	1:1	Kim et al. 1985

---

these inhibitors, DE3 from *Erythrina latissima*, has also been shown
to inhibit tissue plasminogen activator (Joubert et al., 1985). The
cereal inhibitors are isoform mixtures active against subtilisin and
endogenous α-amylases (Warchalewski, 1977a,b; Yoshikawa et al., 1976;

Table 2.   Continued

Species	Inhibitor	Specificity*	Inh/Enz	Selected references
Cesalpinoideae				
Peltophorum africanum	DE1	T-s		Joubert 1981
Mimosideae				
Adenanthera pavonina (carolina tree)	DE1-8	T-s, Ch-s	(DE5)1:1	Richardson et al. 1986 Sudhakar Prabhu and Pattabiranam 1980
Albizzia julibrissin (silk tree)	AII,III BI,II	T-s, Ch-s Ch-s	1:1	Odani et al. 1979
Acacia elata		T-s		Kortt and Jermyn 1981
A. sieberana		T-s, Ch-s	1:1	Joubert 1983
Gramineae				
Hordeum vulgare (barley)	BASI	S-s, αA-s	1:1+1	Svendsen et al.1986
Triticum aestivum (wheat)	WASI	S-s, αA-s	1:1+1	Mundy et al. 1984 Maeda 1986
Secale cereale (rye)	RASI	S-s, αA-s		Weselake et al. 1985 Mosolov and Shulgin 1986

*T, trypsin;   Ch, chymotrypsin;   S, subtilisin;   αA, α-amylase;   s, strong;   w, weak

---

Weselake et al., 1983a,b; Mundy et al., 1983; Hejgaard et al., 1983). More specifically, the α-amylase 2 isozymes from barley, wheat, rye and oats are inhibited, while the α-amylase 1 isozymes are not (Mundy et al., 1984). Amylases from sorghum, rice, saliva, pancreas, *Aspergillus oryzae* or *Bacillus subtilis* were found to be insensitive (Mundy et al., 1984). Not included in Table 2 are certain inhibitors whose classification as Kunitz-type is still uncertain, although their M_rs and amino acid compositions are compatible with such a classification. This is the case for the trypsin inhibitor CPPTI-fm from *Cucurbita pepo* (Pham et al., 1985), the kallikrein inhibitors from potatoes (Hojima et al., 1973), a protease inhibitor from spinach leaves (Satoh et al., 1985), a trypsin inhibitor from white mustard (Menegatti et al., 1985), and the trypsin inhibitors from the tubers of taro (Ogata and Makisumi, 1984). The latter are of particular interest because their dimeric nature has been well established (Ogata and Makisumi, 1985).

```
                1                                                                    50
WBI     E P L L D S E G E L V R N G G T Y Y L L P D R W A L G G G I E A A A T G T E T C P L T V V F S
ELI       V L L D G N G E V V Q N G G T Y Y L L P Q V W A D G G G V Q L A K T G E E T C P L T V V Q S
STI     D F V L D N E G N P L E N G G T Y Y I L S D I T A F - G G I R A A P T G N E R C P L T V V Q S
PAI     D F V L D A E G K F L - N G E I Y Y I L P
AJI1    K E L L D A D G D I L L N G G . Y Y I V
AJI2    S N L L L D T D G N L L E D G G . Y Y I L P A
API     R E L L D V D G N F L R N G G S Y Y I V P A F R G K G G G L E L A R T G S E T C P R T V V Q A
AEI     K G L L D A D G D I L
ASI     . . L L D A D G D L L . S G Y L Y Y I L
BASI    A D P P P V N D T D G H E L R A D A N Y Y V L S A N R A H G G G L T M A P G H G R H C P L F V S Q D
WASI    D P P P V H D T D G N E L R A D A N Y Y V L P A N R A H G G G L T M A P G H G R R C P L F V S G E
RPI     A P P P V Y B T Z S H G L S A B G S Y Y V L P A
```

```
        51                             P₃P₂P₁P′₁P′₂P′₃                                  100
WBI     P N E V S V G E P L R I S S Q L R S G F I P D Y - - - S - L V R I G F A N P P K C A P S P - W W T V
ELI     P N E L S D G K P I R I E S R L R S T F I P D D - - - D - E V R I G F A Y A P K C A P S P - W W T V
STI     R N E L D K G I G T I I S S P Y R I R F I A E G H P L S - L K F D S F A V I M L C V G I P T E W S V
API     P A E Q S R G L P A R L S T P P R I R Y I G P E F Y L T - I E F E E - Q K P P S C L R D S N L Q W K
BASI    P N G Q H D G F P V R I T P Y G V A P S D K I I R - L S T D V R I S F R A Y T T C L Q - S T E W H I
WASI    A D G Q R D G L P V R I A P H G G A P S D K I I R - L S T D V R I S F R A Y T T C V Q - S T E W H I
```

```
        101                                                                           150
WBI     V E D Q P Q Q P S V K L - S E L K S T K F D Y - - - L F K F E K V T S - K F S S Y K L K Y C - A K R
ELI     V E D E Q E G L S V K L - S E D E S T Q F D Y - - - P F K F E Q V S D - K L H S Y K L L Y C - E G K
STI     V E D L P E G P A V K I - G E N K D A M - D G - - - W F R L E R V S D D E F N N Y K L V F C - P Q Q
API     V E E E S Q I - - V K I - A S K E E E Q L F G - - - S F Q I K P Y R D D - - - - Y K L V Y C E P Q Q
BASI    D S E L A A G R R H V I T G P V K D P S P S G R E N A F R I E K Y H G A E V S E Y K L M S C - - - -
WASI    D S E L V S G R R H V I T G P V R D P S P S G R E N A F R I E K Y S G A E V H E Y K L M A C - - - -
```

```
        151                                                     193
WBI     - - - D T C K D I G I Y R D Q - G Y - A R L V V T D E N P L - V V I F - K - K V E S S
ELI     - - H E K C A S I G I N R D Q K G Y - R R L V V T E D N P L T V V L - - K - K D E S S
STI     A E D D K C G D I G I S I D H D D G T R R L V V S K N K P L - V V Q F Q K L D K E S L
AJI1    K D D H C K D L G - S I D D D E N
AJI2    N R D C K D L G I S T D D D N
API     G G R P E C K D L G I S I D D D N N - R R L A V K E G D P L - V V Q F V N A D R E G N
AEI     D D E S C K D L G I S I D D - E N
ASI     S . . E . C Q D L G I . V D D - E
BASI    - - G D W C Q D L G V F R D L K G G A W F L G A T E P Y H V - V V - F K K A P P A
WASI    - - G D S C Q D L G V F R D L K G G A W F L G A T E P Y H V - V V - F K K A P P A
```

The interaction between the Kunitz-type inhibitors and those enzymes against which they show strong inhibition is generally stoichiometric, as summarized in Table 2. All reported trypsin inhibitors form a 1:1 complex with the enzyme and those few that are also effective against chymotrypsin form a 1:1 complex with this enzyme, which can be displaced from the complex by trypsin. Most of the inhibitors that only inhibit chymotrypsin form a 1:1 complex, but certain inhibitors from *Psophocarpus* and *Erythrina* are able to form 1:2 complexes. In this context, it is of interest to note that Bosterling and Quast (1981) demonstrated the ability of STI to bind two molecules of chymotrypsin, an enzyme that is not inhibited by it, while only one molecule of chymotrypsin was bound if the inhibitor had been previously incubated with trypsin. The cereal inhibitors are able to simultaneously bind one molecule of subtilisin and one molecule of α-amylase (Mundy et al., 1983, 1984; Weselake et al., 1983a,b; Halayko et al., 1986).

The inhibitors from species of the Mimosidae subfamily, such as *Acacia*, *Albizzia*, and *Adenanthera* spp., are composed of two disulphide- linked chains, whereas the remaining inhibitors from Leguminoseae and those from Triticeae are single–chained. It seems likely, from their alignment with the single–chain inhibitors, that the two chains are originated from a single chain precursor (see Fig. 1).

Figure.1. (opposite) Alignment of amino acid sequences of members of the soybean trypsin inhibitor (Kunitz) family. WBI is a trypsin inhibitor from winged bean, Psophocarpus tetragonolobus *(Yamamoto et al., 1983); ELI is trypsin inhibitor DE3 from* Erythrina latissima *(Joubert et al., 1985); STI is the Kunitz trypsin inhibitor from soybean (Kim et al., 1985); PAI is a trypsin inhibitor from* Peltophorum africanum *(Joubert, 1981); AJI1 and AJI2 are partial sequences of the two–chained inhibitors AII and BII from* Albizzia julibrissin *(Odani et al., 1979); API is the two–chained trypsin inhibitor DE5 from* Adenanthera pavonina *(Richardson et al., 1986); AEI and ASI are partial sequences of the two–chained trypsin inhibitors from* Acacia elata *(Kortt and Jermyn, 1981) and* Acacia sieberana *(Joubert, 1983), respectively; BASI and WASI are inhibitors of subtilisin and of endogenous α-amylases from barley (Svendsen et al., 1986) and wheat (Maeda, 1986), respectively; RPI is a rice inhibitor (Kato et al., 1972; cited by Mundy et al., 1984). Vertical arrows (▼) indicate the reactive bonds of the trypsin inhibitors. Gaps introduced for the alignment are indicated (-). The N-terminal positions of the second chain of the two–chained inhibitors are indicated by a vertical line (|). Conserved positions are boxed. Unidentified residues are indicated by an asterisk (*).*

Structure and Evolution

Available sequence data for homologues of STI (Kunitz) in different taxa are summarized in Fig. 1, and the homology matrix derived from it is presented in Table 3, where only common segments of the sequences were used to calculate percent homology in the cases in which the sequences were incomplete. As expected, higher homologies are observed within each of the Leguminosae sub–families and among the cereal inhibitors, than between these taxonomic groups, with the notable exception of STI itself, which seems to be significantly closer to the inhibitor from *Peltophorum* (sub–family Caesalpinoideae) than to those of *Erythrina* and *Psophocarpus*, which belong to the Papilionaceae sub–family, together with *Glycine* (soybean). Inhibitors from the most primitive of the Leguminoseae sub–families, Mimosideae, are closer to that of *Peltophorum* and to STI than to the other two inhibitors from the Papilionideae. The cereal inhibitors are significantly closer to those from the Mimosideae than to those from the other two Leguminoseae sub–families. A complete sequence of the *Peltophorum* inhibitor, as well as further sequences from the Caesalpinoideae and the Papilionideae, should help to clarify the evolutionary relationships suggested by the present data.

Table 3. Binary comparisons (% homology) of inhibitors from the STI (Kunitz) family. Homology was calculated for common segments of the sequences that appear in Fig. 1. Abreviations are as in Fig. 1.

INHIBITOR	PAPILIONIDEAE			CESALPINOIDEAE			MIMOSIDEAE			GRAMINEAE		
	WBI	ELI	STI	PAI	AJI1	AJI2	API	AEI	ASI	BASI	WASI	RPI
WBI		66	45	48	40	46	34	37	41	28	26	41
ELI			42	45	33	39	34	33	37	26	26	33
STI				67	51	52	39	43	37	30	30	41
PAI					50	48	52	45	53	33	38	43
AJI1						59	68	79	53	32	32	30
AJI2							68	58	54	37	42	43
API								61	49	26	26	41
AEI									65	32	36	27
ASI										36	36	30
BASI											89	67
WASI												67
RPI												

The reactive bond has been identified in five of the Leguminoseae inhibitors as Arg-Ile or Arg-Ser in exactly homologous positions (Fig. 1). Since no chymotrypsin inhibitor of this group has been sequenced, the apparent conservation of Arg at the P1 position probably reflects this bias. The inhibitors from wheat and barley, which do not inhibit trypsin, respectively have Gly-Ala and Val-Ala in the homologous positions. High variability for the reactive site amino acids have been recognised since early times in proteinase inhibitor research (see Laskowski and Kato, 1980). More recently, this observation has been extended to those amino acids which are in contact with the enzyme, in the case of the ovomucoids (Laskowski et al., 1987), and to a domain of 16 amino acids around the reactive site, in the case of the serpins (Hill and Hastie, 1987). An X-ray crystallographic analysis of the STIa-porcine trypsin complex indicated that STI was in contact with the enzyme at the following positions (see Fig. 1): Asp-Phe (4-5) Asn (16), Pro-Tyr-Arg-Ile-Arg-Phe (65-70), His-Pro (75- 76). Data in Fig.1 are compatible with contact area hypervariability for this group of inhibitors, with the exception of Asn (16), which seems to be fairly conserved, and the already mentioned Arg at the P1 position.

A search for internal repeats was undertaken following the reports of two binding sites for chymotrypsin in STI (Bosterling and Quast, 1981), in WCI2,3 from winged bean (Shibata et al., 1986), and in DE3 from *Erythrina cristagalli* (Joubert and Sharon, 1985). Mostly imperfect short repeats were found and their distribution was not consistent with a straight-forward two-domain structure. Perhaps when inhibitors with a 1:2 stoichiometry are sequenced, a clearer picture will emerge.

A crystallographic study of wheat WASI is in progress (Maeda et al., 1987) and an elucidation of its reactive site with subtilisin, as well as of the contact areas with both subtilisin and endogenous α-amylase would be of great interest. The 3.0 A electron density map is in agreement with predictions of secondary structure based on the amino acid sequence and indicates a similar structure to that of STI (Maeda, private communication). In this context, a survey of the activity against endogenous α-amylases of the inhibitors from the Leguminoseae would help to clarify whether this activity was either lost during the evolution of the lipid-storing species, acquired during the evolution of species with starchy seeds, or still conserved in at least some members of both branches.

Genetics

Genetic variants of STI, designated a, b, and c, have been identified, after screening the USDA germplasm collection by polyacrylamide gel electrophoresis, and found to be allelic forms encoded at a single locus for which a null allele also exists (see Orf and Hymowitz, 1979). Isolation and characterization of the genetic variants was carried out (Freed and Ryan, 1978; Kim et al., 1985): the most frequent allele, STIa (88.8%), differs from the second, STIb (10.9%), at 8 sequence positions, whereas the less frequent one, STIc (0.3%), differs at only one position from STIb, which has led to the speculation that differentiation of the first two alleles was quite ancient and had already been completed in *Glycine soja*, the wild progenitor of the cultivated species *Glycine max* (Kim et al., 1985). The STI locus has been assigned to linkage group 9 in G. max, at 16.2 ± 1.5 map units from the acid phosphatase locus (Hildebrand et al., 1980) and at 15.3 ± 0.9 of the leucine aminopeptidase locus (Kiang and Chiang, 1986).

The gene encoding the barley inhibitor has been assigned to barley chromosome 2, after immunochemical analysis with monospecific antibodies of all barley-wheat addition lines (Hejgaard et al., 1984a), and that for the rye gene has been similarly assigned to chromosome 2R (Hejgaard et al., 1984b).

Physiology

Messenger RNA for STI was isolated and translated *in vitro* by Vodkin (1981). A polypeptide was obtained that had higher apparent M_r than that of the inhibitor obtained from mature seeds, M_r 23,200 vs. M_r 21,500. This finding was confirmed by sequencing a cDNA clone which encoded a protein with a putative leader sequence of 25 amino acids preceding the known sequence of STI (Hoffman et al., 1984). The ultrastructural localization of STI in thin sections was carried out by Horisberger and Tacchini (1982), using specific antibodies and protein-A-gold label. In the cotyledons, STI was found in the cell wall and in most of the protein bodies, but not in the cytoplasm, while in the embryonic axis, some of the label was also associated with the cytoplasm. An investigation of the fate of STI during germination has shown a rather precise proteolytic processing at the carboxyl terminus, which seems to involve a five residue sequence in STIa (Hartl et al., 1986). This would argue against its possible role as a store for sulphur amino acids and would suggest some unknown specific function in germination.

The interaction of barley α-amylase II with BASI has been studied in detail by electrophoresis and by a variety of physical methods (Waselake et al., 1983a; Mundy et al., 1983; Halayko et al., 1986). The inhibitor combines with the enzyme at a molar ratio 2:1, affecting an enzymic tryptophan residue which is essential for productive enzyme-substrate binding. The inhibitor seems to be synthesized in barley endosperm at least up to 30 days after anthesis and its mRNA is not detected in mature dry aleurone, although it appears upon "*in vitro*" incubation and is dramatically induced by treatment with abcisic acid (Mundy and Rogers, 1986). In a study where de-embryonated seeds were incubated with labelled amino acids, the enzyme-inhibitor complex was not significantly labelled (Rodaway, 1978), suggesting a possible regulatory role of the inhibitor during seed development, by inhibiting the enzyme during starch synthesis, and/or, in the event of premature sprouting, by preventing starch degradation (Mundy et al., 1983).

BOWMAN-BIRK INHIBITOR FAMILY

Following the discovery of antitryptic activity in soybeans, two different inhibitors were eventually identified, the already described Kunitz type and a second one, separated by Bowman (1946). A characterization of the second inhibitor was carried out by Birk et al. (1963a,b), hence its present designation as Bowman-Birk inhibitor (BBI), and its covalent structure was established by Odani and Ikenaka (1972, 1973a). Homologues of BBI have been isolated in the seeds of numerous species. A first-hand account of the early research on BBI has been published by Birk (1985) and a thorough review of the BBI family has appeared quite recently (Ikenaka and Norioka, 1986). A succinct description of this family will be presented here and only very recent developments will be treated in some detail.

Distribution and Inhibitory Properties

The Bowman-Birk inhibitor (BBI) is a protein of 71 amino acid residues, with 7 disulphide bridges, that inhibits two serine proteases simultaneously, trypsin and chymotrypsin. The double-headed nature of this inhibitor was elegantly corroborated by Odani and Ikenaka (1973b), who fragmented the molecule with cyanogen bromide and pepsin into the two active domains and a C-terminal tetrapeptide. Homologues of BBI, with two reactive sites and capable of binding two molecules of serine proteases simultaneously, have been described in a number of species. In general, specificity is determined by the residue at P1 position of the reactive site: arginine and lysine for trypsin, tyrosine and phenylalanine for chymotrypsin, and alanine for elastase (see Ikenaka and Norioka, 1986, and Fig. 2). Exceptions are the first reactive sites of the peanut inhibitors (Norioka and Ikenaka, 1983b, 1984) and the second reactive site of the soybean inhibitor C-II (Odani and Ikenaka, 1977), which can bind either trypsin or chymotrypsin, although an Arg residue at the P1 position would predict specificity for trypsin in all the cases. A different exception is that represented by inhibitor II from azuki bean, which can not bind trypsin and chymotrypsin simultaneously, although it is truly double-headed (Yoshikawa and Ogura, 1978).

Inhibitors of the Bowman-Birk type have been described in the seeds of many Leguminoseae: soybean (Odani and Ikenaka, 1972, 1976, 1977), garden bean (Wilson and Laskowski, 1975), azuki bean (Ishikawa et al., 1979; Yoshikawa et al., 1979a; Kiyohara et al., 1981), mung bean (Zhang et al., 1982; Wilson and Chen, 1983), cowpea (Morphy et al., 1985), lima bean (Stevens et al., 1974), *Macrotyloma axillare* (Joubert et al., 1979), *Lonchocarpus capassa* (Joubert, 1984a), *Pterocarpus angolensis* (Joubert, 1982), *Vicia angustifolia* (Shimokawa et al., 1984), peanut (Norioka and Ikenaka, 1983a,b), navy bean (Wagner and Riehm, 1967), kidney bean (Putzai, 1968), black-eyed pea (Gennis and Cantor, 1976), runner bean (Hory and Weder, 1976) and lentil (Weder, 1986). Until recently, the BBI family was thought to be restricted to the Papilionaceae (Fabaceae), the most advanced sub–family of the Leguminoseae. The recent demonstration by Odani et al. (1986) of significant homology of the germ trypsin inhibitors from cereals with BBI extends the distribution of this type of inhibitor to a distant family and suggests that it may be present in other taxa. Besides the typical two–headed inhibitors, a second single–domain class has been identified in cereals which has not been found in any others species (Odani et al., 1986).

```
                          P P P P'P'P'
          1               3 2 1 1 2 3                    44
SBI1          S D H S S S D D E S S K P C C D L C M C T A S M P P - - Q C H C A D I R L N S C
SBI2          S D Q S S S Y D D D E Y S K P C C D L C M C T R S M P P - - Q C S C E D I R L N S C
GBI               Z P S Z S S P P C C B I C V C T A S I P P - - Q C V C T B,I R L B S C
ABI1    S V H H Q D S S D E P S E S S H P C C D L C L C T K S I P P - - Q C Q C A D I R L D S C
ABI2          D S S D E P S E S S E P C C D L C L C T K S I P P - - Q C Q C A D I R L D S C
MBI1            S H D E P S E S S E P C C D S C D C T K S K P P - - Q C H C A N I R L N S C
MBI2    S S H H D S S D E P S E S S E P C C D S C R C T K S I P P - - Q C H C A D I R L N S C
BBI             D D E S S K P C C D Q C A C T K S N P P - - Q C R C S D M R L N S C
BTCI    S G H H Z B S T B Z A S Z S S K P C C R Z C A C T K S I P P - - Z C R C S B V R L N S C
LBI     S G H H E H S T D Z P S Z S S K P C(C,B H(C,A,C)T K S I P P - - Q C R C S D F R L D S C
ABI3    S G H H D E T T D E P S E S S K P C C D Q C S C T K S M P P - - K C R C S D I R L N S C
MAI1      D H H H S T D E P S E S S K P C C D E C A C T K S I P P - - Q C R C T D V R L N S C
MAI2        H E H S S D E S S E S S K P C C D L C T C T K S I P P - - Q C H C N D M R L N S C
VAI             G D D V K S A C C D T C L C T R S Q P P - - T C R C V D V G - E R C
PI1       E A S S S S D D N V C C N G C L C D R R A P P Y F E C V C V D T F - D H C
PI2                     V C C N G C L C D R R A P P Y F E C V C V D T F - D H C
PI3                 A A S D C C S A C I C D R R A P P Y F E C T C G D T F - D H C
LCI     D D D H S D D E P(R)E S E S(S)K P(C,C,S,S,C,-,C,T)R S R P P - - Q C Q C(T)D V R L N(S)C
PAI       K G G A S H P D R Q D K E C C E A C F C T L P Y P T G - T C M C A D V A - D Y C
CPI             S T - T T A C C D S C V C T K S I P P - - Q C R C N D M
WGI1          A A K K R P W K C C D Q A V C T R S I P P - - I C R C M D Q V F Q - C
WGI2            A T R P W K C C D R A I C T K S F P P - - M C R C M D M V E Q - C
```

```
        45             P P P P'P'P'                        88
                       3 2 1 1 2 3
SBI1    H S A C D R C A C T R S M P G Q C R C L D T T D - F - C Y K P - C K S S D E D D D
SBI2    H S D C K S C M C T R S Q P G Q C R C L D T N D - F - C Y K P - C K S R D D
GBI     H S A C K S C M C T R S M P G K C R C L B T T B - Y - C Y K S - C K S B S G Z B B
ABI1    H S A C K S C M C T R S M P G Q C R C L D T H D - F - C H K P - C K S R D K D
ABI2    H S A C K S C M C T R S M P G Q C R C L D T H D - F - C H K P - C K S R D K D
MBI1    H S A C K S C I C T R S M P G K C R C L D T D D - F - C Y K P - C E S M D K D
MBI2    H S A C K S C M C T R S M P G K C R C L D T D D - F - C Y K P - C E S M D K D D D
BBI     H S A C K S C I C A L S Y P A Q C F C V D I T D - F - C Y E P - C K P S E D D K E N
BTCI    H S A C K S C A C T F S I P A Z C F C G B I B B B F - C Y K P - C K S S H S B B B B W N
LBI     H S A C K S C I C T L S I P A Q C V(C,B,B)I T D - F - C Y E P - C K S S H S D D D N N N
ABI3    H S A C K S C A C T Y S I P A K C F C T D I N D - F - C Y E P - C K S S R D D D W D N
MAI1    H S A C S S C V C T F S I P A Q C V C V D M K D - F - C Y A P - C K S S H D D
MAI2    H S A C K S C I C A L S E P A Q C F C V D T T D - F - C Y K S - C H N N A E K D
VAI     H S A C N H C V C N Y S N P P Q C Q C F D T H K - F - C Y K A - C H S E E K E E V I K N
PI1     P A S C N S C V C T R S N P P Q C R C T D K T Q - G R C P V T E C R S
PI2     P A S C N S C V C T R S N P P Q C R C T D K T Q - G R C P V T E C R S
PI3     P A A C N K C V C T R S I P P Q C R C T D R T Q - G R C P L T P C A
LCI     H(S)A C K(S)C(M,C,T,F)S D P G M C S C L D V T D - F - C Y K P - C K S S G D D D(Z,Z)
PAI     P A A C K K C Y C T K S I P P Q C H C A D S A P - N N C Y I P A C E G(F,K)
WGI1    P S T C K A C G P S V G D P S R R V C Q D Q Y V
WGI2    A A T C K K C G P A T S D S S R R V C E D • Y
```

286

Table 4. Binary comparisons (% homology) of members of the Bowman-Birk inhibitor family. Homology was calculated for common segments of the sequences that appear in Fig.2. Abreviations for inhibitors are as in Fig. 2. Tribes were abreviated as follows: Gly, Glycineae;Pha,Phaseoleae; Vic, Vicieae; Sti,Sti losantheae; Dal, Dalbergiae;Gra, Gramineae.

GROUP			I							II						III	IV							
	INHIBITOR / TRIBE		SBI1	SBI2	GBI	ABI1	ABI2	MBI1	MBI2	BBI	BTCI	LBI	ABI3	MAI1	MAI2	VAI	PI1	PI2	PI3	LCI	PAI	CPI	WGI1	WGI2
I	Gly	SBI1		77	72	73	75	72	75	67	65	63	63	62	64	47	38	35	41	66	39	50	31	31
	Gly	SBI2			69	72	75	69	69	67	59	59	61	60	61	49	38	35	39	63	36	50	35	33
	Pha	GBI				77	78	74	79	57	64	69	65	65	67	44	42	40	41	66	37	61	37	33
	Pha	ABI1					97	79	83	62	67	69	64	66	65	48	39	35	39	61	36	61	35	35
	Pha	ABI2						82	88	63	69	71	66	68	67	50	41	38	41	65	38	61	35	35
	Pha	MBI1							93	63	67	68	68	67	67	45	35	33	36	64	37	58	31	33
	Pha	MBI2								61	70	69	67	68	68	46	37	32	37	64	37	65	35	35
II	Gly	BBI									75	79	75	75	74	45	34	34	36	65	39	61	37	41
	Pha	BTCI										83	76	79	67	44	35	30	37	63	38	61	38	35
	Pha	LBI											77	81	72	44	38	33	41	59	40	65	38	39
	Pha	ABI3												74	63	43	35	31	37	59	37	58	37	37
	Pha	MAI1													68	48	37	33	39	67	39	65	37	37
	Pha	MAI2														46	35	31	36	58	33	69	35	39
III	Vic	VAI															43	43	43	42	40	54	36	33
IV	Sti	PI1																87	75	37	37	36	27	27
	Sti	PI2																	75	31	34	32	27	27
	Sti	PI3																		37	46	35	29	31
	Dal	LCI																			35	54	33	33
	Dal	PAI																				33	27	31
	Vic	CPI																					54	58
	Gra	WGI1																						66
	Gra	WGI2																						

Figure 2. (opposite). Alignment of amino acid sequences of proteinase inhibitors of the Bowman Birk family. SBI1 and SBI2 are respectively inhibitors CII and DII from soybean (Odani and Ikenaka, 1976, 1977); GBI is inhibitor II' from garden bean (Wilson and Laskowski, 1975); ABI1 and ABI2 are inhibitors I- A,A' and IIa,c, respectively, from azuki bean (Kiyohara et al., 1981; Ishikawa et al., 1985); MBI1 and MBI2 are two inhibitors from mung bean (Zhang et al., 1982; Wil son and Chen, 1983); BBI is the Bowman–Birk inhibitor from soybean (Odani and Ikenaka, 1972); BTCI is an inhibitor from cowpea (Morphy et al., 1985); LBI is inhibitor IV-IV' from lima bean (Stevens et al., 1974); ABI3 is inhibitor I-II from azuki bean (Ishikawa et al., 1979; Yoshikawa et al., 1979a); MAI1 and MAI2 are respectively inhibitors DE3 and DE4 of Macrotyloma axillare (Joubert et al., 1979); VAI is an inhibitor from Vicia angustifolia (Shimokawa et al. , 1984); PI1, PI2 and PI3 are respectively inhibitors AII, BIII, and BII from peanuts (Norioka and Ikenaka, 1983a,b); LCI is inhibitor DE4 from Lonchocarpus capassa (Joubert, 1984a); PAI is inhibitor DE1 from Pterocarpus angolensis (Joubert, 1982); CPI is an inhibitor from chickpea (Belew and Eaker, 1976); WGI1 and WGI2 are inhibitors from wheat germ (Odani et al., 1986). Vertical arrows (▼) indicate the reactive bonds. Gaps are indicated (-)

Structure and Evolution

Inhibitors of the BBI family have been classified into four groups according to their sequence homology and a phylogenetic tree has also been deduced (see Ikenaka and Norioka, 1986). In Fig. 2 the sequences included in the alluded classification have been aligned with the following additional ones: cowpea inhibitor (Morphy et al., 1985), chickpea inhibitor (Belew and Eaker, 1976), two sequences from species of the tribe Dalbergieae, namely *Lonchocarpus capassa* (Joubert, 1984a) and *Pterocarpus angolensis* (Joubert, 1982), and two trypsin inhibitors from wheat germ (Odani et al., 1986). The percentage of homology for the length of sequence compared in each case is presented for all binary comparisons in Table 4. From these data the following aspects merit attention:-

i) Although quite diverged, the cereal inhibitors are more closely related to each other than to any other of the inhibitors.

ii) The inhibitor from chick pea is about equidistant to the cereal inhibitors (54%–58% homology), to group I (50%–65%), to group II (61%–69%), and to the only other sequenced inhibitor from the tribe Vicieae (Group III, 54%).

iii) The inhibitor from cowpea fits into group II.

iv) The two inhibitors from the tribe Dalbergieae are quite different from each other (35% homology); while LCI–DE4 is close and about equidistant to group I (61%–66%) and to group II (58%–67%), PAI–DE1 is quite apart from all the other inhibitors (33%–46%).

v) Inhibitors from the same species, i.e. soybean, belong to more than one group.

These observations clearly indicate that a good part of the divergence of the multigene families encoding inhibitors of the BBI–type occurred before speciation of the Leguminoseae and that the apparent lack of correlation between the sequence data and the taxonomic classification is due to non–homology of the loci being compared, which makes the above–mentioned phylogenetic tree difficult to interpret in evolutive terms. Besides sequencing new inhibitors from new species, a clear priority from an evolutionary point of view would be to determine all the different sequences of BBI–type inhibitors within a number of species and to investigate the genome organization of the corresponding multi–gene families.

An ancestral monovalent inhibitor that would have generated the double–headed ones by internal gene duplication has been postulated (Tan and Stevens, 1971; Odani and Ikenaka, 1976). In this context, the recent finding of single–headed inhibitors in wheat germ further supports the hypothesis and suggests that these are the most primitive among the known members of this family. Then, the cereal double–headed inhibitors would be the closest to the ancestral double–headed inhibitor, being the less diverged from the single–headed one, and the inhibitor from chickpea would be the most primitive among the dicot inhibitors, being the closest to the wheat ones. However, the single–domain cereal inhibitors could be the result of post–translational processing of double–headed precursor or could be encoded by genes originated through deletion of the second domain sequence. Molecular cloning of the pertinent genes and further protein sequences might allow to discriminate between these alternatives.

Very recently, the structure of the BBI-type inhibitors AII from peanut has been determined by X-ray crystallography at a 3.3 Å resolution (Suzuki et al., 1987). The structures of the two domains are similar and are related by an intramolecular approximate twofold rotation axis, while the two independent protease binding sites protrude from the molecular body on opposite sides. Based on this three-dimensional structure, the evolutionary scheme proposed by Odani and Ikenaka (1976) has been further refined (Suzuki et al., 1987), postulating a second partial duplication, that would have extended the C- terminal and in turn replaced the N-terminal residues of the first domain, and a change of the pairing partners of the disulphide bridges, as represented in Fig. 3.

Other recent structural studies have been carried out with inhibitor BIII from peanut, using 1H NMR (Koyama et al., 1986), and with the complex of bovine trypsin and inhibitor ABI from azuki beans (Tsunogae et al., 1986). The first study indicated that cleavage of the two reactive sites resulted in conformational changes that exclusively affected the disulphide loops containing the reactive sites. From the second work it was concluded that, similarly to other families of inhibitors, both the primary and the secondary contact regions ("front side" and "back side" of the ring) were essential for the inhibitory interaction.

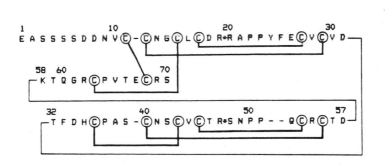

Figure. 3. The primary sequence of PI1 from peanut, aligned to highlight the three–dimensional domain structure and internal homology, according to Suzuki et al. (1987). The reactive sites are marked by asterisks and thick lines show disulphide bridges.

Genetics and Molecular Cloning

There is a surprising lack of genetic studies concerning the intraspecific variation and genome organization of the multi-gene families encoding the BBI family of inhibitors. As already pointed out, such studies would greatly help our understanding of the evolution of this protein family. The molecular cloning and analysis of a gene coding for BBI has been reported by Larkins and co-workers (Hammond et al., 1984). They constructed and characterized a cDNA clone corresponding to BBI and, using this clone as a probe, determined that the BBI gene was present in only one or two copies per genome. They did not find cDNA clones for other members of the family. Sequence analysis of a genomic clone revealed that it was similar, but not identical, to the cDNA sequence and that it had no introns. Additional cDNA and genomic sequences will have to be analysed to explain the significance of the observed variation between the two clones. It is of interest to note that while the BBI cDNA probe did not hybridize with other genes of the BBI family in soybean, it was able to do so with homologous genes from mimosa and redbud. Based on the DNA sequence, it seems that the initial translation product must contain a short leader sequence. An estimate of divergence time of 310 million years was made for the first and second domains of BBI based on the divergence of their nucleotide sequences.

Physiology

The BBI and its homologues do not appear to accumulate in any plant tissue other than the developing seed (Hwang et al., 1978). BBI mRNA accumulates early during the mid-maturation stage and reaches a steady state later in development (Foard et al., 1982; Hammond et al., 1984). The quantity of protein accumulated is of the same order as that for the Kunitz inhibitor.

During germination, at least part of the inhibitor is rapidly released from the seed within the first 8h of imbibition (Hwang et al., 1978; Wilson, 1980; Tan-Wilson and Wilson, 1982; Horisberger and Tacchini, 1982). The rapid release suggests its elution from a barrier-free pool, which is in line with the fact that BBI is detected in the intercellular space prior to germination but is absent from this space in 4-day old seedlings. Besides this release, it has now been documented in azuki bean (Yoshikawa et al., 1979b), soybean (Madden et al., 1985), and mung bean (Lorensen et al., 1981; Wilson and Chen, 1983) that this type of inhibitor undergoes a specific and extensive proteolytic processing during the early stages of germination and seedling growth. The significance of these phenomena is unclear, but they seem to exclude a reserve role for this protein family.

CEREAL TRYPSIN/α-AMYLASE INHIBITOR FAMILY

This protein family includes a wide range of components whose structural relationships, often unsuspected, have only recently been fully demonstrated. Inhibitors of heterologous α-amylases were first described in wheat endosperm by Kneen and Sandstedt (1943). Years later, work on the main wheat albumins (Fish and Abbot, 1969; Ewart, 1969; Feillet and Nimmo, 1970; Sodini et al., 1970; Cantagalli et al., 1971, and others), which were found to be identical to some of the inhibitors, led to the realization that the inhibitors represent a major part of the albumin and globulin fraction of the endosperm. The first trypsin inhibitor of this family was isolated from barley endosperm by Mikola and Suolinna (1969), but its relationship to the α-amylase inhibitors was discovered only recently (Odani et al., 1982, 1983a). The CM- proteins from wheat, barley, and rye, which were so designated because of their solubility in chloroform:methanol mixtures (Garcia-Olmedo and Garcia-Faure, 1969; Garcia-Olmedo and Carbonero, 1970; Rodriguez-Loperena et al., 1975; Salcedo et al., 1978b, 1982; Paz-Ares et al., 1983a), were eventually found to be members of the trypsin/α-amylase inhibitor family, an observation that led to the discovery of new α-amylase and trypsin inhibitors (Shewry et al., 1984; Barber et al., 1986a,b; Sanchez-Monge et al., 1986b). Finally, homologous relationships of the cereal inhibitors were established with the 2S storage proteins from castor bean (*Ricinus communis*) (Odani et al., 1983c) and the Kazal secretory trypsin inhibitor from bovine pancreas (Odani et al., 1983b), thus showing that this protein super-family is distributed beyond the plant kingdom.

Distribution and Inhibitory Properties

Inhibitors of heterologous α-amylases have been described in a number of species, although the available information for some of them is not as complete as that for those of wheat and barley.

Fractionation of a partially purified NaCl extract by gel filtration allowed a classification of the wheat inhibitors into three classes - monomeric, dimeric and tetrameric - which showed different specificities: the dimeric inhibitor seemed to inhibit mainly α-amylase from human saliva, while all three inhibitors were active against α-amylase from the insect *Tenebrio molitor* (Petrucci et al., 1974; for a review see Buonocore et al., 1977). Purification and characterization of individual inhibitors from wheat was undertaken at several laboratories (Shainkin and Birk, 1970; Saunders and Lang, 1973; Silano et al., 1973; Petrucci et al., 1976, 1978; O'Donnell and McGeeney, 1976; O'Connor and McGeeney, 1981a; Maeda et al., 1982) and two types were characterized, respectively designated 0.28 and 0.19. The first type represented monomeric variants of about M_r 12,000 that were more active against insect α-amylases than against the human salivary or pancreatic ones. Those of the second type were dimeric of about M_r 24,000 and were more effective against insect enzymes. The dimers could be dissociated into M_r 12,000 subunits that were initially thought to be different (Silano et al., 1973), although more recent evidence suggests that they might be mixtures of homodimers (Maeda et al., 1985). Both types seem to have one molecule of carbohydrate per subunit, which might be important for their inhibitory activity (Silano et al., 1977; Petrucci et al., 1978), and to share a number of physicochemical characteristics (Silano et al., 1973; Silano and Zahnley, 1978). Their interaction with the enzyme was also studied and stoichiometries of 2:1 for 0.28 and 1:1 for 0.19 were proposed. The sugar moieties were suspected to be essential for the formation of the complex, which would explain the reversion of the inhibition by maltose (Silano et al., 1977; and Buonocore et al., 1980).

The wheat tetrameric inhibitor has been less actively investigated. O'Connor and McGeeney (1981a) isolated an inhibitor with an apparent M_r of 63,000, which dissociated into subunits of M_r 14,000 and 15,000 with no carbohydrate, while Buonocore et al. (1985) determined an apparent M_r of 48,000 by equilibrium sedimentation and separated four electrophoretic bands, all of which seemed to be inhibitory. According to these authors, this inhibitor was active against both insect and human α–amylases but not against those from bacteria (*Bacillus subtilis*) or from fungi (*Aspergillus oryzae*). Amino acid analyses and circular dicroism spectra of the subunits indicated that they were similar to those of the monomeric and dimeric inhibitors. Recent reconstitution experiments have shown that proteins CM2, CM3 and CM16 from tetraploid wheat and the same proteins plus CM1 and CM17 from hexaploid wheat are components of the tetrameric inhibitors (Sanchez-Monge, unpublished), thus identifying the subunits of the inhibitors described by Buonocore et al. (1985).

An extensive survey of the inhibitory properties of the three size classes of inhibitors versus α–amylases from 18 insect, 23 marine, and 17 avian and mammalian species was carried out by Silano et al. (1975). It was found that the monomeric class was mostly effective against the insect enzymes, especially against those of grain predators, the M_r 24,000 class was more effective against marine, avian and mammalian α–amylases, and no clear pattern was observed for the tetrameric class. These findings probably reflect the properties of the predominant components within each class, and not the properties of specific purified components. Thus Orlando et al. (1983) found inhibition of the *B.subtilis* α–amylase by components of the 0.19 family, and the monomeric fraction from *Aegilops speltoides* also inhibits the human enzyme (Bedetti et al., 1974).

The existence of α–amylase inhibitors corresponding to the three size classes has been recently demonstrated in barley, and the previously characterized proteins CMa, CMb and CMd have been identified as subunits of the tetrameric inhibitor (Sanchez-Monge et al., 1986b). These three proteins have been shown to be truly homologous to those of the monomeric and dimeric inhibitors from wheat and only one of them, CMa, has been found to be active by itself (Barber et al., 1986a,b; Sanchez-Monge et al., 1986b). Reconstitution experiments indicated that all binary mixtures were active, including those between inactive subunits, and that the mixture of the three subunits had the highest specific activity (Sanchez–Monge et al., 1986b). The barley tetrameric inhibitor was active against the α–amylase from *T.molitor* but showed no effect against salivary α–amylase. Evidence has also been obtained for the presence of a tetrameric inhibitor with similar subunits in the wild barley *Hordeum chilense* (Sanchez–Monge et al., 1987).

A barley protein designated CMb' has been recently characterized as the subunit of a homodimeric α–amylase inhibitor that seems to be active against α–amylase from *T. molitor* but not against the salivary one (Sanchez–Monge, unpublished).

The only reported inhibitor from this family which has been found to be bifunctional was isolated from ragi (*Eleusine coracana*) by Shivarat and Pattabiranam (1980, 1981). This inhibitor is able to inhibit α–amylase and trypsin independently and is able to form a ternary complex with the two enzymes. The α–amylases of porcine pancreas, human pancreas and human saliva were inhibited in the ratio 5:5:1. A probable isoform of this inhibitor was later isolated by Manjunath et al. (1983).

Oligomeric inhibitors of α–amylase have been identified in *Phaseolus vulgaris* (Pick and Wober, 1978; Lajolo and Finardi– Filho, 1985). There is no evidence that the *Phaseolus* inhibitor is related to the cereal ones, whereas the amino acid composition of that from black bean is quite untypical of this group.

Other α–amylase inhibitors, whose assignment to this family is still uncertain, are heterodimers with the subunits linked by disulphide bridges; they have been found in rye (Granum, 1978), pearl millet (Chandrasekher and Pattabiranam, 1985), sorghum (Moideen Kutty and Pattabiranam, 1986) and *Echinocloa frumentacea* (Moideen Kutty and Pattabiranam, 1985). Finally, a 24,000 M_r heterodimer has been described in *Setaria italica* that is made up of non–covalently linked subunits of M_r 12,000 and 16,000 (Nagaraj and Pattabiranam, 1985). More structural information about these inhibitors will be needed for their proper classification. It is not unlikely that at least some of them might have to be included in a new class.

Besides the trypsin inhibitor from ragi, which is bifunctional, other trypsin inhibitors from this group have been identified in barley, maize, rye, rice, and probably in wheat and sorghum.

The barley inhibitor was first identified by Mikola and Suolinna (1969), who isolated a M_r 14,100 protein from endosperm active against trypsin and inactive against chymotrypsin, papain, subtilopeptidase A, pepsin, bacterial or fungal proteinases, as well as against the endogenous proteinases from green malt. Antibodies raised against this protein did not react with inhibitors from barley embryo or with wheat endosperm extract; cross–reactivity with rye endosperm extract was observed (Mikola and Kirsi, 1972). A similar inhibitor was purified by Boisen (1976) from a different genetic stock. The inhibitor was sequenced by Odani et al. (1983a), who did not find activity against elastase or α–amylases from various sources.

Inhibitors which are probably related to that from barley have been characterized in wheat endosperm (Shyamala and Lyman, 1964; Boisen and Djurtoft, 1981a; Mitsunaga et al., 1982). At least four different species seem to be present in this tissue, which inhibit trypsin at a 1:1 ratio and chymotrypsin in a non-stoichiometric manner. They have similar heat stability and isoelectric points to that of barley, and their reported amino acid compositions are compatible with interspecific homology, although they are moderately divergent.

A trypsin inhibitor which seems to be closely related to that of barley has been characterized in rye (Polanowski, 1974; Boisen and Djurtoff, 1981b; Chang and Tsen, 1981a,b). This inhibitor is also weakly active against chymotrypsin.

The trypsin inhibitor from maize has been well characterized. It was first isolated from the opaque–2 mutant as a M_r 11,000 protein which inhibited trypsin by forming a 1:1 complex and was inactive against chymotrypsin (Swartz et al., 1977). The complete amino acid sequence of this inhibitor was obtained and found to be homologous to the barley one (Mahoney et al., 1984). Similar or identical inhibitors have been identified in different types of maize (Johnson et al., 1980) and in teosinte (Corfman and Reeck, 1982). An of the Hageman factor fragment for which three variants were isolated from maize by Hojima et al. (1980a) is probably identical to the trypsin inhibitor.

An inhibitor from rice bran of about M_r 14,500 which forms a 2:1 complex with trypsin (Tashiro and Maki, 1979; Maki et al., 1980) and one from sorghum (Filho, 1974) seem to be also of the barley type.

No inhibitory properties have been described for the 2S storage proteins from *Ricinus* (Sharief and Li, 1982), *Brassica* (Crouch et al., 1983), *Lupinus* (Lilley and Inglis, 1986) or *Bertholletia* (Ampe et al., 1986).

Structure and Evolution

Total or partial amino acid sequences have been obtained for different members of this protein family either by direct protein sequencing or indirectly from cloned cDNAs (Fig. 4). Percentages of homology calculated for all possible binary comparisons are presented in Table 5. A higher number of members have been characterized in wheat and barley than in any other species. Evolutionary implications of the observed homologies are best understood if they are considered together with the "*in vitro*" activities of the proteins and the chromosomal locations of their corresponding genes in both species. The distribution among chromosomes of the multi–gene family encoding this group of proteins has been investigated through the analysis of wheat aneuploids and barley-wheat addition lines. In the case of wheat and barley, a number of loci can be postulated based on the general observation that there is greater homology between a given

Table 5.   Binary comparisons (% homology) of members of the cereal α-amylase/trypsin inhibitor family. Homology was calculated for common segments of the sequences that appear in Fig. 4. Abreviations are as in Fig. 4.

PROTEINS / SPECIES		CM16	CM17	CMb	CM3	CMd	CM2	CM1	CMa	CMc	C13	0.53	0.19	0.28	C44	C23	C38	CMe	RBI	MTI	PA	2SC	2SR	2SB	2SL	KB
WHEAT	CM16		78	83	45	40	45	41	38	33	35	27	27	27	32	22		38	43	37	19	17	16	27	13	
WHEAT	CM17			70	37	36	37	37	35	33	37	26	26	30	30	25		37	37	33	20	19	18	19	19	
BARLEY	CMb				52	47	45	41	38	33	35	30	30	30	29	26		45	47	43	22	17	16	23	13	
WHEAT	CM3					70	50	39	43	38	34	41	41	30	26	32		46	45	48	16	17	13	27	24	
BARLEY	CMd						47	37	40	29	30	30	23	22	35			43	39	35	22	16	13	27	19	
WHEAT	CM2							86	86	72	40	38	34	37	39	32		54	52	41	22	21	13	27	17	
WHEAT	CM1								82	69	43	34	31	33	35	32		53	48	41	19	21	13	23	17	
BARLEY	CMa									78	41	31	28	30	32	32		50	48	38	19	24	13	23	21	
BARLEY	CMc										34	31	28	30	32	35		45	43	33	22	17	13	27	20	
BARLEY	C13											22	23	23	22	30	23	36	34	38	19	15	12	16	16	17
WHEAT	0.53												94	58	45	26	27	28	29	30	19	19	15	17	21	11
WHEAT	0.19													56	44	26	28	30	29	30	19	20	16	18	23	11
WHEAT	0.28														63	24	28	30	29	30	22	19	15	17	19	10
BARLEY	C44															24	22	23	26	26	27	17	14	16	16	8
BARLEY	C23																32	42	49	47	22	18	13	19	16	16
BARLEY	C38																	34	35	31	7	16	15	17	16	15
BARLEY	CMe																		56	51	27	23	14	18	18	22
RAGI	RBI																			64	22	19	14	20	18	21
MAIZE	MTI																				22	16	12	15	18	24
PEA	PA																					11	14	12	8	
CA.BEAN	2SC																						35	30	28	8
RAPE	2SR																							21	22	9
BR.NUI	2SB																								22	6
LUPIN	2SL																									14
COW	KB																									

protein from one genome and the appropriate one from a different genome than between that protein and any other encoded in the same genome. The proposed loci are listed together with the *in vitro* activities of the corresponding proteins in Table 6. These data indicate that this dispersed multi–gene family has originated both by translocation and by intrachromosomal duplication, and that most if not all of the dispersion must have occurred prior to the branching–out of the barley genome from the diploid genomes included in allohexaploid wheat. The clearest case of an intrachromosomal duplication is that of the Cma and Cmc loci in chromosome 1 of barley, whose corresponding proteins are closer to each other than to any other encoded in the same genome. Nevertheless, protein CMa shows an even higher similarity to proteins CM1 and CM2, which are respectively encoded in chromosomes 7D and 7B of wheat and, as CMa, are subunits of the tetrameric α–amylase inhibitors. In contrast, CMc is a trypsin inhibitor which shows a much greater divergence from CMe, the other trypsin inhibitor encoded in chromosome 3, than from CMa (45% homology vs 78%).

Figure 4. (opposite) Alignment of amino acid sequences of members of the cereal α–amylase/trypsin inhibitors family. Wheat CM–proteins CM1, CM2, CM3, CM16 and CM17 are subunits of tetrameric α–amylase inhibitors (Shewry et al., 1984; Barber et al., 1986a); barley CM–proteins CMa, CMb and CMd are subunits of tetrameric α–amylase inhibitors (Barber et al., 1986b) and CMc and CMe are trypsin inhibitors (Shewry et al., 1984; Barber et al., 1986a; Odani et al., 1983a; Lazaro et al., 1985); 0.19 and 0.53 are wheat dimeric α–amylase inhibitors (Maeda et al., 1985) and 0.28 is a wheat monomeric α–amylase inhibitor (Kashlan and Richardson, 1981); C13, C23, C38 and C44 are amino acid sequences deduced from barley cDNA clones pUP13, pUP23, pUP38 and pUP44, respectively, the latter corresponds to a dimeric α–amylase inhibitor (Paz–Ares et al., 1986; Lazaro, unpublished); RBI is an α–amylase/trypsin inhibitor from ragi, Eleusine coracana (Campos and Richardson, 1983); MTI is a trypsin inhibitor from maize (Mahoney et al., 1984); PA is a sulphur–rich pea albumin (Higgins et al., 1986); 2SC, 2SR, 2SB and 2SL are 2S storage proteins from Ricinus communis (Sharief and Li, 1982), Brassica napus (Crouch et al., 1983), Bertholletia excelsa (Ampe et al., 1986), and Lupinus angustigolius (Lilley and Inglis, 1986), respectively; KB is the bovine Kazal inhibitor (Greene and B artelt, 1969). Gaps introduced for the alignment are indicated (-). Vertical arrows (▼) indicate the reactive bonds of the trypsin inhibitors. The three sequence blocks approximately correspond to the three domains defined by Kreis et al. (1985). The N–terminal positions of the second chain of the two chained 2S globulins are indicated by a vertical line (|). Conserved positions are boxed. Unidentified residues are indicated by an asterisk (*).

Over a dozen different sequences of this family have been detected as abundant proteins and/or mRNAs in the endosperm of barley (Salcedo et al., 1984; Paz–Ares et al., 1986 and unpublished), which is a diploid species, whereas the number of variants per genome in wheat seems to be lower, possibly as a result of diploidization (Aragoncillo et al., 1975; Sanchez–Monge et al., 1986a).

Because of the complexity of this protein family, as has been ascertained in wheat and barley, the evolutionary implications of the limited sequence information available in other species must be drawn with caution. Thus, the bifunctional inhibitor from *E. coracana* has the highest homology with the maize trypsin inhibitor, followed by barley trypsin inhibitor CMe, the barley protein encoded by cDNA clone pUP-23, and some of the wheat and barley subunits of the tetrameric α–amylase inhibitors, but proteins with higher percentage of interspecific homology may yet be found in these species (Table 5). Protein PA from pea seeds is quite distant from the wheat and barley proteins, showing the highest homology with barley trypsin inhibitor CMe and with the barley protein encoded by clone pUP44. It should be pointed out that in this and subsequent cases of weak homology (20%-30% range), most of the homology is due to highly conserved or invariant positions, most notably the cysteines (Fig. 4).

```
      1                                                           53
CM16     I - G N E - D C T P W M S T L - I T P L P S C R - D Y V E Q Q A C
CM17         N E - D C T P W T S T L - I * P L P * C R - N Y V * * Q A C
CMb      V - G S E - D C T P W T A T P - I T P L P S C R - D Y V E Q Q A C
CM3      S - G S - - - C V P G V A F R - T N L L P H C R - D Y V L Q Q T C
CMd      A - A A A T D C S P G V A F P - T N L L G V A R - D Y V L Q Q T C
CM2      T - G P Y - - C Y P G M G L P - S N F L E G C R - E Y V A Q Q T C
CM1      T - G P Y - - C Y A G M S L P - I N P L E G C R - E Y V A S Q * C
CMe      T - G Q Y - - C Y A G M G L P - S N P L Q G C R - E Y V A Z Q T C
CMc      T - S I Y T - C Y E G M G L P - V N P L Q G C R - F Y V A * Q T C
C13  E R D Y G E - Y - C R V G K S I P - I N P L P A C R - E Y I T - R R C A - - V G - - D Q Q V P - D V - -
0.53     S - G P W M - C Y F G Q A F Q - V P A L P G C R - P L L - K L Q C - - - N G - - - - S Q V - P E - -
0.19     S - G P W M - C Y P G Q A F Q - V P A L P G C R - P L L - R L Q C - - - N G - - - - S Q V - P E - -
0.28     S - G P W S W C N P A T G Y K - V S A L T G C R - A M V - K L Q C - - - V G - - - - S Q V - P E - -
C44  R S D N S G P W M W H C D P E M G H K - V S P L T R C R - A L V - K L E C - - - - V G - - - - N R V - P E - -
C23  A A T L E S V K D E C Q L G V D F P - H N P L A T C H - T Y V I K R V C G - - R G - - P - S R P - M L - -
C38                                        L P E W M T S A E L N Y P G - - - - Q P - Y L - -
CMe      F - G D S - - C A P G D A L P - H N P L R A C R - T Y V V S Q I C H - - Q G - - P R L L T - S D - -
RBI      S V G T S - - C I P G M A I P - H N P L D S C R - W Y V A K R A C G - - V G - - P R L A T - Q E - -
MII      S A G T S - - C V P G W A I P - H N P L P S C C - W Y V T S R R C G - - I G P R P R L P W - P E - -
PEA      I S C N G V C S P F D I P P C G S P L C R C I P A G L V I G N C R H P Y G
2SC      P S Q Q G - - C R G Q I Q - E - Q Q N L R Q C Q - E Y I - K Q Q V - - - S G - Q - G P R R   Q E - -
2SA  A G P F R I P K C R K E F Q - Q - A Q H L R A C Q - Q W L H K Q A M Q - - S G - G - G P Q Q R P P L - -
2SB              E E C R E Q M Q R Q - Q M - L S H C R - M Y R Q Q M E E S P R R G - - - - - - - M E - -
2SL  F R S S E Q S - - C K R Q L Q - Q - V N - L R H C E - N H I - D Q R I Q - - - - - - S S Q E - S E E S

                                                   [ Q Q Q E E E E D R H K ]
```

```
      54                                                         106
C13  L K Q Q - C C R E L S D L - P E S C R C D A L S I L V - Q - - - - - G - - V I T E D G - S R V G - - R M E
0.53 A V L R D C C Q Q L A D I - S E W P R C G A L - Y S M - L - - - - D S - - M Y K E H G - V S E G Q A G T G
0.19 A V L R D C C Q Q L A H I - S E W C R C G A L - Y S M - L - - - - D S - - M Y K E H G - A Q E G Q A G T G
0.28 A V L R D C C Q Q L A D I N N G W C R C G D L - S S M - L - - - - R A - - V Y Q E L G - V R E G K E - V -
C44  D V L R D C C Q E V A N I S N E W C R C G D L - G S M - L - - - - R S - - V Y A A L G - V G G G P E E V -
C23  V K E R - C C R E L A A V - P D H C R C E A L R I L M - D - - - - - G - - V R T P E G R V V E G - R - L G
C38  A K L Y - C C Q E L A E T - P Q Q C R C E A L R T S M A L P V P P Q P - - V D P S T G - - N V G Q S G L M
CMe  M K R R - C C D E L S A I - P A Y C R C E A L R I I M - Q - - - - - G - - V V T W Q G - A F E G - A Y F K
RBI  M K A R - C C R Q L E A I - F A Y C R C E A V R I L M - D - - - - - G - - V V T P S G - Q H E G - R L L Q
MII  L K R R - C C R E L A D I - P A Y C R C T A L S I L M - D - - - - - G A I P P G P D A - Q L E G - - A L E
2SC  R S L R G C C D H L Q D M - Q S Q C R C E G L R Q A - - - - - - - I Q Q - Q L Q G - - Q N V F E A F R
2SA  - - L Q Q C C N E L H Q E - P L C V C P T L K G A S K A - - V K Q Q I Q Q Q G Q Q Q G K Q Q M V S R I Y Q
2SB  P A M S E L L E Q L E G M - Q E S L R L L E L H H A M H H - - H M Q Q - - E E M Q P R G E Q H R R H H - - R
2SL  E E L D Q C C E Q L N E L N S Q R C Q C R A L Q Q I Y E - - - - - - S Q S E - Q C E G R Q Q E Q Q L E G E
KB                                                                       N I L - G R E A - K C T N
```

```
      107                                                        158
C13  A V P R C D G E R I H S M G S Y L T A - Y S E C N - - P H - N P - - G T P R G D C - V L - F G - G G I S
0.53 A F P S C R R E V V K L T A - A S I T - A V - C R L - P I V V D A S G D G A Y V K D V - - A A Y P D A
0.19 A F P S C R R E V V K L T A - A S I T - A V - C R L - P I V V D A S G D G A Y V C K D V - - A A Y P D A
0.28 - L P G C R V E V M K L T A - A S V P - E V - C K V - P I P N P - S G D R A G V C Y G D W - C A Y P D V
C44  - F P G C Q K D V M K L L V - A G V P - A L - C N V - P I P N E A A G T R - G V C Y - - W - S A S T D T
C23  D R R D C P R E E Q P A F A A T L V T - A A E C N L S S V - Q E - P G V R - L V L L A D G
C38  D L P G C P R E M Q R D F V R L L V A - P G D C N L A T I - - H - - N - V R Y - C P A V - E Q - P L W I
CMe  D S P N C P R E R D T S V A A N L V T - P Q E C N L A T I - - H - - G - S A Y - C P E L Q P G Y - G
RBI  D L P G C P R Q V Q R A F A P K L V T - E V E C N L A T I - - H - - G - G P F - C L S L - L G - A G E
MII  D L P G C P R A V G Q G F A A T L V T - E A E C N L E T I - - S
2SC  T A A N L P S M C G - - - - - V S P T - Q - - C R F
2SA  T A T H L P K V C N - - - - - I P Q V - S U - C P F Q K T M P G
2SB  M A E N L P S R C N - - - - - L S P M - R - - C P M G G S
2SL  L E K - L P R I C G - - - - - F G P L - R R - C N I N P D E E
KB   E V N G C P R I Y N P V C G T D G V T Y S N E C L L C M E - - N - - K - E R Q - T P V L I Q K S - G P C
```

There is considerable divergence within the group of 2S storage proteins, which have no known inhibitory activity, and between this group and the rest of the sequenced proteins. The intra–group homology is in the 20%–35% range, whereas homology with the nearest cereal components is in the 18%–27% range (Table 5).

Table 6. Chromosomal locations of genes and inhibitory activities of the α-amylase/ trypsin inhibitors from wheat and barley.

Chromosome group	Genome*	Species	Protein grouped by loci	Inhibitory activity	References**
3	B	W	I (0.53)		a,b,c
	D	W	III (0.19)	α-amylase (dimeric)	a,b,c
?	H	B	CMb' (c44)		d
	H	B	CMe	trypsin	e,f
4	A	W	CM16		a,b
	D	W	CM17	α-amylase (tetrameric)	a,b
	H	B	CMb		e
	A	W	CM3		a,b
	H	B	CMd	α-amylase (tetrameric)	e
6	D	W	II (0.28)	α-amylase (monomeric)	b,c,g
7	B	W	CM2		a,b,h
	D	W	CM1	α-amylase (tetrameric)	a,b,h
1*	H	B	CMa		e
	H	B	CMc	trypsin	e
?	H	B	C13,C23,C38	?	i,j

* Hexaploid wheat (T. aestivum): genomes AABBDD and barley (H. vulgare): genomes HH. Barley chromosome 1 homologous to chromosome group 7 of wheat.

** a) Aragoncillo et al. (1975); b) Fra-Mon et al. (1984); c) Sanchez-Monge et al. (1986a); d) Sánchez-Monge and Lazaro (unpublished); e) Salcedo et al. (1984); f) Hejgaard et al. (1984); g) Pace et al. (1978); h) García-Olmedo and Carbonero (1970); i) Paz-Ares et al. (1986); j) Lázaro (unpublished).

The Kazal secretory trypsin inhibitor from bovine pancreas shows significant homology only with the trypsin inhibitors from maize, ragi and barley CMe (Table 5). The three domains proposed for barley trypsin inhibitor CMe by Kreis et al. (1985) are approximately indicated in Fig. 4. The reactive site is just at the right-hand border of the first domain and the sequence -Pro-Arg-Leu-, which corresponds to the -P2-P1-P'1- residues, has been conserved in the three proteins with anti-trypsin activity which have been completely sequenced (Fig. 4). Otherwise, the region around the trypsin reactive site appears as extremely variable throughout this family, with numerous deletions and/or insertions.

A weak but significant homology has been found between the alluded domains of these proteins and similar domains present in prolamins such as α-gliadin, B1-hordein, ˜γ˜-secalin, HMW-prolamin, and other cereal prolamins, in which these domains are separated by stretches of repeated sequences (Kreis et al., 1985). The sequence information accumulated since that finding further supports the proposed homology as better fitting comparisons can be made.

The sequences in Fig. 4 were searched for internal repeats because the homology of the Kazal inhibitors with the C- terminal part of these proteins, and the double-headed nature of the bifunctional inhibitor from ragi and of the trypsin inhibitor from rice bran, suggested a possible duplication. Only in the case of the monomeric α-amylase inhibitor (0.28) a weak homology was found between the N-terminal and the C- terminal domains:

```
18
L T G C R A M V   K L Q C V G S Q V P E
.  . . .    .  . .          .  . . .  11/20
L P G C R K E V M K L T A A   S   V P E
                                     96
```

Genetics

Apart from the studies already discussed regarding the chromosomal locations of genes encoding proteins of this group, research on other aspects of their genetic control and regulation has also been carried out. The considerable sequence divergence found for this protein family within a genome is in sharp contrast with the low variability which has been observed for the different members of the group. In a survey of tetraploid and hexaploid wheat cultivars, no variants were found for proteins CM1 and CM2, and only a rare allelic variant of protein CM3 was present in a small group of closely related cultivars (Garcia-Olmedo and Garcia-Faure, 1969; Garcia-Olmedo and Carbonero, 1970; Rodriguez-Loperena et al., 1975; Salcedo et al., 1978a). Similarly, in a wide-ranging survey of *Hordeum vulgare* cultivars and *H. spontaneum* accessions, proteins CMa and CMc were invariant in the cultivated species and presented rare variants in the wild one, protein CMd was invariant in both species, and proteins CMb and CMe had allelic variants in both species (Salcedo et al., 1984; Molina-Cano et al., 1987). In the case of the trypsin inhibitor CMe, two variants, designated CMe2/CMe2', appeared in over 40% of the samples and were jointly inherited and codominantly expressed with respect to CMe (Salcedo et al., 1984).

Other surveys have been concerned with the variation of the inhibitory activity against trypsin (Kirsi, 1973; Kirsi and Ahokas, 1983) and against α-amylases (Bedetti et al., 1974). The considerable variation in the levels of activity indicated that these were probably due not only to changes of the specific activity among variants but also to extreme variations in the expression levels of the corresponding genes.

The regulation of the expression of genes from this family has been studied in some detail in a number of cases. Salcedo et al. (1978a) demonstrated that the net accumulation of protein CM3, now known to be a subunit of the tetrameric α-amylase inhibitor in wheat, and that of its allelic variant CM3', varied linearly with gene dosage but the amount of CM3' for a given dosage was about half of that of CM3. The expression level cosegregated with the structural variation indicating that it was dependent on cis-acting genetic elements. A gene-dosage compensation effect was described by Aragoncillo et al. (1978) for proteins CM16 and CM17, which are also equivalent subunits of the tetrameric α-amylase inhibitor in wheat. Although the net output of each of the proteins varied linearly with gene dosage, the amount of protein at a given dosage of its structural gene was significantly higher if the chromosome carrying the gene for the other protein was absent. This suggests some degree of coordinated expression mediated by trans-acting genetic elements of the genes encoding subunits of the tetramers.

The effects of high-lysine mutations on the expression levels of different genes from this family have been investigated (Salcedo et al., 1984; Lazaro et al., 1985). The most notable effect concerns the gene for the trypsin inhibitor CMe, whose product is not detected in mutant Riso 1508 and is at a much lower level in Hiproly, as has been confirmed recently (Salcedo, unpublished). In both cases, the mutations affect loci in chromosome 7, while the structural gene for the inhibitor is in chromosome 3 (see Lazaro et al., 1985). Just the opposite effect has been described for the high-lysine mutation opaque-2 of maize, which greatly increases the corresponding trypsin inhibitor (Halim et al., 1973).

Physiology

Tissue localization of the barley trypsin inhibitor(s) was found to be restricted to the starchy endosperm and the aleurone layer, as they were not detected in husks, coleoptiles, leaves or roots (Kirsi and Mikola, 1971). Similarly, Buonocore et al. (1977) concluded that the α-amylase inhibitors were endosperm-specific. Synthesis of this protein family seems to precede that of the bulk of reserve proteins and starch. Thus, Kirsi (1973) detected the barley trypsin inhibitor at about 5 days after anthesis (daa) and found that most of it was accumulated between 10 and 23 daa. He did not find a response to nitrogen fertilization. The α-amylase inhibitors were detected at 8 daa (Pace et al., 1978). In a study of in vivo and in vitro synthesis of these proteins in barley, Paz-Ares et al. (1983b) concluded that synthesis took place between 10 and 30 daa, with a peak between 15 and 20 daa. These proteins were synthesized by membrane-bound polysomes as precursors of higher apparent M_r (13,000-21,000) than the mature proteins (12,000-16,000). The largest in vitro product (21,000) was found to be the precursor of subunit CMd. Accordingly, a putative leader sequence has been found in those cDNA clones of this family that were large enough to include the 5'-end of the mRNA sequence (Paz-Ares et al., 1986).

The subcellular location of the different members of this family has not been investigated. Indirect evidence of association with starch granules has been reported (see Buonocore et al., 1977) and indeed at least some of the inhibitors show affinity for insoluble starch (O'Connor and McGeeney, 1981b). However, Paz-Ares et al. (1983b) following different homogenization and subcellular fractionation procedures consistently found these proteins in the supernatant, indicating that their association with the particulate fraction was labile if it existed at all. On the other hand, Zawistowska and Bushuk (1986) recovered CM-proteins in gluten preparations, where they seemed to have a considerable amount of bound polar lipids.

During germination, the anti-trypsin activity disappears rather abruptly after remaining stable for 2-3 days (Kirsi and Mikola, 1971). Similarly, rapid disappearance with the onset of germination have been observed for the α-amylase inhibitor (Pace et al., 1978). These observations suggest that these inhibitors do not play a specific role during germination.

POTATO INHIBITORS I AND II FAMILIES

The families of the non-homologous potato inhibitors I and II will be considered together because the two inhibitors play a joint role in the systemic response to wounding (see Ryan, 1984). They were first discovered in potato tubers, where they accumulate throughout development and represent a substantial fraction of the soluble protein, and later they were found to accumulate in wounded tomato and potato leaves (Green and Ryan, 1972; Plunkett et al., 1982; Sanchez-Serrano et al., 1986; Cleveland et al., 1987).

Distribution and Inhibitory Properties

Inhibitor I is an oligomer of M_r 41,000, made of protomers of M_r 8,100, with one disulphide bond and one reactive site each, which inhibits chymotrypsin-like enzymes (Ryan, 1984). Besides potato, tomato and other Solanaceae (Gurusiddaiah et al., 1972; Lee et al., 1986), homologues of this inhibitor have been found in plant species outside this family, such as broad bean (Svendsen et al., 1984) and barley (Svendsen et al., 1980, 1982; Jonassen and Svendsen, 1982), and in a lower animal, the leech (Seemuller et al., 1980).

The inhibitor from broad bean (*Vicia faba*) has no cysteine and is active against microbial serine proteases, including subtilisin, with which it forms a 1:1 complex, and is inactive against chymotrypsin and trypsin (Svendsen et al., 1984). The subtilisin inhibitors from *Vigna unguiculata* and *Phaseolus vulgaris* seem to be closely related to that from *Vicia faba* as judged from the amino acid compositions (Vartak et al., 1980; Seidl et al., 1978, 1982). It is very probable that the constitutive trypsin inhibitors from buckwheat, reported by Kiyohara and Iwasaki (1985), are also members of this protein family.

Two inhibitors, CI-1 and CI-2, have been well characterized in barley. Both of them inhibit chymotrypsin and subtilisin, but CI-1 has a single reactive site for the two enzymes and CI- 2 has one reactive site for chymotrypsin and two for subtilisin (Jonassen and Svendsen, 1982). Neither of the inhibitors has cysteines.

Eglin, the inhibitor from the leech, has no cysteines and inhibits elastase and cathepsin G (Seemuller et al., 1980). A more distantly related inhibitor of this family has been identified in yeast (Maier et al., 1979).

Inhibitor II is a dimer of M_r 23,000, made of two protomers of M_r 12,000, with five disulphide bonds and two reactive sites per monomer, which respectively inhibit chymotrypsin and trypsin (Ryan, 1984). The inhibitors from this family in tomato and potato are related to two smaller trypsin and chymotrypsin inhibitors from potato tubers, PCI-I and PTI-I (Hass et al., 1982) and with another proteinase inhibitor from egg plant (Richardson, 1979). Wound-inducible antitrypsin activity has also been detected in species such as alfalfa, tobacco, strawberry, cucumber, squash, clover, broadbean and grape (Walker-Simmons and Ryan, 1977). It is yet unknown if these inhibitors are homologues of potato inhibitors I or II and in some cases, such as that of the alfalfa inhibitors (Brown and Ryan, 1984), it is suspected that they might belong to entirely different groups.

Structure and Evolution

Available sequence information of the potato inhibitor I family is summarized in Fig. 5. In potatoes, a mixture of 10 or more isoinhibitors are present and variation is observed at 16 out of the 84 positions of the mature protein (Richardson, 1974; Richardson and Cossins, 1974; Graham et al., 1985a,b, Cleveland et al., 1987). The tomato sequence presented 80% homology with respect to the potato isoinhibitors (Fig. 5). As expected, lower homologies were found between the tomato inhibitor and those from barley (CI-1C, 25%; CI-2, 33%) and broad bean (35%). A remarkably low divergence (35% homology) was found for the animal inhibitor Eglin, whereas the homology of the yeast inhibitor (≈16%) was rather weak (Fig. 5). The reactive sites seem to be located at exactly homologous positions (Richardson et al., 1977), except for CI-1C from barley, which has a second subtilisin site (Met30-Ser31) besides the standard one (Jonassen and Svendsen, 1982). In other inhibitor families, the three-dimensional structure is stabilized by disulphide bridges, which seem to be essential for their activity and, accordingly, they are highly conserved throughout evolution. Members of potato

```
        1                                                    42
II-I                         L M C E G K Q M W P E L I G V P T K L A K E I I G K E
PI-I                         K E F E C D G K L Q W P E L I G V P T K L A K E I I E K Q
                                   K       N   S                      6
CI-2  (Z,V,S,S)K K P E G V N T G A G D R H N L K T E W P E L V G K S V E E A K K V I L - Q
CI-1C          Y P E P T E G S I G A S G A K T S W P E V V G M S A E K A K E I I L - R
VSI                              R T S W P E L V G V S R E E A R K I K E - -
LIE                          T E F G S E L K - S F P E V V G K T V D Q A R E Y F T - L
YIB                          T K N F I V T L K K N T P D V E A K K F L D S V H H A G - G

        43                                                           87
II-I    N P S I T N I P I L L S G S P I T L D Y L C D R V R L F D N I L G F V V Q - M P V V T
PI-I    N S L I T N V H I L L N G S P V T L D Y R C N R V R L F D N I L G N V V E - I P V V G
             S   Q   K         M               D       Y   D       L   R L A
CI-2    D K P E A Q I I V L P V G T I V T M E Y R I D R V R L F V D K L D N I A E - V P R V G
CI-1C   D K P N A Q I E V I P V D A M V P L N F N P N R V F V L V H,K,A,T,T V A,Z,- V,S R V G
VSI     E K P E A E I Q V V P Q E S F V T A D Y K F Q R V R L Y V D E S N K V V R -(A,A,P) I G
LIE     H Y P Q Y N V Y F L P E G S P V T L D L R Y N R V R V F Y N P G T N V V N H V P H V G
                                               S
YIB     S I L H K F D I I K G Y T I K V P D V L H L N K L K E K H N D V I E N V E E D K E V H T N
              V   E
```

Figure. 5. Alignment of amino acid sequences of members of the potato inhibitor I family. TI-I is from tomatoes (Graham et al., 1985a); PI-I is from potatoes (Melville and Ryan, 1972; Cleveland et al., 1987); CI-2 and CI-1C are from barley (Svendsen et al., 1982); VSI is from broadbean (Svendsen et al.,1984); LIE is the inhibitor Eglin from leech (Seemuller et al., 1980); YIB is from yeast (Maier et al., 1979). Vertical arrows (▼) indicate the reactive bonds. Conserved positions are boxed.

Inhibitor I family either have one disulphide bond or none and it does not seem to be essential for their activity. Indeed, the crystal and molecular structure of barley inhibitor CI-2, which lacks cysteine, and of its complex with subtilisin Novo have been determined and few conformational changes have been found between the free and the complexed inhibitor (McPhalen et al., 1985 and McPhalen and James, 1987).

Although members of the inhibitor II family have been well characterized (Plunkett et al., 1982), direct sequencing methods have not allowed the determination of complete amino acid sequences, which have now been deduced from the corresponding cDNA clones in tomato (Graham et al., 1985b) and in potato (Sanchez–Serrano et al., 1986). Homology between the deduced sequences of tomato and potato is high in the region between positions 29 and 154, where only 19 differences appear (Fig. 6). The preceding signal sequence is 6 amino acids shorter (positions 21 to 28) in the tomato. As shown in Fig. 6, a high degree of homology was observed between the putative inhibitor II from potato, the two smaller proteinase inhibitor peptides from potato tuber, PCI-I and PTI, and a similar inhibitor from eggplant (Richardson, 1979; Hass et al., 1982; Sanchez–Serrano et al., 1986). These smaller peptides could be derived from different isoforms of

303

inhibitor II by further processing, as suggested by the fact that PCI-I is completely homologous to the appropriate segment of one of the two sequences deduced from potato cDNA (Sanchez-Serrano et al., 1986). As pointed out by Graham et al. (1985b), the full- length inhibitors have a duplicated-domain structure (Fig. 6). Based on the known position of the single reactive site of the eggplant inhibitor (Richardson, 1979), the putative reactive sites of the larger inhibitors have been postulated at the homologous positions of the two domains (Fig. 6), which would explain the known specificities of the different inhibitors (Graham et al., 1985b).

```
                  1        SIGNAL    SEQUENCE                              44
         TPI-II   M A V H K E V N F V A Y L L I V L G M F L - - - - - - Y V D A K A C T R E C G N L G F G
         PPI-II   M D V H K E V N F V A Y L L I V L G L L V L V S A M E H V D A K A C T L E C G N L G F G

                  45                                                      88
         TPI-II   I C P R S E G S P L N P I C I N C C S G Y K G C N Y Y N S F G K F I C E G E S D P K R P
         PPI-II   I C P R S E G S P E N R I C T N C C A G Y K G C N Y Y S A N G A F I C E G Q S D P K K P
                                          P
         PCI-I                      P I C T N C C A G Y K G C N Y Y S A N G A F I C E G Q S D P K Y P
         PTI                        R I C T N C C A G Y K G C N Y Y S A N G A F I C E G E S D P K N P
         EPI                        L C T N C C A G R K G C S Y F S E D G T F I C K G E S N P E N P
                                    I         N

                  89                                                      132
         TPI-II   N A C T F N C D P N I A Y S R C P R S Q G K S L I Y P T G C T T C C T G Y K G C Y Y F G
         PPI-II   K A C P L N C D P H I A Y S K C P R S E G K S L I Y P T G C T T C C T G Y K G C Y Y F G
                                                                                                       S
         PCI-I    K A C P L N C D P H I A Y S K C P R S
         PTI      N V C P R N C D T N I A Y S K C L R
         EPI      K A C P R N C D G R I A Y G I C P L S

                  133                              154
         TPI-II   K D G K F V C E G E S D E P K A N M Y P V M
         PPI-II   K N G K F V C E G E S D E P K A N M G P A M
```

Figure. 6. Alignment of amino acid sequences of members of the potato inhibitor II family. TPI-II is from tomatoes (Graham et al., 1985b); PPI-II is from potatoes (Sanchez-Serrano et al., 1986); PCI-I and PTI are from potatoes (Hass et al., 1982); EPI is from eggplant (Richardson, 1979). Vertical arrows (▼) indicate reactive bonds. Horizontal arrows (►) indicate internal duplications. Conserved positions are boxed.

Genetics and Molecular Cloning

The chromosomal location of genes encoding the chymotrypsin inhibitors in chromosome 5 of barley and the 20-fold increase in their expression as a result of the high-lysine mutation in the locus lys of chromosome 7 have been reported (Jonassen, 1980; Boisen et al., 1981; Hejgaard et al., 1984a). As already discussed, the same high-lysine mutation exerts the opposite effect on the gene encoding barley trypsin inhibitor CMe (Lazaro et al., 1985). The gene for an inhibitor from rye which cross-reacts with antibodies raised against the CI inhibitors from barley has been located in chromosome 1R (Hejgaard et al., 1984b).

Although the variability of the concentration of the inhibitors in potato tubers has been investigated (Ryan et al., 1977), no other formal genetic studies seem to have been carried out concerning these two inhibitor families. In contrast, considerable progress in the molecular cloning of these genes has been achieved in different species: cDNA clones for type-I inhibitors from tomato (Graham et al., 1985a) and barley (Williamson et al., 1987) and for type-II inhibitors from tomato (Graham et al., 1985b) and potato (Sanchez-Serrano et al., 1986) are now available, together with genomic clones for type-I from tomato (Lee et al., 1986) and potato (Cleveland et al., 1987) and for inhibitor-II from potato (Keil et al., 1986).

Comparison of the amino acid sequences, deduced from nucleotide sequences, with the N-terminal sequences of the native inhibitor-I proteins from tomato and potato indicates that in both cases a precursor is coded which must yield the mature protein after processing an N-terminal pre- and pro- sequences. The putative pre- or transit sequences are 23 residues long in both cases, whereas the pro-sequences would have 19 residues, 9 of which are charged in tomato and 16 residues, 6 of which are charged, in potato (Graham et al., 1985a,b, Cleveland et al., 1987). The nucleotide sequences of several cDNAs corresponding to inhibitor CI-2 from barley have an open reading frame that corresponds exactly to the mature protein sequence. Although there is another ATG codon further 69 nucleotides upstream, the sequence between the two ATG codons would code for a typical signal peptide. An in-frame TAA stop codon is consistently present and in vitro translation starts in the downstream ATG (Williamson et al., 1987). The open reading frames in cDNAs corresponding to inhibitor-II in tomato and potato code for precursors of 148 and 154 amino acids, respectively. The N-terminal sequences of these precursors have all the features of typical leader sequences. The tomato leader sequence is six amino acids shorter, which accounts for the difference in length of the precursors (Graham et al., 1985b; Sanchez-Serrano et al., 1986).

Inhibitor-I genes have two introns both in tomato and in potato, whereas a potato gene for inhibitor II has only one intron (Lee et al., 1986; Cleveland et al., 1987; Keil et al., 1986). The cDNA and genomic clones of the inhibitors have been searched for common features that could be related to the regulation of the systemic wound response. The two tomato proteinase inhibitor mRNAs share a variant polyadenylation signal (AATAAG) and a conserved 3^l–untranslated 10–base palindromic region, but these are not present in the potato mRNAs (Graham et al., 1985a,b; Cleveland et al., 1987; Sanchez– Serrano et al., 1986). An imperfect direct repeat 100 bp long was found in the 5^l–flanking region of the inhibitor I gene from tomato that was shared by other *Lycopersicon* species with inducible inhibitors, but not by the potato genes for inhibitors I and II (Lee et al., 1986; Cleveland et al., 1987; Keil et al., 1986). Homology of the 3^l–untranslated regions of the proteinase inhibitor II genes from potato and an extensin gene have been also pointed out by Sanchez–Serrano et al. (1987), but again this homology is not shared by all the inducible genes.

Physiology

Certain species such as tomatoes, respond to wounding by insects or other severe mechanical damage within a few hours by accumulating proteinase inhibitors in leaves throughout the plant, including non–wounded parts (Green and Ryan, 1972; Plunkett et al., 1982;

```
           1                                                36
CMTI-I          R V C P R I L M E C K K D S D C L A E C V C L E H - G Y - C G
CMTI-III        R V C P R I L M K C K K D S D C L A E C V C L E H - G Y - C G
CMTI-IV   H E E R V C P R I L M K C K K D S D C L A E C V C L E H - G Y - C G
CPGTI-I         R V C P K I L M E C K K D S D C L A E C I C L E H - G Y - C G
CPTI-II         R V C P K I L M E C K K D S D C L A E C I C L E H - G Y - C G
CPTI-III  H E E R V C P K I L M E C K K D S D C L A E C I C L E H - G Y - C G
CSTI-IIb        M V C P K I L M K C K H D S D C L L D C V C L E D I G Y - C G V S
CSTI-IV         M M C P R I L M K C K H D S D C L P G C V C L E H I E Y - C G
CM-1            G I C P R I L M E C K R D S D C L A Q C V C K R Q - G Y - C G
CM-3            A I C P R I L V E C K R D S D C P A Q C I C K R Q - G Y - C G
TKTI                 L L M P V K P N D D R V I G C W C I S R - G Y L C G C M

          37              50
TKTI       P C K L N D D S L C G R K G
```

Figure. 7. Alignment of amino acid sequences of the squash family of serine proteinase inhibitors. CMTI-I, CMTI-III and CMTI-IV are from Cucurbita maxima; *CPGTI-I, CPTI-II and CPTI-III are from* C. pepo; *CSTI-IIb and CSTI-IV are from* Cucumis sativus (Wieczorek et al., 1985). *Inhibitors CM-1 and CM-3 are from* Momordica repens (Joubert, 1984b) *and TKTI is from* Trichosanthes kirilowii (Tan et al., 1984). *Conserved positions are boxed. Vertical arrows (▼) indicate the reactive bonds.*

Graham et al., 1986). The induction of the inhibitor proteins is mediated by a putative wound signal, called the proteinase inhibitor inducing factor (PIIF), whose activity has been shown to be associated with fragments of the plant cell wall, apparently released at the wound site by endogenous endopolygalacturonases (Bishop et al., 1984; Ryan, 1984). Following a single wound on a lower leaf of a tomato plant, mRNAs for the two inhibitors began to accumulate in leaves 2 to 4 hours later, and the proteins were detected with a further 2 hours lag. The levels of the mRNAs reached a maximum at about 8 hours after wounding and then decayed with apparent half–lifes of 10 hours; consecutive wounds had an accumulative effect. Synthesis of the inhibitors was found to be regulated at the level of transcription (Graham et al., 1986).

The wound–induced response has been investigated in transgenic tobacco plants in which either the complete inhibitor–II gene from potato (Sanchez–Serrano et al., 1987) or a fusion of this gene with that of chloramphenicol acetyl transferase (Thornburg et al., 1987) had been introduced. The fact that these genes were wound–induced in the transgenic plants indicated that tobacco has the capacity to regulate the expression of these inducible genes. The sequences necessary and sufficient for wound–inducibility were located within 1kbp upstream of the genes.

SQUASH TRYPSIN/HAGEMAN FACTOR INHIBITOR FAMILY

A new family of very small inhibitor (29–32 residues, 3 disulphide bridges) that are active against trypsin and Hageman factor have been described in the seeds of the Cucurbitaceae (Hojima et al., 1980b; Polanowski et al., 1980; Joubert, 1984b). Members of this family have been isolated and sequenced from squash (*Cucurbita maxima*) by Wilusz et al. (1983), from zucchini (*C. pepo* var. Giromontia), summer squash (*C. pepo*), and cucumber (*Cucumis sativus*) by Wieczorek et al. (1985), and from *Momordica repens* by Joubert (1984b). Families of isoinhibitors have been found in all of these species, which seem to result both from multiple structural genes and from proteolytic processing: while some of the sequences are clearly different from each other, in other cases the heterogeneity is only due to the presence or absence of N–terminal peptides (Fig. 7). A double–headed inhibitor from a Chinese medical herb, *Trichosanthes kirilowii* (Cucurbitaceae), which has 41 amino acid residues (Tan et al., 1984), is over 30% homologous to the shorter squash inhibitors and 26% homologous to soybean Bowman–Birk family.

Disulphide bond positions have been proposed, based on the similarity of the squash inhibitors with wheat–germ agglutinin (Siemion et al., 1984). The reactive site is either Arg–Ile or Lys–Ile in agreement with their trypsin specificity (Joubert, 1984b; Wieczorek et al., 1985). In spite of their small size, they are quite strong inhibitors. Up to now these inhibitors have been found only within the Cucurbitaceae and it will be of interest to investigate their range of distribution and variation.

HOMOLOGY OF PROTEIN Z WITH α1-ANTITRYPSIN AND RELATED PROTEINS

Protein Z is a major barley endosperm albumin of M_r 43,000 which is present in both free and bound forms in the mature grain, where it seems to act as a storage protein and contains a substantial part of the grain lysine (Hejgaard, 1976, 1982; Hejgaard and Boisen, 1980; Giese and Hejgaard, 1984). Nucleotide and amino acid sequencing of cDNA clones have established the 180 C-terminal residues of this protein (Fig. 8). This sequence has been found to be 26–32% homologous with human α1-antitrypsin, human α1-antichymotrypsin, human antithrombin III, mouse contrapsin and chicken ovalbumin (Hejgaard et al., 1985). The sequence homology and the specific cleavage by trypsin at a bond corresponding to the reactive sites of the inhibitors suggest a possible inhibitory activity for protein Z, although no activity against microbial or pancreatic serine proteinases has been found (Hejgaard et al., 1985).

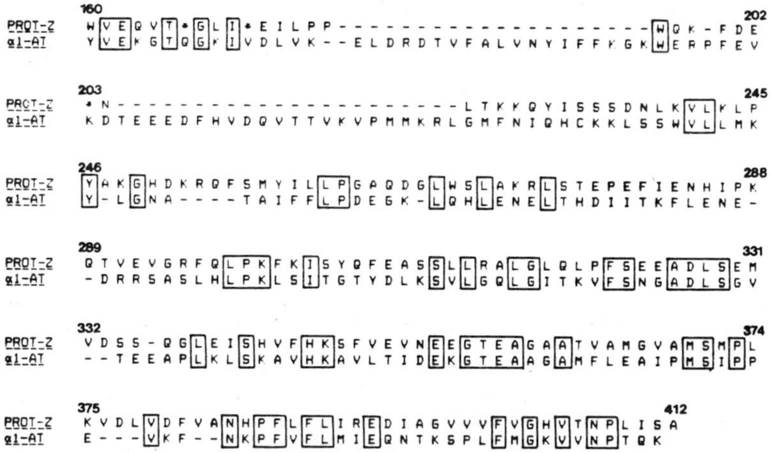

Figure. 8. Comparison of a partial sequence from barley protein Z with α1-antitrypsin (Hejgaard et al., 1985). Conserved positions are boxed.

308

RAGI I-2 / MAIZE BIFUNCTIONAL INHIBITOR FAMILY

Besides the already described bifunctional inhibitor from ragi (*Eleusine coracana*), a second unrelated inhibitor which was active against porcine pancreatic α-amylase, was characterized in that species (Shivaraj and Pattabiraman, 1980; Campos and Richardson, 1984). A M_r 10,000 protein was also purified from barley and found to be homologous to the ragi inhibitor I-2 (Svensson et al., 1986; Mundy and Rogers, 1986). This protein was considered as a probable amylase/protease inhibitor (PAPI) both because of this relationship with the ragi inhibitor and because of its weak but significant homology with Bowman-Birk type proteinase inhibitors (Mundy and Rogers, 1986), although no target enzyme was identified for it (Svensson et al., 1986). Expression of the gene encoding PAPI, which has been studied using a cDNA probe, takes place primarily in aleurone tissue during late stages of grain development. PAPI mRNA is present at high levels in aleurone tissue of dessicated grain, as well as in incubated aleurone layers prepared from rehydrated grain, PAPI protein being secreted into the incubation medium (Mundy and Rogers, 1986).

The sequence of a maize protein which is a potent in vitro inhibitor of bovine trypsin and insect α-amylase has been recently determined and found to be partially homologous with barley PAPI and ragi I-2 inhibitor (Richardson et al., 1987). The new maize bifunctional inhibitor showed extensive homology with the sweet protein thaumatin II, present in the fruits of *Thaumatococcus danielli*, and with a pathogenesis-related (PR) protein induced in tobacco plants following infection with tobacco mosaic virus. This striking observation suggest a possible inhibitory function both for thaumatin II and the PR- protein (Richardson et al., 1987). Sequence alignments showing the above described relationships are presented in Fig. 9.

INHIBITORS OF METALLOCARBOXYPEPTIDASES

Inhibitors of metallo-carboxypeptidases (CPI) were identified both in potato tubers (Rancour and Ryan, 1968) and tomato fruits (Hass and Ryan, 1980a), and were later found to accumulate in wounded leaves (Graham and Ryan, 1981). These inhibitors are polypeptides of M_r 4,200 which are unusually heat stable. They are active against most animal carboxypeptidases and inactive against all plant and most microbial carboxypeptidases tested (Hass et al., 1981). Inhibition occurs by a competitive mechanism and results in the rapid removal of the carboxy-terminal amino acid, whose cleavage does not lead to elimination of the inhibitory activity (Hass and Ryan, 1980b).

A

```
        1                                                              44
PAPI   M A R A Q V L L M A A A L V L M L T A A P R A A V A L N - C G Q V D - S K M K P C L T Y
RAI-2                                              A I S - C G Q V S - S A I G P C L A Y
MBI                                      T A A - R I - W A R T G C - Q F D A S G R G S C R T G
                                         46

        45                                                             88
PAPI   V Q G - G P G P S G E C C N G V R D L H N Q A Q S S G D R Q T V C N C - L K G I A R G I
RAI-2  A R G A G A A P S A S C Q S G V R S L N A A A R T T A D R R A A C N C S L K S A A S R V
MBI    D C G - G
                73

        89                                                             123
PAPI   H N L N L N N A A S I P S K C N V N V P Y T I S P D - - I D C S R I Y
RAI-2  S G L N A G K A S S I P G R C G V R L P Y A I S A S - - I D C S R V N N
MBI    N L D F F D - I S I L D G - - F N V P Y S F L P D G G S G C S R
        97                                                             125
```

B

```
        1                                                              46
MBI    A V F T V V N Q C P F T V W A A S V P V - - - - - G G G R Q L N R G E S W R I T A P A G T T
TPR    A T F D I V N Q C T Y T V W A A A S P - - - - - - G G G R Q L N S G Q S W S I N V N P G T V
THA    A T F E I V N R C S Y T V W A A A S K G D A A L D A G G R Q L N S G E S W T I N V E P G T K

        47                                                             92
MBI    A A R I W A R T G C Q F D A S G R G S C R T G D C G G V V Q C T G Y G R A P N T L A E Y A L
TPR    Q A R I W G R T N C N F D G S G R G N C E T G D C N G M L E C Q G Y G K P P N T L A E F A L
THA    G G K I W A R T D C Y F D D S G R G I C R T G D C G G L L Q C K R F G R P P T T L A E F S L

        93                                                             138
MBI    K Q F N N L D F F D I S I L D G F N V P Y S F L P D G G S G C S R G P R C A V D V N A R C P
TPR    N Q - P N Q D F V D I S L V D G F N I P M E F S P T N G - G C - R N L R C T A P I N E Q C P
THA    N Q Y G K - D Y I D I S N I K G F N V P M D F S P T T R - G C - R G V R C A A D I V G Q C P

        139                                                            184
MBI    A E L R - Q D G V C N N A C P V F K K D E Y C C V G S A A N N C H P T N Y S R Y F K G Q C P
TPR    A Q L K - I Q G G C N N P C T V I K T N E F C C - T N G P G S C G P T D L S R F F K A R C P
THA    A K L K A P G G G C N D A C T V F Q T S E Y C C - - - T T G K C G P T E Y S R F F K R L C P

        185                                              214
MBI    D A Y S Y P K D D A T S T F T C P A G T N Y K V V F C P
TPR    D A Y S N Y P Q D D P P S L F T C P P G T N Y R V V F C P
THA    D A F S Y V L D K P T - T V T C P G S S N Y R V T F C P T A
```

Figure. 9. a) Comparison of the complete sequences of PAPI, a probable α-amylase/proteinase inhibitor from barley (Mundy and Rogers, 1986), and RAI-2, the α-amylase inhibitor I-2 from ragi (Campos and Richardson, 1984), with homologous segments of MBI, a bifunctional α-amylase/trypsin inhibitor from maize (Richardson et al., 1987). b) Alignment of the complete sequence of MBI with those of thaumatin (THA) and a pathogenesis-related protein induced by tobacco mosaic virus (TPR), as reported by Richardson et al., 1987). Conserved positions are boxed.

Sequences of the highly homologous tomato and potato isoinhibitors have been aligned in Fig. 10. The structure of CPI-II from potato has been deduced by X-ray diffraction analysis at 2.5Å resolution of a complex of the inhibitor with carboxypeptidase A (Rees and Lipscomb, 1982). CPI-II, which has no helical structures or β-pleated sheets, binds to the enzyme like an extended substrate. The Val38-Gly39 peptide bond is cleaved and Gly39 gets trapped in the binding pocket of the enzyme. In the case of the tomato inhibitor, which lacks an amino acid at position 39, inhibition also takes place, indicating that this amino acid is not required for activity.

A significant increase of CPI was observed in the upper, unwounded leaves 38h following wounding of the lower leaves from potato plants and this response increased as the plants grew older (Hollander-Czytko et al., 1985). The levels of CPI accumulation in upper, unwounded leaves were increased by multiple wounds in a similar manner to that described for proteinase inhibitors I and II of potatoes. Finally, the inhibitor was also found to accumulate in the vacuoles of unwounded, wound-induced leaves (Hollander-Czytko et al., 1985).

INHIBITORS OF THIOL-PROTEINASES

Protein inhibitors of thiol-proteinases, such as papain, ficin, bromelain, etc., have been identified in different plant species, but only a limited number of studies have been devoted to them and it is not yet possible to classify them into structural families.

A family of isoinhibitors from pineapple stems that competitively inhibit papain, ficin, and endogenous bromelain, is probably among the best characterized of this group (Reddy et al., 1975). They consist of two chains with 41 and 10-11 amino acid residues respectively (Fig. 11). It has been proposed that the two chains, which are linked by disulphide bridges, are generated from single-chain precursors by excision of a "bridge" peptide which links the amino-terminal of the A chain to the carboxy-terminal of the B chain (Reddy et al., 1975).

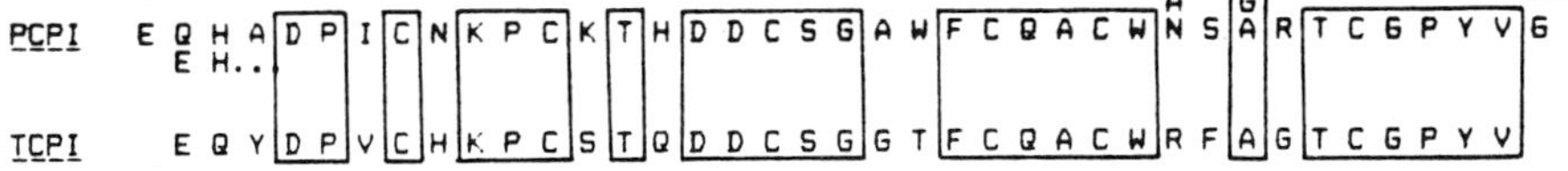

Figure. 10. Comparison of carboxypeptidase inhibitors from potatoes (PCPI) and tomatoes (TCPI) (see Hass and Ryan, 1981). PCPI has a variant that differs at the N-terminal as indicated in the figure. Conserved positions are boxed.

An inhibitor of M_r 24,000 from the seeds of the Bauhinia tree is active against a wide range of proteinases: papain, ficin, bromelain, trypsin, chymotrypsin and pepsin (Goldstein et al., 1973).

Two inhibitors from maize endosperm, with M_rs of about 13,000 and 9,500 respectively, have been found to be active against papain, ficin and bromelain (Abe et al., 1980), whereas an inhibitor isolated from dehulled rice seeds, which has an M_r of about 12,000, was only active against the first two enzymes (Abe and Arai, 1985). An unstable papain inhibitor has been purified from the grain of foxtail millet (Tashiro and Maki, 1986).

INHIBITORS OF CARBOXYL PROTEINASES

A polyvalent, heat stable inhibitor of M_r 27,000 which is active against cathepsin D, a carboxyl proteinase, has been isolated from potatoes (Keilova and Tomasek, 1976a,b). This inhibitor simultaneously inhibits trypsin and chymotrypsin and has no effect on hog and chicken pepsin, rennin, or cathepsin E. The inhibition of cathepsin D and trypsin is reversible and competitive, with Ki values of 3.8 x 10M and 8.6 x 10M, respectively.

Broad specificity proteinase inhibitors have been also characterized in cultured cells of *Scopolia japonica* (Solanaceae; Sakato et al., 1975). The inhibitors were shown to be polypeptides of M_r 4,000-6,000. They are potent inhibitors of pepsin, trypsin, chymotrypsin, plasmin and kallikrein, and had no effect on papain. These inhibitors form stable complexes with trypsin and chymotrypsin in an equimolar ratio. The inhibitory mechanism for both enzymes is non-competitive.

```
        1                                                                    41
A-1  D E Y K C Y C A D T Y S D C P G F C K K C K A E F G K Y I C L D L I S P N D C V K

        1                 11
B-2  T A C S E C V C P L Q
```

Figure.11. Sequences of the A and B chains of an inhibitor of thiol-proteinase from pineapple stem (Reddy et al., 1975).

INHIBITORS AND PLANT METABOLISM

Most of the described inhibitors are only active against heterologous enzymes and it is hard to envisage their involvement in metabolic processes of their respective species, with the possible exception of a storage role which has been postulated for them in view of their special accumulation in reserve tissues and the dependence of the concentration of some of them on nitrogen fertilization. Although such a role cannot be excluded, and might be even probable in some cases (especially in tissues like the cereal endosperm that are completely degraded during germination), in other cases a storage function is not likely because they are lost from the seed by diffusion (Tan–Wilson and Wilson, 1982) or undergo a specific proteolytic processing that seems to be different from the non–specific one of the true reserve proteins (Hartl et al., 1986; Yoshikawa et al., 1979b; Madden et al., 1985; Lorensen et al., 1981; Wilson and Chen, 1983).

Only those few inhibitors that have been shown to affect endogenous proteases and amylases are likely to play specific roles in plant metabolism.

Proteinase inhibitors that inhibit endogenous proteinases are found in barley (Mikola and Enari, 1970; Kirsi and Mikola, 1971), rice (Horiguchi and Kitagishi, 1971), wheat (Preston and Kruger, 1976), maize (Reed and Penner, 1976; Abe et al., 1980), mung beans (Baumgartner and Chrispeels, 1976), Scots pine (Salmia and Mikola, 1980; Salmia, 1980), and alfalfa (Gonnelli et al., 1985). At least some of these inhibitors might be specific for thiol–proteinases (see Ryan, 1981). They are usually present in small amounts and rapidly disappear during early stages of germination. The disappearance of inhibitory activity is accompanied by a rise of the endopeptidase activity, but the kinetics of the two processes, both in mung bean (Baumgartner and Chrispeels, 1976) and in Scots pine (Salmia, 1980), suggest that they are not causally related. Furthermore, since the endogenous proteinase and the inhibitor have different subcellular locations, these authors have postulated that the inhibitors do not control the breakdown of storage proteins by the enzyme associated with the protein bodies, but probably play a protective function with respect to other cellular components. As previously indicated, a possible role in the prevention of starch degradation has been proposed for the endogenous α–amylase inhibitors (Mundy et al., 1983), but as in the case of the endogenous proteinase inhibitors, no clear metabolic role has been fully demonstrated for them.

INHIBITORS AND PLANT PROTECTION

Several lines of evidence indicate that possibly a majority of the plant inhibitors that are active against heterologous proteinases and α–amylases play a protective role against the attacks of animal predators, insects, fungi, bacteria, and viruses. The inhibition of hydrolytic enzymes from these organisms may be directly toxic or even lethal to them, or could affect their fitness in such a way that different feeding habits might evolve. The action of some inhibitors on insect and microbial proteases has been repeatedly reported (Birk et al., 1963b; Applebaum et al., 1964a; Applebaum and Konijn, 1966; Peng and Black, 1976; Yoshikawa et al., 1976; Mosolov et al., 1976, 1979). Similarly, α–amylases of different origins have been tested against plant inhibitors (Applebaum et al., 1964b; Silano et al., 1975; Powers and Culbertson, 1982; Orlando et al., 1983; Baker, 1987). Particularly significant in the present context was the observation that insects which were able to feed on a tissue like the cereal endosperm, which has a high concentration of α–amylase inhibitors, had significantly higher α–amylase activities than those that were not able to do so (Silano et al., 1975). More recently, Gatehouse et al. (1986) have studied the effects on larvae development of inhibitor preparations from wheat endosperm that were strong inhibitors of digestive α–amylases from larvae of *Tribolium confusum*, a storage pest of wheat products, and *Collosobruchus maculatus*, a storage pest of legume seeds. While very high inhibitor concentrations were required to affect survival of larvae from the first species, quite low concentrations were effective against those of the second. The authors concluded that a detoxification mechanism must have been functioning in the first case, besides the considerably higher α–amylase activity, to account for the low sensitivity.

A more direct link of proteinase inhibitors with plant protection is represented by the already described systemic response which has been investigated by Ryan and co–workers. Increased levels of inhibitors are induced not only by the mechanical lesions caused by insects (see Ryan, 1984) but also following infection by fungi: Peng and Black (1976) demonstrated that while an incompatible race of *Phytophthora infestans* was able to elicit higher inhibitor levels in tomato plants, a compatible one produced either a small transitory increase or no increase. The systemic response was two times greater with the incompatible race than with compatible ones.

The effects of proteinase inhibitors on insects have been investigated in a number of cases. A possible involvement of these inhibitors in the protection of barley against grasshoppers was postulated because leaf extracts of the less sensitive variety tested had a considerably higher anti–chymotrypsin activity (Weiel and Hapner, 1976). Growth retardation and delayed pupation of corn borer larvae have been observed in feeding experiments with the Kunitz soybean trypsin inhibitor, while the maize trypsin inhibitor had no effect (Steffens et al., 1978). These and similar studies suggest but do not demonstrate a protective role for the inhibitors. More

314

conclusive evidence has been obtained for the resistance of the cowpea (*Vigna unguiculata*) to the bruchid beetle *Callosobruchus maculatus*. The only resistant variety found in a screening of over five thousand varieties had a higher level of proteinase inhibitors, whose antimetabolic nature was demonstrated in feeding trials (Gatehouse et al., 1979). In comparison, soybean trypsin inhibitor and lima been trypsin inhibitor were relatively ineffective (Gatehouse and Boulter, 1983). A convincing demonstration of the protective potential of the cowpea inhibitor has been obtained by cloning the corresponding gene and expressing it in tobacco under a constitutive promoter. According to a recent news item (Newmark, 1987), the transformed tobacco plants have been tested against bud- and army worms, and in both cases, they fail to grow and eventually die. This approach will allow in the near future testing the possible protective properties of different inhibitors.

An exciting new development concerns virus resistance. It seems that resistance to the cowpea mosaic virus in cultivar Arlington depends on the inhibition by a plant inhibitor of the thiol-proteinase encoded by the virus, thus preventing proteolytic processing of the viral polyprotein (see Ponz and Bruening, 1986). The recently reported homology of a PR-protein, induced by infection with tobacco mosaic virus, with a bifunctional inhibitor from maize (Richardson et al., 1987) is also of considerable interest in the present context.

INHIBITORS AND NUTRITION

Due to their abundance in many plant tissues, which are used as food and feed components, and to their relative stability versus heat, enzymatic degradation, or extreme pH, the potential incidence of proteinase and α-amylase inhibitors on human health and on animal production has been extensively investigated, leading to some controversies which are not quite settled yet.

The effect of trypsin inhibitors has been recently reviewed by Liener (1986), who has pointed out the need for further research to remove the remaining uncertainty concerning possible risks to humans. The pancreas of rats and chicks fed on raw soybeans is stimulated to produce more trypsin and becames enlarged as a result of hypertrophy and hyperplasia. The secretory activity is subject to inhibition by intraluminal trypsin through the inhibition of the release of the hormone cholecystokinin (CCK) from the intestinal mucosa. The inhibition of trypsin by the inhibitor results in an increased release of CCK (Green and Lyman, 1972; Liddle et al., 1984). Most of the information relating to the nutritional effects of trypsin inhibitors has come from experiments with animals and while the above described effects appear also in mouse, hamster and young guinea pig, besides the rat and the chick, other animals such as the dog, pig and calf do not display this sensitivity (see Liener, 1986). For this reason, extrapolation to man is not easy, although it has been shown that raw soybeans can inhibit completely trypsin and chymotrypsin in human pancreatic juice (Krogdahl and Holm, 1981). The Kunitz inhibitor, which is heat labile, is inactivated by incubation with gastric juice, while the Bowman– Birk inhibitor is both heat stable and insensitive to gastric juice, and therefore, more likely to be present in processed soybean–containing foods and to survive passage through the stomach. The infusion of Bowman–Birk inhibitor into the duodenum of human subjects has been shown to significantly increase the enzyme content of the pancreatic juice (Goodale et al., 1985). A reported outbreak of gastrointestinal illness in individuals who had consumed underprocessed soy protein extender in a tuna fish salad would be a good illustration of the potential risk for humans (Gunn et al., 1980). In contrast, a potentially beneficial side–effect has been postulated for these inhibitors with the claim that diets containing soybeans or soybean protease inhibitors prevent the appearance of tumors in mouse skin treated with carcinogens, of breast tumors in rats subjected to ionizing radiations, and of spontaneous liver cancer in C3H mice (see Troll et al., 1984).

Inhibitors of salivary and pancreatic α–amylases in the diet could potentially affect starch digestion. Indeed, Puls and Keup (1973) proposed the use of the wheat α–amylase inhibitor as an antihyperglycaemic agent, based on tests carried out in rats, dogs, and humans. A large number of commercial preparations of α–amylase inhibitors or "starch blockers" obtained from kidney beans reached the market. Eventually, these products were discredited and their distribution discontinued on the basis of their ineffectiveness in humans (Carlson et al., 1983; Bo–Linn et al., 1982). A biochemical analysis of commercial starch blockers subsequently showed not only that these preparations were not particularly enriched in α–amylase inhibitors with respect to the crude meal, but that they contained trypsin inhibitors and lectins, as well as enough endogenous, inhibitor–insensitive α–amylase to offset their alleged effectiveness as inhibitors of starch digestion (Liener et al., 1984). More recent studies, using better characterized inhibitors, seem to confirm significant effects in rats and humans. An α–amylase inhibitor, prepared from black kidney beans (*Phaseolus vulgaris*), given to young rats by stomach tubing mixed with a starch meal was able to reduce in a dose–dependent manner the hyperglycaemia resulting from starch utilization (Lajolo et al., 1984). Similarly, a purified white bean inhibitor was found to be stable in human gastrointestinal secretions, to slow dietary starch digestion in vitro, and to rapidly inactivate amylase in the human intestinal lumen (Layer et al., 1985). The potential usefulness of α–amylase inhibitors in the treatment of diabetes and obesity cannot be inferred from these studies but deserves reconsideration.

ACKNOWLEDGEMENTS

Assistance from D. Lamoneda, and Grants No. 2022/83 and No. 0193/85 from the Comision Asesora de Investigacion Cientifica y Tecnica are gratefully acknowledged.

REFERENCES

ABE, M. and ARAI, S. (1985) Purification of a cysteine proteinase inhibitor from rice, *Oryza sativa.* Agric. Biol. Chem. 49:3349–3350.

ABE, M., ARAI, S., KATO, H. and FUJIMAKI, M. (1980) Thiol-protease inhibitor ocurring in endosperm of corn. Agric. Biol. Chem. 44:685–686.

AMPE, C., VAN DAME, J., CASTRO, L.A.B., SAMPAIO, M.J.A.M., VAN MONTAGU, M. and VANDEKERCKHOVE, J. (1986) The amino acid sequence of the 2S sulphur-rich proteins from seeds of Brazil nut (*Bertholletia excelsa* H.B.K.). Eur. J. Biochem. 159:597–604.

APPLEBAUM, S.W., BIRK, Y., HARPAZ, I. and BONDI, A. (1964a) Comparative studies on proteolytic enzymes of *Tenebrio molitor* L. Comp. Biochem. Physiol. 2:85–103.

APPLEBAUM, S.W., HARPAZ, I. and BONDI, A. (1964b) Amylase secretion in the larvae of *Prodenia litura.* Comp. Biochem. Physiol. 13:107–111.

APPLEBAUM, S.W. and KONIJN, A.M. (1966) The presence of a *Tribolium* protease inhibitor in wheat. J. Insect. Physiol. 12: 665–669.

ARAGONCILLO, C., RODRIGUEZ-LOPERNA, M.A., CARBONERO, P. and GARCIA-OLMEDO, F. (1975) Chromosomal control of non-gliadin proteins from the 70% ethanol extract of wheat endosperm. Theor. Appl. Genet. 45:322–326.

ARAGONCILLO, C., RODRIGUEZ-LOPERENA, M.A., SALCEDO, G., CARBONERO, P. and GARCIA-OLMEDO, F. (1978) Influence of homoeologous chromosomes on gene-dosage effects in allohexaploid wheat (*Triticum acstivum* L.). Proc. Natl. Acad. Sci. USA 75:1446–1450.

BAILLARGEON, M.W., LASKOWSKI, M., NEVES, D.E., PORUBCAN, M.A., SANTINI, R.E. and MARKLEY, J.L. (1980) Soybean trypsin inhibitor (Kunitz) and its complex with trypsin. Carbon-13 nuclear magnetic resonance studies of the reactive site arginine. Biochemistry 19:5703–5710.

BAKER, J.E. (1987) Purification of isoamylases from the rice wevil, *Sitophilus oryzae,* by high-performance liquid chromatography and their interaction with partially-purified amylase inhibitors from wheat. Insect. Biochem. 17:37–44.

BARBER, D., SANCHEZ-MONGE, R., GARCIA-OLMEDO, F., SALCEDO, G. and MENDEZ, E. (1986a) Evolutionary implications of sequential homologies among members of the trypsin/α-amylase inhibitor family (CM-proteins) in wheat and barley. Biochim. Biophys. Acta 873:147–151.

BARBER, D., SANCHEZ-MONGE, R., MENDEZ, E., LAZARO, A., GARCIA-OLMEDO, F. and SALCEDO, G. (1986b) New α-amylase and trypsin inhibitors among the CM-proteins of barley (*Hordeum vulgare*). Biochim. Biophys. Acta 869:115–118.

BAUMGARTNER, B. and CHRISPEELS, M.J. (1976) Partial characterization of a protease inhibitor which inhibits the major endopeptidase present in the cotyledons of mung beans. Plant Physiol. 58:1–6

BEDETTI, C., BOZZINI, A., SILANO, V. and VITTOZZI, L. (1974) Amylase protein inhibitors and the role of *Aegilops* species in polyploid wheat speciation. Biochim. Biophys. Acta 362: 299–307.

BELEW, M. and EAKER, D. (1976) The trypsin and chymotrypsin inhibitors in chickpeas. Identification of the trypsin-reactive site, partial amino acid sequence and further physico-chemical properties of the major inhibitor. Eur. J. Biochem. 62:499–508.

BIRK, Y. (1985) The Bowman–Birk inhibitor. Trypsin and chymotrypsin-inhibitor from soybeans. Int. J. Peptide Prot. Res. 25:113–131.
BIRK, Y., GERTLER, A. and KHALEF, S. (1963a) A pure trypsin inhibitor from soya beans. Biochem. J. 87:281–284.
BIRK, Y., GERTLER, A. and KHALEF, S. (1963b) Separation of a *Tribolium*-protease inhibitor from soybeans on a calcium phosphate column. Biochim. Biophys. Acta 67:326–328.
BISHOP, P., PEARCE, G., BRYANT, J.E. and RYAN, C.A. (1984) Isolation and characterization of the proteinase inhibitor inducing factor from tomato leaves. J. Biol. Chem. 259:13172–13177.
BOISEN, S. (1976) Characterization of the endosperm trypsin inhibitor in barley. Phytochemistry 15:641.
BOISEN, S. (1983) Protease inhibitors in cereals. Occurrence, properties, physiological role and nutritional influence. Acta Agric. Scand. 33:369–381.
BOISEN, S., ANDERSEN, C.Y. and HEJGAARD, J. (1981) Inhibitors of chymotrypsin and microbial serine proteases in barley grains. Isolation, partial characterization and immunochemical relationships of multiple molecular forms. Physiol. Plant. 52:167–176.
BOISEN, S. and DJURTOFT, R. (1981a) Trypsin inhibitor from wheat endosperm: purification and characterization. Cereal Chem. 58:460–463.
BOISEN, S. and DJURTOFT, R. (1981b) Trypsin inhibitor from rye endosperm: purification and properties. Cereal Chem. 58:194–198.
BO–LINN, G.W., SANTA ANA, C.A., MORAWSKI, S.G. and FORDTRAN, J. S. (1982) Starch blockers – their effect on calorie absorption from a high starch meal. N. Engl. J. Med. 307: 1413–1416.
BOSTERLING, B. and QUAST, U. (1981) Soybean trypsin inhibitor is double headed. Biochim. Biophys. Acta 657:58–72.
BOWMAN, D.E. (1946) Differentiation of soybean antitrypsin factors. Proc. Soc. Exp. Biol. Med. 63:547–550.
BROWN, W.E. and RYAN, C.A. (1984) Isolation and characterization of a wound-induced trypsin inhibitor from alfalfa leaves. Biochemistry 23:3418–3422.
BUONOCORE, V., DE BIASI, M.-G., GIARDINA, P., POERIO, E. and SILANO, V. (1985) Purification and properties of an α-amylase tetrameric inhibitor from wheat kernel. Biochim. Biophys. Acta 831:40–48.
BUONOCORE, V., GRAMENZI, F., PACE, W., PETRUCCI, T., POERIO, E. and SILANO, V. (1980) Interaction of wheat monomeric and dimeric protein inhibitors with α-amylase from yellow mealworm (*Tenebrio molitor* L. Larva). Biochem. J. 187:637–645.
BUONOCORE, V., PETRUCCI, T. and SILANO, V. (1977) Wheat protein inhibitors of α-amylase. Phytochemistry 16:811–820.
CAMPOS, F.A.P. and RICHARDSON, M. (1983) The complete amino acid sequence of the bifunctional α-amylase/trypsin inhibitor from seeds of ragi (Indian finger millet; *Eleusine coracana* Goertn.). FEBS Lett. 152:300–304.
CAMPOS, F.A.P. and RICHARDSON, M. (1984) The complete amino acid sequence of α-amylase inhibitor I-2 from seeds of ragi (Indian finger millet; *Eleusine coracana*). FEBS Lett. 167: 221–225.
CANTAGALLI, P., DI GIORGIO, G., MORISI, G., POCCHIARI, F. and SILANO, V. (1971) Purification and properties of three albumins from *T. aestivum* seeds. J. Sci. Food Agric. 22:256 –259.
CARLSON, G.L., LI, B.U.K., BASS, P. and OLSEN, W.A. (1983) A bean

α-amylase inhibitor formulation (starch blocker) is ineffective in man. Science 219:393–395.

CHANDRASEKHER, G. and PATTABIRAMAN, T.N. (1985) Natural plant enzyme inhibitors. XIX. Purification and properties of an α-amylase inhibitor from bajra (*Peunisetum typhoideum*). Ind. J. Biochem. Biophys. 20:241–245.

CHANG, C.-R. and TSEN, C.C. (1981a) Isolation of trypsin inhibitors from rye, triticale, and wheat samples. Cereal Chem. 58:207–210.

CHANG, C.-R. and TSEN, C.C. (1981b) Characterization and heat stability of trypsin inhibitors from rye, triticale, and wheat samples. Cereal Chem. 58:211–213.

CLEVELAND, T.E., THORNBURG, R.W. and RYAN, C.A. (1987) Molecular characterization of wound inducible inhibitor I gene from potato and the processing of its mRNA and protein. Plant Mol. Biol. 8:199–207.

CORFMAN, R.S. and REECK, G.R. (1982) Immunoadsorbent isolation of trypsin inhibitors from corn and teosinte seeds. Biochim. Biophys. Acta 715:170–174.

CROUCH, M.L., TENBARGE, K.M., SIMON, A.E. and FERL, R. (1983) cDNA clones for *Brassica napus* seed storage proteins: evidence from nucleotide sequence analysis that both subunits of napin are cleaved from a precursor polypeptide. J. Mol. Appl. Genet. 2:273–283.

EWART, J.A.D. (1969) Isolation and characterization of a wheat albumin. J. Sci. Food Agric. 20:730–733.

FEILLET, P. and NIMMO, C.C. (1970) Soluble proteins of wheat. III. Isolation and characterization of two albumins, ALB 13A and ALB 13B, from flour. Cereal Chem. 47:447–464.

FILHO, J.X. (1974) Trypsin inhibitors in sorghum grains. J. Food Sci. 39:422–423.

FISH, W.W. and ABBOT, D.C. (1969) Isolation and characterization of a water-soluble wheat-flour protein. J. Sci. Food Agric. 20:723–730.

FOARD, D., GUTAY, P., LADIN, B., BEACHY, R. and LARKINS, B. (1982) In vitro synthesis of the Bowman-Birk and related soybean protease inhibitors. Plant Mol. Biol. 1:227–243.

FRA-MON, P., SALCEDO, G., ARAGONCILLO, C. and GARCIA-OLMEDO, F. (1984) Chromosomal assignment of genes controlling salt-soluble proteins (albumins and globulins) in wheat and related species. Theor. Appl. Genet. 69:167–172.

FREED, R.C. and RYAN, D.S. (1978) Changes in Kunitz trypsin inhibitor during germination of soybeans: an immunoelectrophoresis assay system. J. Food Sci. 43:1316–1319.

GARCIA-OLMEDO, F. and CARBONERO, P. (1970) Homoeologous protein synthesis controlled by homoeologous chromosomes in wheat. Phytochemistry 9:1495–1497.

GARCIA-OLMEDO, F. and GARCIA-FAURE, R. (1969) A new method for the estimation of common wheat (*T.aestivum* L.) in pasta products. Lebens. Wiss. Technol. 2:94–96.

GATEHOUSE, A.M.R. and BOULTER, D. (1983) Assessment of the antimetabolic effects of trypsin inhibitors from cowpea (*Vignia unguiculata*) and other legumes on development of the bruchid beetle *Callosobruchus maculatus*. J. Sci. Food Agric. 34:345–350.

GATEHOUSE, A.M.R., FENTON, K.A., JEPSON, I. and PAVEY, D.J. (1986) The effects of α-amylase inhibitors on insect storage pests: inhibition of α-amylase in vitro and effects on development in vivo. J. Sci. Food Agric. 37:727–734.

GATEHOUSE, A.M.R., GATEHOUSE, J.A., DOBIE, P., KILMINSTER, A.M. and BOULTER, D. (1979) Biochemical basis of insect resistance in *Vignia unguiculata*. J. Sci. Food Agric. 30:948–958.

GENNIS, L.S. and CANTOR, C.R. (1976) Double–headed inhibitors from black–eyed peas. I. Purification of two new protease inhibitors and the endogenous protease by affinity chromatography. J. Biol. Chem. 251:734–740.

GIESE, H. and HEJGAARD, J. (1984) Synthesis of salt–soluble proteins in barley. Pulse–labelling study of grain filling in liquid–cultured detached spikes. Planta 161:172–177.

GOLDSTEIN, Z., TROP, M. and BIRK, Y. (1973) Multifunctional proteinase inhibitor from *Bauhinia* seeds. Nature New Biol. 246:29–30.

GONNELLI, M., CIONI, P., ROMAGNOLI, A., GABELLIERI, E., BALESTRERI, E. and FELICIOLI, R. (1985) Purification and characterization of two leaf polypeptide inhibitors of leaf protease from alfalfa (*Medicago sativa*). Arch. Biochem. Biophys. 238:206–212.

GOODALE R.L., LIENER, I.E., DESHMUKH, A., FRICK, T., SATTERBERD, T., WARD, G. and BORNER, J.W. (1985) Do trypsin inhibitors affect feedback control of human pancreatic secretion?. Dig. Dis. Sci. 30:972 (abs. cited in Liener 1986).

GRAHAM, J.S., HALL, G., PEARCE, G. and RYAN, C.A. (1986) Regulation of synthesis of proteinase inhibitors I and II mRNAs in leaves of wounded tomato plants. Planta 169:399–405.

GRAHAM, J.S., PEARCE, G., MERRYWEATHER, J., TITANI, K., ERICSSON, L. and RYAN, C.A. (1985a) Wounded–induced proteinase inhibitors from tomato leaves. I. The cDNA–deduced primary structure of pre–inhibitor I and its post–translational processing. J. Biol. Chem. 260:6555–6560.

GRAHAM, J.S., PEARCE, G., MERRYWEATHER, J., TITANI, K., ERICSSON, L. and RYAN, C.A. (1985b) Wounded–induced proteinase inhibitors from tomato leaves. II. The cDNA–deduced structure of pre–inhibitor II. J. Biol. Chem. 260:6561–6564.

GRAHAM, J.S. and RYAN, C.A. (1981) Accumulation of metallocarboxy-peptidase inhibitor in leaves of wounded potato plants. Biochem. Biophys. Res. Commun. 101:1164–1170.

GRANUM, P.E. (1978) Purification and characterization of an α–amylase inhibitor from rye (*Secale cereale*) flour. J. Food Biochem. 2:103.

GREEN, G.M. and LYMAN, R.L. (1972) Feedback regulation of pancreatic enzyme secretion as a mechanism for trypsin inhibitor–induced hypersecretion in rats. Proc. Soc. Exp. Biol. Med. 140:6–12.

GREEN, T. and RYAN, C.A. (1972) Wound–induced proteinase inhibitor in plant leaves: a possible defense mechanism against insects. Science 175:776–777.

GREENE, L.J. and BARTELT, D.C. (1969) The structure of the bovine pancreatic secretory trypsin inhibitor – Kazal's inhibitor. J. Biol. Chem. 244:2646–2657.

GUNN, R.A., TAYLOR, P.R. and GANGAROSA, E.J. (1980) Gastrointestinal illness associated with consumption of soy protein extender. J. Food Protect. 43:525–529.

GURUSSIDAIAH, S., KUO, T. and RYAN, C.A. (1972) Immunochemical comparison of chymotrypsin inhibitor I among several genera of the Solanaceae. Plant Physiol. 50:627–631.

HALAYKO, A.J., HILL, R.D. and SVENSSON, B. (1986) Characterization of the interaction of barley α–amylase II with an endogenous α–amylase inhibitor from barley kernels. Biochim. Biophys. Acta 873:92-101.

HALIM, A.H., WASSOM, C.E. and MITCHELL, H.L. (1973) Trypsin inhibitor in corn (*Zea mays* L.) as influenced by genotype and moisture stress. Crop. Sci. 13:405-407.

HAMMOND, R., FOARD, D. and LARKINS, B. (1984) Molecular cloning and analysis of a gene coding for the Bowman-Birk protease inhibitor in soybean. J. Biol. Chem. 259:9883-9890.

HARTL, P.M., TAN-WILSON, A.L. and WILSON, K.A. (1986) Proteolysis of Kunitz soybean trypsin inhibitor during germination. Phytochemistry 25:23-26.

HASS, G.M., AGER, S.P., LE THOURNEAU, D., DERR-MAKUS, J.E. and MAKUS, D.J. (1981) Specificity of the carboxypeptidase inhibitor from potatoes. Plant Physiol. 67:754-758.

HASS, G.M., HERMODSON, M.A., RYAN, C.A. and GENTRY, L. (1982) Primary structures of two low molecular weight proteinase inhibitors from potatoes. Biochemistry 21:752-756.

HASS, G.M. and RYAN, C.A. (1980a) Carboxypeptidase inhibitor from ripened tomatoes: purification and properties. Phytochemistry 19:1329-1333.

HASS, G.M. and RYAN, C.A. (1980b) Cleavage of the carboxypeptidase inhibitor from potatoes by carboxypeptidase B. Biochem. Biophys. Res. Commun. 97:1481-1486.

HASS, G,M, and RYAN, C.A. (1981) Carboxypeptidase inhibitor from potatoes. Methods Enzymol. 80:778-791.

HEJGAARD, J. (1976) Free and protein-bound β-amylases of barley grain. Characterization by two-dimensional immunoelectrophoresis. Physiol. Plant. 38:293-299.

HEJGAARD, J. (1982) Purification and properties of protein Z - a major albumin of barley endosperm. Physiol. Plant. 54:174-182.

HEJGAARD, J., BJORN, S.E. and NIELSEN, G. (1984a) Localization to chromosomes of structural genes for the major protease inhibitors in barley grains. Theor. Appl. Genet. 68:127-130.

HEJGAARD, J., BJORN, S.E. and NIELSEN, G. (1984b) Rye chromosomes carrying structural genes for the major grain protease inhibitors. Hereditas 101:257-259.

HEJGAARD, J. and BOISEN, S. (1980) High-lysine proteins in Hiproly barley breeding: identification, nutritional significance and new screening methods. Hereditas 93:311-320.

HEJGAARD, J., RASMUSSEN, S.K., BRANDT, A. and SVENDSEN, I. (1985) Sequence homology between barley endosperm protein Z and protease inhibitors of the α1-antitrypsin family. FEBS Lett. 180:89-94.

HEJGAARD, J., SVENDSEN, I. and MUNDY, J. (1983) Barley α amylase/subtilisin inhibitor II. N-terminal amino acid sequence and homology with inhibitors of the soybean trypsin inhibitor (Kunitz) family. Carlsberg Res. Commun. 48:91-94.

HIGGINS, T.J.V., CHANDLER, P.M., RANDALL, P.J., SPENCER, D., BEACH, L.R., BLAGROVE, R.J., KORTT, A.A. and INGLIS, A.S. (1986) Gene structure, protein structure, and regulation of the synthesis of a sulfur-rich protein in pea seeds. J. Biol. Chem. 261:11124-11130.

HILDEBRAND, D.F., ORF, J.H. and HYMOWITZ, T. (1980) Inheritance of an acid phosphatase and its linkage with the Kunitz trypsin inhibitor in seed protein of soybeans. Crop. Sci. 20:83-85.

HILL, R.E. and HASTIE, N.D. (1987) Accelerated evolution in reactive centre regions of serine protease inhibitors. Nature 326:96-99.

HOFFMAN, L.M., SENGUPTA–GOPALAN, C. and PAAREN, H.E. (1984) Structure of soybean Kunitz trypsin inhibitor mRNA determined from cDNA by using oligodeoxynucleotide primers. Plant Mol. Biol. 3:111–117.

HOJIMA, Y., MORIWAKI, C. and MORIYA, H. (1973) Studies of kallikrein inhibitors from potatoes. IV. Isolation of two inhibitors and determination of molecular weights and amino acid compositions. J. Biochem. 73:923–932.

HOJIMA, Y., PIERCE, J.V. and PISANO, J.J. (1980a) Hageman factor fragment inhibitor in corn seeds: purification and characterization. Thromb. Res. 20:149–162.

HOJIMA, Y., PIERCE, J.V. and PISANO, J.J. (1980b) Plant inhibitors of serine proteinases: Hageman factor fragment, kallireins, plasmin, thrombin, factor Xa, trypsin and chymotrypsin. Thromb. Res. 20:163–171.

HOLLANDER–CZYTKO, H., ANDERSEN, J.K. and RYAN, C.A. (1985) Vacuolar localization of wound-induced carboxypeptidase inhibitor in potato leaves. Plant Physiol. 78:76–79.

HORIGUCHI, T. and KITAGISHI, K. (1971) Studies on rice seed protease. V. Protease inhibitor in rice seed. Plant Cell Physiol. 12:907–915.

HORISBERGER, M. and TACCHINI, M. (1982) Ultrastructural localization of Kunitz trypsin inhibitor in soybeans using gold granules labelled with protein A. Experientia 38:726.

HORY, H.D. and WEDER, J.K. (1976) Trypsin- und chymotrypsin-inhibitorem in leguminosem. VII. Partialsequenzen des trypsin–chymotrypsin–inhibitors PCI 3 aus *Phaseolus coccineus*. Z. Lebensm. Unters.–Forsch. 162:349–356.

HUNKAPILLER, M.W., FORGAC, M.D., YU, E.H. and RICHARDS, J.H. (1979) ^{13}C NMR studies of the binding of soybean trypsin inhibitor to trypsin. Biochem. Biophys. Res. Commun. 87:25–31.

HWANG, D.L., YANG, W.K. and FOAARD, D.E. (1978) Rapid release of protease inhibitors from soybeans: immunochemical quantitation and parallels with lectin. Plant Physiol. 61:30–34.

IKENAKA, T. and NORIOKA, S. (1986) Bowman–Birk family serine proteinase inhibitors. In: Barrett, A.J. and Salvesen, G. (eds.) Proteinase Inhibitors. Elsevier, pp.361–374.

ISHIKAWA, C., NAKAMURA, S., WATANABE, K. and TAKAHASHI, K. (1979) The amino acid sequence of azuki bean proteinase inhibitor I. FEBS Lett. 99:97–100.

ISHIKAWA, C., WATANABE, K., SAKATA, N., NAKAGAKI, C., NAKAMURA, S. and TAKAHASHI, K. (1985) Azuki bean (*Vigna angularis*) protease inhibitors: isolation and amino acid sequences. J. Biochem. 97:55–70.

JOHNSON, R., MITCHELL, H.L., WASSOM, C.E. and REECK, G.R. (1980) Comparison of trypsin inhibitors from several types of corn. Phytochemistry 19:2749–2750.

JONASSEN, I. (1980) Characteristics of hiproly barley II. Quantification of two proteins contributing to its high lysine content. Carlsberg Res. Commun. 45:59–68.

JONASSEN, I. and SVENDSEN, I. (1982) Identification of the reactive sites in two homologous serine proteinase inhibitors isolated from barley. Carlsberg Res. Commun. 47: 199–203.

JOUBERT, F.J. (1981) Purification and some properties of a proteinase inhibitor (DE-1) from *Peltophorum africanum* (Weeping wattle) seed. Hoppe–Seyler's Z. Physiol. Chem. 362:1515–1521.

JOUBERT, F.J. (1982) Purification and some properties of a proteinase inhibitor (DE-1) from *Pterocarpus angolensis* (Kiaat) seed. S. Afr. J.

Chem. 35:72-76.

JOUBERT, F.J. (1983) Purification and properties of the proteinase inhibitors from *Acacia sieberana* seed. Phytochemistry 22:53-57.

JOUBERT, F.J. (1984a) Proteinase inhibitors from *Lonchocarpus capassa* (apple-leaf) seed. Phytochemistry 23:957-961.

JOUBERT. F.J. (1984b) Trypsin isoinhibitors from *Momordica repens* seeds. Phytochemistry 23:1401-1406.

JOUBERT, F.J., HEUSSEN, C. and DOWDLE, B.D. (1985) The complete amino acid sequence of trypsin inhibitor DE-3 from *Erythrina latissima* seeds. J. Biol. Chem. 260:12948-12953.

JOUBERT, F.J., KRUGER, H., TOWNSHEND, G.S. and BOTES, D.P. (1979) Purification, some properties and the complete primary sequences of two protease inhibitors (DE-3 and DE-4) from *Macrotyloma axillare* seed. Eur. J. Biochem. 97:85-91.

JOUBERT, F.J. and SHARON, N. (1985) Proteinase inhibitors from *Erythrina corolladendron* and *E. cristagalli* seeds. Phytochemistry 24:1169-1179.

KASHLAN, N. and RICHARDSON, M. (1981) The complete amino acid sequence of a major wheat protein inhibitor of α-amylase. Phytochemistry 20:1781-1784.

KEIL, M., SANCHEZ-SERRANO, J., SCHELL, J. and WILLMITZER, L. (1986) Primary structure of proteinase inhibitor II gene from potato (*Solanum tuberosum*). Nucl. Acid. Res. 14:5641-5650.

KEILOVA, H. and TOMASEK, V. (1976a) Isolation and properties of cathepsin D inhibitor from potatoes. Collection Czechoslov. Chem. Commun. 41:489-497.

KEILOVA, H. and TOMASEK, V. (1976b) Further characteristics of cathepsin D inhibitor from potatoes. Collection Czechoslov. Chem. Commun. 41:2440-2447.

KIANG, Y.T. and CHIANG, Y.C. (1986) Genetic linkage of a leucine amino peptidase locus with the Kunitz trypsin inhibitor locus in soybeans. J. Heredity 77:128-129.

KIM, S.H., HARA, S., IIASA, S., IKENAKA, T., TODA, H., KITAMURA, K. and KAIZUMA, N. (1985) Comparative study on amino acid sequences of Kunitz-type soybean trypsin inhibitors Tia, Tib and Tic. J. Biochem. 98:435-448.

KIRSI, M. (1973) Formation of proteinase inhibitors in developing barley grains. Physiol. Plant. 29:141-144.

KIRSI M. and AHOKAS, H. (1983) Trypsin inhibitor activities in the wild progenitor of barley. Phytochemistry 22:2739-2740.

KIRSI, M. and MIKOLA, J. (1971) Occurrence of proteolytic inhibitors in various tissues of barley. Planta 96:281-291.

KIYOHARA, T. and IWASAKI, T. (1985) Chemical and physicochemical characterization of the permanent and temporary trypsin inhibitors from buckwheat. Agric. Biol. Chem. 49:589-594.

KIYOHARA, T., YOKOTA, K., MASAKI, Y., MATSUI, P., IWASAKI, T. and YOSHIKAWA, M. (1981) The amino acid sequences of proteinase inhibitors I-A and I-A' from azuki beans. J. Biochem. 90:721-728.

KNEEN, E. and SANDSTEDT, R.M. (1943) An amylase inhibitor from certain cereals. J. Am. Chem. Soc. 65:1247-1252.

KOIDE, T. and IKENAKA, T. (1973a) Studies on soybean trypsin inhibitors. 1. Fragmentation of soybean trypsin inhibitor (Kunitz) by limited proteolysis and by chemical cleavage. Eur. J. Biochem. 32:401-407.

KOIDE, T. and IKENAKA, T. (1973b) Studies on soybean trypsin inhibitors. 3. Amino acid sequence of the carboxyl terminal region and the complete amino acid sequence of soybean trypsin inhibitor (Kunitz). Eur. J. Biochem. 32:417–431.
KOIDE, T., TSUNAZAVA, S. and IKENAKA, T. (1973) Studies on soybean trypsin inhibitors. 2. Amino acid sequence around the reactive site of soybean trypsin inhibitor. Eur. J. Biochem. 32:408–416.
KORTT, A.A. and JERMYN, M.A. (1981) *Acacia* proteinase inhibitors. Purification and properties of the trypsin inhibitors from *Acacia elata* seed. Eur. J. Biochem. 115:551–557.
KOWALSKI, D. and LASKOWSKI, M. (1976a) Chemical–enzymatic replacement of Ile64 in the reactive site of soybean trypsin inhibitor (Kunitz). Biochemistry 15:1300–1309.
KOWALSKI, D. and LASKOWSKI, M. (1976b) Chemical–enzymatic insertion of an amino acid residue in the reactive site of soybean trypsin inhibitor (Kunitz). Biochemistry 15:1309–1315.
KOWALSKI, D., LEARY, T.R., McKEE, R.E., SEALOCK, R.W., WANG, D. and LASKOWSKI, M. (1974) Replacements, insertions, and modifications of amino acid residues in the reactive site of soybean trypsin inhibitor (Kunitz). In: Fritz, H., Tschesche, H., Green, L.J. and Truscheit, E. (eds.) Bayer Symposium V. Proteinase inhibitors. Springer–Verlag Berlin, pp. 311–324.
KOYAMA, S., KOBAYASHI, Y., NORIOKA, S., KYOGOKU, Y. and IKENAKA, T. (1986) Conformational studies by 1H nuclear magnetic resonance of the trypsin–chymotrypsin inhibitor B-III from peanuts and its enzymatically modified derivative. Biochemistry 25:8076–8082.
KREIS, M., FORDE, B.G., RAHMAN, S., MIFLIN, B.J. and SHEWRY, P.R. (1985) Molecular evolution of the seed storage proteins of barley, rye and wheat. J. Mol. Biol. 183:499–502.
KROGDAHL, A. and HOLM, H. (1981) Soybean proteinase inhibitor and human proteolytic enzymes: selective inactivation of inhibitors by treatment with human gastric juice. J. Nutr. 111:2045–2051.
KUNITZ, M. (1945) Crystallization of a trypsin inhibitor from soybean. Science 101:668–669.
KUNITZ, M. (1946) Crystalline soybean trypsin inhibitor. J. Gen. Physiol. 29:149–154.
KUNITZ, M. (1947a) Crystalline soybean trypsin inhibitor. 2. General properties. J. Gen. Physiol. 30:291–307.
KUNITZ, M. (1947b) Isolation of a crystalline protein compound of trypsin and soybean trypsin inhibitor. J. Gen. Physiol. 30:311–320.
KUNITZ, M. (1949) The kinetics and thermodynamics of reversible denaturation of crystalline soybean trypsin inhibitor. J. Gen. Physiol. 32:241–263.
LAJOLO, F.M. and FINARDI-FILHO, F. (1985) Partial characterization of the amylase inhibitor of black beans (*Phaseolus vulgaris*) variety Rico 23. J. Agric. Food Chem. 33:132–138.
LAJOLO, F.M., NANCINI-FILHO, J. and MENEZES, E.W. (1984) Effect of a bean (*Phaseolus vulgaris*) α–amylase inhibitor on starch utilization. Nutr. Rep. Int. 30:45–54.
LASKOWSKI, M. and KATO, I. (1980) Protein inhibitors of proteinases. Ann. Rev. Biochem. 49:593–626.
LASKOWSKI, M., KATO, I., ARDELT, W., COOK, J., DENTON, A., EMPIE, M.W., KOHR, W.J., PARK, S.J., PARKS, K., SCHATZLEY, B.L., SCHOENBERGER, O.L., TASHIRO, M., VICHOT, G., WHATLEY, H.E., WIECZOREK, A. and WIECZOREK, M. (1987) Ovomucoid third domains from

100 avian species: isolation, sequences and hypervariability of enzyme-inhibitor contact residues. Biochemistry 26:202-221.

LAYER, P., CARLSON, G.L. and DIMAGNO, P. (1985) Partially purified white bean amylase inhibitor reduces starch digestion in vitro and inactivates intraduodenal amylase in humans. Gastroenterology 88:1895-1902.

LAZARO, A., BARBER, D., SALCEDO, G., MENDEZ, E. and GARCIA OLMEDO, F. (1985) Differential effects of high-lysine mutations on the accumulation of individual members of a group of proteins encoded by a disperse multigene family in the endosperm of barley (*Hordeum vulgare* L.). Eur. J. Biochem. 149:617-623.

LEE, J.S., BROWN, W.E., GRAHAM, J.S., PEARCE, G., FOX, E.A., DREHER, T.W., AHERN, K.G., PEARSON, G.D. and RYAN, C.A. (1986) Molecular characterization and phylogenetic studies of a wound-inducible proteinase inhibitor I gene in *Lycopersicon* species. Proc. Natl. Acad. Sci. USA 83:7277-7281.

LIDDLE, R.H., GOLDFINE, I.D. and WILLIAMS, J.A. (1984) Bioassay of plasma cholecystokinin in rats: effects of food, trypsin inhibitor, and alcohol. Gastroenterology 87:542-549.

LIENER, I.E. (1986) Trypsin inhibitors: concern for human nutrition or not?. J. Nutrition 116:920-923.

LIENER, I.E., DONATUCCI, D.A. and TARCZA, J.C. (1984) Starch blockers: a potential source of trypsin inhibitors and lectins. Am. J. Clin. Nutr. 39:196-200.

LILLEY, G.G. and INGLIS, A.S. (1986) Amino acid sequence of conglutin δ, a sulfur-rich seed protein of *Lupinus angustifolius* L. Sequence homology with the C-III α-amylase inhibitor from wheat. FEBS Lett. 195:235-241.

LORENSEN, E., PREVOSTO, R. and WILSON, K. (1981) The appearence of new active forms of trypsin inhibitor in germinating mung bean (*Vigna radiata*) seeds. Plant Physiol. 68:88-92.

MADDEN, M., TAN-WILSON, A. and WILSON, K. (1985) Proteolysis of soybean Bowman-Birk trypsin inhibitor during germination. Phytochemistry 24:2811-2815.

MAEDA, K. (1986) The complete amino acid sequence of the endogenous α-amylase inhibitor in wheat. Biochim. Biophys. Acta 871:250-256.

MAEDA, K., KAKABAYASHI, S. and MATSUBARA, H. (1985) Complete amino acid sequence of an α-amylase inhibitor in wheat kernel (0.19 inhibitor). Biochim. Biophys. Acta 828:213-221.

MAEDA, K., SATO, M., KATO, Y., TANAKA, N., HATA, Y., KATSUBE, Y. and MATSUBARA, H. (1987) Crystallographic study of endogenous α-amylase inhibitor from wheat. J. Mol. Biol. in press.

MAEDA, K., TAKAMORI, Y. and OKA, O. (1982) Isolation and properties of an α-amylase inhibitor (0.53) from wheat (*Triticum aestivum*). Agric. Biol. Chem. 46:2873-2875.

MAHONEY, W.C., HERMODSON, M.A., JONES, B., POWERS, D.D., CORFMAN, R.S. and REECK, G.R. (1984) Amino acid sequence and secondary structural analysis of the corn inhibitor of trypsin and activated Hageman factor. J. Biol. Chem. 259:8412-8416.

MAIER, K., MULLER, R., TESCH, R., WITT, I. and HOLZER, H. (1979) Amino acid sequence of yeast proteinase B inhibitor 1. Comparison with inhibitor 2. Biochem. Biophys. Res. Commun. 91:1390-1398.

MAKI, Z., TASHIRO, M., SUGIHARA, N. and KANAMORI, M. (1980) Double-headed nature of a trypsin inhibitor from rice bean. Agric. Biol. Chem. 44:953-955.

MANJUNATH, N.H., VEERABHADRAPPA, P.S. and VIRUPAKSHA, T.K. (1983) Isolation and characterization of a trypsin inhibitor from finger millet. Phytochemistry 22:2349–2357.

McPHALEN, C.A. and JAMES, M.N.G. (1987) Crystal and molecular structure of the serine proteinase inhibitor CI-2 from barley seeds. Biochemistry 26:261–269.

McPHALEN, C.A., SVENDSEN, I., JONASSEN, I. and JAMES, M.N.G. (1985) Crystal and molecular structure of chymotrypsin inhibitor 2 from barley seeds in complex with subtilisin NOVO. Proc. Natl. Acad. Sci. USA 82:7242–7246.

MELVILLE, J.C. and RYAN, C.A. (1972) Chymotrypsin inhibitor I from potatoes: large scale preparation and the characterization of its subunit components. J. Biol. Chem. 247:3445–3453.

MENEGATTI, E., PALMIERI, S., WALDE, P. and LUISI, P.L. (1985) Isolation and characterization of a trypsin inhibitor from white mustard (Sinapis alba L.). J. Agric. Food Chem. 33:784–789.

MIKOLA, J. and ENARI, T.M. (1970) Changes in the contents of barley proteolytic inhibitors during malting and mashing. J. Inst. Brew. 76:182–188.

MIKOLA, J. and KIRSI, M. (1972) Differences between endospermal and embryonal trypsin inhibitors in barley, wheat and rye. Acta Chem. Scand. 26:787–795.

MIKOLA, J. and SOULINNA, E.M. (1969) Purification and properties of a trypsin inhibitor from barley. Eur. J. Biochem. 9:555–560.

MITSUNAGA, T., KIMURA, Y. and SHIMIZU, M. (1982) Isolation and partial characterization of a trypsin inhibitor from wheat endosperm. J. Nutr. Sci. Vitaminol. 28:419–429.

MOIDEEN KUTTY, A.V. and PATTABIRAMAN, T.N. (1985) Isolation and characterization of an amylase inhibitor from Echinocloa frunentacea grains. Ind. J. Biochem. Biophys. 22:155–160.

MOIDEN KUTTY, A.V. and PATTABIRAMAN, T.N. (1986) Isolation and characterization of an amylase inhibitor from sorghum seeds, specific for human enzymes. J. Agric. Food Chem. 34:552–557.

MOLINA–CANO, J.L., FRA–MON, P., SALCEDO, G., ARAGONCILLO, C., ROCA DE TAGORES, F. and GARCIA–OLMEDO, F. (1987) Morocco as a possible domestication center for barley: biochemical and agromorphological evidence. Theor. Appl. Genet. 73:531–536.

MORPHY, L., MIZUTA, K., IKEMOTO, H. and VENTURA, M.M. (1985) Complete amino acid sequence of the trypsin and chymotrypsin inhibitor from Vignia unguiculata (L) walp seeds. An. Acad. Brasil. Cienc. 57:387.

MOSOLOV, V.V., LOGINOVA, M.D., FEDURKINA, N.V. and BENKEN, I.I. (1976) The biological significance of proteinase inhibitors in plants. Plant Sci. Lett. 7:77–80.

MOSOLOV, V.V., LOGINOVA, M.D., MALOVA, E.L. and BENKEN, I.I. (1979) A specific inhibitor of Colletotrichum lindemunthianum protease from kidney bean (Phaseolus vulgaris) seeds. Planta 144:265–269.

MOSOLOV, V.V. and SHULGIN, M.N. (1986) Protein inhibitors of microbial proteinases from wheat, rye and triticale. Planta 167:595–600.

MUNDY, J., HEJGAARD, J. and SVENDSEN, I. (1984) Characterization of a bifunctional wheat inhibitor of endogenous α–amylase and subtilisin. FEBS Lett. 167:210–214.

MUNDY, J. and ROGERS. J.C. (1986) Selective expression of a probable α–amylase/protease inhibitor in barley aleurone cells: comparison to the barley α–amylase/subtilisin inhibitor. Planta 169:51–63.

MUNDY, J., SVENDSEN, I. and HEJGAARD, J. (1983) Barley α-amylase/ subtilisin inhibitor I. Isolation and characterization. Carlsberg Res. Commun. 48:81-90.

NAGARAJ, R.H. and PATTABIRAMAN, T.N. (1985) Isolation of an amylase inhibitor from *Setaria italica* grains by affinity chromatography on Blue-sepharose and its characterization. J. Agric. Food Chem. 33:646-650.

NEWMARK, P. (1987) Trypsin inhibitor confers pest resistance. Biotechnology 5:426.

NORIOKA, S. and IKENAKA, T. (1983a) Amino acid sequence of a trypsin-chymotrypsin inhibitor, B-III, of peanut (*Arachis hypogaea*). J.Biochem. 93:479-485.

NORIOKA, S. and IKENAKA, T. (1983b) Amino acid sequences of trypsin-chymotrypsin inhibitors (A-I, A-II, B-I and B-II) from peanut (*Arachis hypogaea*): A discussion on the molecular evolution of legume Bowman-Birk type inhibitors. J. Biochem. 94:589-599.

NORIOKA, S. and IKENAKA, T. (1984) Inhibition mechanism of a peanut trypsin-chymotrypsin inhibitor B-III: Determination of the reactive sites for trypsin and chymotrypsin. J. Biochem. 96:1155-1164.

O'CONNOR, C.M. and McGEENEY, K.F. (1981a) Isolation and characterization of four inhibitors from wheat flour which display differential inhibition specificities for human salivary and human pancreatic α-amylases. Biochim. Biophys. Acta 658:387-396.

O'CONNOR, C.M. and McGEENEY, K.F. (1981b) Interaction of human α-amylases with inhibitors from wheat flour. Biochim. Biophys. Acta 658:397-405.

ODANI, S. and IKENAKA, T. (1972) Studies on soybean trypsin inhibitors. IV. Complete amino acid sequence and the antiproteinase sites of Bowman-Birk soybean proteinase inhibitor. J. Biochem. 71:839-848.

ODANI, S. and IKENAKA, T. (1973a) Studies on soybean trypsin inhibitors. VIII. Disulfide bridges in soybean Bowman-Birk inhibitor. J. Biochem. 74:697-715.

ODANI, S. and IKENAKA, T. (1973b) Scission of soybean Bowman-Birk proteinase inhibitor into two small fragments having either trypsin or chymotrypsin inhibitory activity. J. Biochem. 74:857-860.

ODANI, S. and IKENAKA, T. (1976) The amino acid sequences of two soybean double-headed proteinase inhibitors and evolutionary consideration on the legume proteinase inhibitors. J. Biochem. 80:641-643.

ODANI, S. and IKENAKA, T. (1977) Studies on soybean trypsin inhibitors. XI. Complete amino acid sequence of a soybean trypsin-chymotrypsin- elastase inhibitor, C-II. J. Biochem. 92:1523-1531.

ODANI, S., KOIDE, T. and ONO, T. (1982) Sequence homology between barley trypsin inhibitor and wheat α-amylase inhibitors. FEBS Lett. 141:279-282.

ODANI, S., KOIDE, T. and ONO, T. (1983a) The complete amino acid sequence of barley trypsin inhibitor. J. Biol. Chem. 258:7998-8003.

ODANI, S., KOIDE, T. and ONO, T. (1983b) A possible evolutionary relationship between plant trypsin inhibitor, α-amylase inhibitor, and mammalian pancreatic secretory trypsin inhibitor (Kazal). J. Biochem. 93:1701-1704.

ODANI, S., KOIDE, T. and ONO, T. (1986) Wheat germ trypsin inhibitors. Isolation and structural characterization of single-headed and double-headed inhibitors of the Bowman-Birk type. J. Biochem. 100:975-983.

ODANI, S., KOIDE, T., ONO, T. and OHNISHI, K. (1983c) Structural relationship between barley (*Hordeum vulgare*) trypsin inhibitor and castor-bean (*Ricinus communis*) storage protein. Biochem. J. 213-543-545.

ODANI, S., ODANI, S., ONO, T. and IKENAKA, T. (1979) Proteinase inhibitors from a Mimosideá legume, *Albizzia julibrissin*. Homologues of soybean trypsin inhibitor (Kunitz). J. Biochem. 86:1795-1805.

O'DONELL, M.D. and McGEENEY, K.F. (1976) Purification and properties of an α-amylase inhibitor from wheat. Biochim. Biophys. Acta 422:159-169.

OGATA, F. and MAKISUMI, S. (1984) Isolation and characterization of trypsin inhibitors from tubers of taro, *Colocasia antiquorum* var. nymphaifolia?. J. Biochem. 96:1565-1574.

OGATA, F. and MAKISUMI, S. (1985) Subunit compositions of trypsin inhibitors from tubers of taro, *Colocasia antiquorum* var. nymphaifilia?. J. Biochem. 97:589-597.

ORF, J.H. and HYMOWITZ, T. (1979) Genetics of the Kunitz trypsin inhibitor: an antinutritional factor in soybeans. J. Am. Chem. Soc. 56:722-726.

ORLANDO, A.R., ADE, P., DI MAGGIO, D., FANELLI, C. and VITTOZZI, L. (1983) The purification of a novel amylase from *Bacillus subtilis* and its inhibition by wheat proteins. Biochem. J. 209:561-564

PACE, W., PARLAMENTI, R., RAB, A., SILANO, V. and VITTOZZI, L. (1978) Protein α-amylase inhibitors from wheat flour. Cereal Chem. 55:244-254.

PAZ-ARES, J., HERNANDEZ-LUCAS, C., SALCEDO, G., ARAGONCILLO, C., PONZ, F. and GARCIA-OLMEDO, F. (1983a) The CM-proteins from cereal endosperm: immunochemical relationships. J. Exp. Bot. 34:388-395.

PAZ-ARES, J., PONZ, F., ARAGONCILLO, C., HERNANDEZ-LUCAS, C., SALCEDO, G., CARBONERO, P. and GARCIA-OLMEDO, F. (1983b) In vivo and in vitro synthesis of CM-proteins (A-hordeins) from barley (*Hordeum vulgare* L.). Planta 157:74-80

PAZ-ARES, J., PONZ, F., RODRIGUEZ-PALENZUELA, P., LAZARO, A., HERNANDEZ-LUCAS, C., GARCIA-OLMEDO, F. and CARBONERO, P. (1986) Characterization of cDNA clones of the family of trypsin/α-amylase inhibitors (CM-proteins) in barley (*Hordeum vulgare* L.). Theor. Appl. Genet. 71:842-846.

PENG, J.H. and BLACK, L.L. (1976) Increased proteinase inhibitor activity in response to infection of resistant tomato plants by *Phytophthora infestans*. Phytopathology 66: 958-963.

PETRUCCI, T., RAB, A., TOMASI, M. and SILANO, V. (1976) Further characterization studies of the α-amylase protein inhibitor of gel electrophoretic mobility 0.19 from the wheat kernel. Biochim. Biophys. Acta 420:288-297.

PETRUCCI, T., SANNIA, G., PARLAMENTI, R. and SILANO, V. (1978) Structural studies of wheat monomeric and dimeric protein inhibitors of α-amylase. Biochem. J. 173:229-235.

PETRUCCI, T., TOMASI, M., CANTAGALLI, P. and SILANO, V. (1974) Comparison of wheat albumin inhibitors of α-amylase and trypsin. Phytochemistry 13:2487-2495.

PHAM, T.-C., LELUK, J., POLANOWSKI, A. and WILUSZ, T. (1985) Purification and characterization of the trypsin inhibitor from *Cucurbita pepo* var. *patissonina* fruits. Biol. Chem. Hoppe-Seyler

366:939-944.

PICK, K.-H. and WOBER, G. (1978) Proteinaceous α-amylase inhibitor from beans (*Phaseolus vulgaris*). Purification and partial characterization. Hoppe-Seyler's Z. Physiol. Chem. 359:1371-1377.

PLUNKETT, G., SENEAR, D.F., ZUROSKE, G. and RYAN, C.A. (1982) Proteinase inhibitor I and II from leaves of wounded tomato plants: purification and properties. Arch. Biochem. Biophys. 213:463-472.

POLANOWSKI, A. (1974) Trypsin inhibitor from starchy endosperm of rye seeds. Acta Soc. Bot. Pol. 43:27-37.

POLANOWSKI, A., WILUSZ, T., NIENARTOWICZ, B., CIESLAR, E., SLOMINSKA, A. and NOWAK, K. (1980) Isolation and partial amino acid sequence of the trypsin inhibitor from the seeds of *Cucurbita maxima*. Acta Biochim. Pol. 27:371-382.

PONZ, F. and BRUENING, G. (1986) Mechanisms of resistance to plant viruses. Ann. Rev. Phytopathol. 24:355-381.

POWERS, J.R. and CULBERTSON, J.D. (1982) In vitro effect of bean amylase inhibitor on insect amylases. J. Food Protec. 45:655-657.

PRESTON, K.R. and KRUGER, J.E. (1976) Localization and activity of proteolytic enzymes in developing wheat kernels. Can. J. Plant Sci. 56:217-223.

PULS, W. and KEUPS, V. (1973) Influence of an α-amylase inhibitor (BAY d 7791) on blood glucose, serum insulin and NEFA in starch loading tests in rats, dogs and man. Diabetologia 9:97-102.

PUTZAI, A. (1968) General properties of a protease inhibitor from the seeds of kidney bean. Eur. J. Biochem. 5:252-259.

RANCOUR, J.M. and RYAN, C.A. (1968) Isolation of a carboxypeptidase B inhibitor from potatoes. Arch. Biochem. Biophys. 125:380-382.

READ, R.J. and JAMES, M.N.G. (1986) Introduction to the protein inhibitors: X-ray crystallography. In: Barrett,A.J. and Salvesen, G. (eds.). Proteinase Inhibitors. Research monographs in cell tissue physiology. Vol. 12. Elsevier, Amsterdam, New York, Oxford, pp. 301-336.

REDDY, M.N., KEIM, P.S., HEINRIKSON, R.L. and KEZDY, F.J. (1975) Primary structural analysis of sulfhydril protease inhibitors from pineapple stem. J. Biol. Chem. 250:1741-1750.

REED, C. and PENNER, D. (1976) Peptidases and trypsin inhibitor in the developing endosperm of opaque-2 and normal corn. Agronomy J. 70:337-340.

REES, D.C. and LIPSCOMB, W.N. (1982) Refined crystal structure of the potato inhibitor complex of carboxypeptidase A at 2.5Å resolution. J. Mol. Biol. 160:475-498.

RICHARDSON, M. (1974) Chymotryptic inhibitor I from potatoes: the amino acid sequence of subunit A. Biochem. J. 137:101-112.

RICHARDSON, M. (1979) The complete amino acid sequence and the trypsin reactive (inhibitory) site of the major proteinase inhibitor from the fruits of the aubergine (*Solanum melongena* L.). FEBS Lett. 104:322-326.

RICHARDSON, M. (1981) Protein inhibitors of enzymes. Food Chem. 6:235-253.

RICHARDSON, M., BARKER, R.D.J., McMILLAN, R.T. and COSSINS, L. (1977) Identification of the reactive sites of chymotrypsin inhibitor I from potatoes. Phytochemistry 16:837-839.

RICHARDSON, M., CAMPOS, F.A.P., XAVIER-FILHO, J., MACEDO, M.L.R., MAIA, G.M.C. and YARWOOD, A. (1986) The amino acid sequence and reactive (inhibitory) site of the major trypsin isoinhibitor (DE5)

isolated from seeds of the Brazilian carolina tree (*Adenanthera pavonina* L.). Biochim. Biophys. Acta 872:134-140.
RICHARDSON, M. and COSSINS, L. (1974) Chimotryptic inhibitor I from potatoes: the amino acid sequences of subunits B, C and D. FEBS Lett. 45:11-13 (corrigendum in FEBS Lett. 52:161 (1975)).
RICHARDSON, M., VALDES-RODRIGUEZ, S. and BLANCO-LABRA, A. (1987) A possible function for thaumatin and a TMV-induced protein suggested by homology to a maize bifunctional enzyme inhibitor. Nature 327:432-434.
RODAWAY, S.J. (1978) Composition of α-amylase from aleurone layers of grains of Himalayan barley. Phytochemistry 17:385-389.
RODRIGUEZ-LOPERENA, M.A., ARAGONCILLO, A., CARBONERO, P. and GARCIA-OLMEDO, F. (1975) Heterogeneity of wheat endosperm proteolipids (CM-proteins). Phytochemistry 14:1219-1223.
RYAN, C.A. (1973) Proteolytic enzymes and their inhibitors in plants. Ann. Rev. Plant Physiol. 24:173-196.
RYAN, C.A. (1981) Proteinase inhibitors. In: Marcus, A.(ed.) The Biochemistry of Plants Vol.6, Academic Press, New York, San Francisco, London, pp. 351-370.
RYAN, C.A. (1984) Defense responses of plants. In: Verma, D.P.S. and Hohn, T.H. (eds.) Plant gene research: genes involved in microbe plant interactions, Springer Verlag, Wien,New York, pp. 375-386.
RYAN, C.A., KUO, T., PEARCE, G. and KUNKEL, R. (1977) Variability in the concentration of three heat-stable inhibitor proteins in potato tubers. Am. Potato J. 53:433-440.
SAKATO, K., TANAKA, H. and MISAWA, M. (1975) Broad-specificity proteinase inhibitors in *Scopolia japonica* (Solanaceae) cultured cells. Isolation, physicochemical properties and inhibition kinetics. Eur. J. Biochem. 55:211-219.
SALCEDO, G., ARAGONCILLO, A., RODRIGUEZ-LOPERENA, M.A., CARBONERO, P. and GARCIA-OLMEDO, F. (1978a) Differential allelic expression at a locus encoding an endosperm protein in tetraploid wheat (*T.turgidum*). Genetics 89:147-156.
SALCEDO, G., FRA-MON, P., MOLINA-CANO, J.L., ARAGONCILLO, A. and GARCIA-OLMEDO, F. (1984) Genetics of CM-proteins (A-hordeins) in barley. Theor. Appl. Genet. 68:53-59.
SALCEDO, G., RODRIGUEZ-LOPERENA, M.A. and ARAGONCILLO, A. (1978b) Relationships among low MW hydrophobic proteins from wheat endosperm. Phytochemistry 17:1491-1494.
SALCEDO, G., SANCHEZ-MONGE, R. and ARAGONCILLO, A. (1982) The isolation and characterization of low molecular weight hydrophobic salt-soluble proteins from barley. J. Exp. Bot. 33:1325-1331.
SALMIA, M.A. (1980) Inhibitors of endogenous proteinases in Scots pine seeds: fractionation and activity changes during germination. Physiol. Plant. 48:266-270.
SALMIA, M.A. and MIKOLA, J. (1980) Inhibitors of endogenous proteinases in the seeds of Scots pine, *Pinus sylvestris*. Physiol. Plant 48:126-130 SANCHEZ-MONGE, R., BARBER, D., MENDEZ, E., GARCIA-OLMEDO, F. and SALCEDO, G. (1986a) Genes encoding α-amylase inhibitors are located in the short arms of chromosomes 3B, 3D and 6D of wheat (*Triticum aestivum* L.). Theor. Appl. Genet. 72:108-113.
SANCHEZ-MONGE, R., FERNANDEZ, J.A. and SALCEDO, G. (1987) Subunits of tetrameric α-amylase inhibitors of *Hordeum chilense* are encoded by genes located in chromosomes 4Hch and 7Hch. Theor. Appl. Genet. (in press).

330

SANCHEZ–MONGE, R., GOMEZ, L., GARCIA–OLMEDO, F and SALCEDO, G. (1986b) A tetrameric inhibitor of insect α–amylase from barley. FEBS Lett. 207:105–109.
SANCHEZ–SERRANO, J.J., KEIL, M., O'CONNOR, A., SCHELL, J. and WILLMITZER, L. (1987) Wound–induced expression of a potato proteinase inhibitor II gene in transgenic tobacco plants. EMBO J. 6:303–306.
SANCHEZ–SERRANO, J.J., SCHMIDT, R., SCHELL, J. and WILLMITZER, L. (1986) Nucleotide sequence of proteinase inhibitor II encoding cDNA of potato (*Solanum tuberosum*) and its mode of expression. Mol. Gen. Genet. 203:15–20.
SATOH, S., SATOH, E., WATANABE, T. and FUJII, T. (1985) Isolation and characterization of a protease inhibitor from Spinach leaves. Phytochemistry 24:419–423.
SAUNDERS, R.M. and LANG, J.A. (1973) α–amylase inhibitor in *T. aestivum*: purification and physical–chemical properties. Phytochemistry 12:1237–1241.
SEALOCK, R.W. and LASKOWSKI, M. (1969) Enzymatic replacement of arginyl by a lysyl residue in the reactive site of soybean trypsin inhibitor. Biochemistry 8:3703–3710.
SEEMULLER, U., EULITZ, M., FRITZ, H. and STROBL, A. (1980) Structure of the elastase–cathepsin G inhibitor of the Leech *Hirudo medicinalis*. Hoppe–Seyler's Z. Physiol. Chem. 361:1841–1846.
SEIDL, D.S., ABREU, H. and JAFFE, W.G. (1978) Purification of a subtilisin inhibitor from black bean seeds. FEBS Lett.92: 245–250.
SEIDL, D.S., ABREU, H. and JAFFE, W.G. (1982) Partial characterization of a subtilisin inhibitor from black bean seeds. Int. J. Pep. Prot. Res. 19:153–157.
SHAINKIN, R. and BIRK, Y. (1970) α–amylase inhibitors from wheat. Isolation and characterization. Biochim. Biophys. Acta 221:502–513.
SHARIEF, F.S. and LI, S.S.-L. (1982) Amino acid sequence of small and large subunits of seed storage protein from *Ricinus communis*. J. Biol. Chem. 257:14753–14759.
SHEWRY, P.R., LAFIANDRA, D., SALCEDO, G., ARAGONCILLO, A., GARCIA–OLMEDO, F., LEW, E.J.-L., DIETLER, M.D. and KASARDA, D.D. (1984) N–terminal amino acid sequences of chloroform/ methanol–soluble proteins and albumins from endosperm of wheat, barley and related species. FEBS Lett. 175:359–363.
SHIBATA, H., HARA, S., IKENAKA, T. and ABE, J. (1986) Purification and characterization of proteinase inhibitors from winged bean (*Psophocarpus tetragonolobus*) seeds. J. Biochem. 99:1147–1155.
SHIMOKAWA, Y., KUROMIZU, K., ARAKI, T., OHATA, J. and ABE, O. (1984) Primary structure of *Vicia angustifolia* proteinase inhibitor. Eur. J. Biochem. 143:677–684.
SHIVARAJ, B. and PATTABIRAMAN, T.N. (1980) Natural plant enzyme inhibitors. Part VIII. Purification and properties of two α–amylase inhibitors from ragi (*Eleusine coracana*) grains. Ind. J. Biochem. Biophys. 17:181–185.
SHIVARAJ, B. and PATTABIRAMAN, T.N. (1981) Natural plant enzyme inhibitors. Characterization of an unusual α–amylase/ trypsin inhibitor from ragi (*Eleusine coracana* Geartn). Biochem. J. 193:29–36.
SHYAMALA, G. and LYMAN, R.L. (1964) The isolation and purification of a trypsin inhibitor from wholewheat flour. Can. J. Biochem. 42:1825–1832.

SIEMION, I.Z., WILUSZ, T. and POLANOWSKI, M. (1984) On the genetic and structural similarities between the squash seeds polypeptide trypsin inhibitors and wheat germ agglutinin. Mol. Cell. Biochem. 60:159–161.

SILANO, V., FURIA, M., GIANFREDA, L., MACRI, A., PALESCANDOLO, R., RAB, A., SCARDI, V., STELLA, E. and VALFRE, F. (1975) Inhibition of amylases from different origins by albumins from the wheat kernel. Biochim. Biophys. Acta 391:170–178.

SILANO, V., POCCHIARI, F. and KASARDA, D.D. (1973) Physical characterization of α–amylase inhibitors from wheat. Biochim. Biophys. Acta 317:139–148.

SILANO, V., POERIO, E. and BUONOCORE, V. (1977) A model for the interaction of wheat monomeric and dimeric protein inhibitors with α–amylase. Mol. Cell. Biol. 18:87–91.

SILANO, V. and ZAHNLEY, J.C. (1978) Association of *Tenebrio molitor* L. α–amylase with two protein inhibitors – one monomeric, one dimeric – from wheat flour. Differential scanning calorimetric comparison of heat stabilities. Biochim. Biophys. Acta 533:181–185.

SODINI, G., SILANO, V., DE AGAZIO, M., POCCHIARI, F., TENTORI, L. and VIVALDI, G. (1970) Purification and properties of a *Triticum aestivum* specific albumin. Phytochemistry 9:1167–1172.

STEFFENS, R., FOX, F.R. and KASSEL, B. (1978) Effect of trypsin inhibitors on growth and metamorphosis of corn borer larvae *Ostrinia nubilalis*. J. Agric. Food Chem. 26:170–175.

STEVENS, F.C., WUERY, S. and KRAHU, J. (1974) Structure-function relationship in lima bean proteinase inhibitor. In: Fritz, H., Tschesche, H., Greene, L.J. and Truscheit, E. (eds.) Proteinase inhibitors, Bayer–Symposium V, Springer Verlag, pp. 344–354.

SUDHAKAR PRABHU, K. and PATTABIRAMAN, T.N. (1980) Natural plant enzyme inhibitors. IX. Isolation and characterization of a trypsin/chymotrypsin inhibitor from Indian red wood (Adenanthera pavonina) seeds. J. Sci. Food Agric. 31:967–980.

SUZUKI, A., TSUNOGAE, Y., TANAKA, I., YAMANE, T., ASHIDA, T., NORIOKA, S., HARA, S. and IKENAKA, T. (1987) The structure of Bowman–Birk type protease inhibitor A-II from peanut (*Arachis hypogaea*) at 3,3Å resolution. J. Biochem. 101:267–274.

SVENDSEN, I., BOISEN, S. and HEJGAARD, J. (1982) Amino acid sequence of serine protease inhibitor CI-1 from barley. Homology with barley inhibitor CI-2, potato inhibitor I, and leech eglin. Carlsberg Res. Commun. 47:45–53.

SVENDSEN, I., HEJGAARD, J. and CHAVAN, J.K. (1984) Subtilisin inhibitor from seeds of broad bean (*Vicia faba*); purification, amino acid sequence and specificity of inhibition. Carlsberg Res. Commun. 49:493–502.

SVENDSEN, I., HEJGAARD, J. and MUNDY, J. (1986) Complete amino acid sequence of the α–amylase/subtilisin inhibitor from barley. Carlsberg Res. Commun. 51:43–50.

SVENDSEN, I., JONASSEN, I., HEJGAARD, J. and BOISEN, S. (1980) Amino acid sequence homology between a serine protease inhibitor from barley and potato inhibitor I. Carlsberg Res. Commun. 45:389–395.

SVENSSON, B., ASANO, K., JONASSEN, I., POULSEN, F.M., MUNDY, J. and SVENDSEN, I. (1986) A 10 kd seed protein homologous with an α–amylase inhibitor from Indian finger millet. Carlsberg Res. Commun. 51:493–500.

SWARTZ, M.J., MITCHELL, H.L., COX, D.J. and REECK, G.R. (1977) Isolation and characterization of trypsin inhibitor from opaque-2 corn seeds. J. Biol. Chem. 252:8105–8107.

SWEET, R.M., WRIGHT, H.T., JANIN, J., CHOTINA, C.H. and BLOW, D.M. (1974) Crystal structure of the complex of porcine trypsin with soybean trypsin inhibitor (Kunitz) at 2.6 Å resolution. Biochemistry 13:4212–4228.

TAN, C. and STEVENS, F. (1971) Amino acid sequence of lima bean protease inhibitor component IV. 2. Isolation and sequence determination of the chymotryptic peptides and the complete amino acid sequence. Eur. J. Biochem. 1 8:515–523.

TAN, F., ZHANG, G., MU, J., LIN, N. and CHI, C. (1984) Purification, characterization and sequence determination of a double-headed trypsin inhibitor peptide from *Trichosanthes kirilowii* (a chinese medical herb). Hoppe-Seyler's Z. Physiol. Chem. 365:1211–1217.

TAN-WILSON, A.L. and WILSON, K.A. (1982) Nature of proteinase inhibitors released from soybeans during imbibition and germination. Phytochemistry 21:1547–1551.

TASHIRO, M. and MAKI, Z. (1979) Purification and characterization of a trypsin inhibitor from rice bran. J. Nutr. Sci. Vitaminol. 25:255–264.

TASHIRO, M. and MAKI, Z. (1986) Isolation of protein inhibitors of papain, trypsin, and α-amylase in the grain of foxtail millet. Agric. Biol. Chem. 50:2955–2957.

THORNBURG, R.W., AN, G., CLEVELAND, T.E., JOHNSON, R. and RYAN, C.A. (1987) Wound-inducible expression of a potato inhibitor II-chloramphenicol acetyltransferase gene fusion in transgenic tobacco plants. Proc. Natl. Acad. Sci. USA 84:744–748.

TROLL, W., FRENKEL, K. and WIESNER, R. (1984) Protease inhibitors as anticarcinogens. J. Nat. Cancer Inst. 73:1245–1250.

TSUNOGAE, Y., TANAKA, I., YAMANE, T., KIKKAWA, J., ASHIDA, T., ISHIKAWA, C., WATANABE, K., NAKAMURA, S. and TAKAHASHI, K. (1986) Structure of the trypsin-binding domain of Bowman-Birk type protease inhibitor and its interaction with trypsin. J. Biochem. 100:1637–1646.

VARTAK, H.G., RELE, M.V. and JAGANNATHAN, V. (1980) Proteinase inhibitors from *Vignia unguiculata* subsp. *cylindrica*. III. Properties and kinetics of inhibitors of papain, subtilisin and trypsin. Arch. Biochem. Biophys. 204:134–140.

VODKIN, L.O. (1981) Isolation and characterization of messenger RNAs for seed lectin and Kunitz trypsin inhibitor in soybeans. Plant Physiol. 68:766–771.

WAGNER, L.P. and RIEHM, J.P. (1967) Purification and partial characterization of a trypsin inhibitor isolated from the navy bean. Arch. Biochem. Biophys. 121:672–677.

WALKER-SIMMONS, M. and RYAN, C.A. (1977) Wound-induced accumulation of trypsin inhibitor activities in plant leaves. Plant Physiol. 59:437–439.

WARCHALEWSKI, J.R. (1977a) Isolation and purification of native α-amylase inhibitors from winter wheat. Bull. Acad. Pol. Ser. Sci. Biol. 25:725–729.

WARCHALEWSKI, J.R. (1977b) Isolation and purification of native α-amylase inhibitors from malted winter wheat. Bull. Acad. Pol. Ser. Sci. Biol. 25:731–735.

WEDER, J.P. (1986) Inhibition of human proteinase by grain legumes. Adv. Exp. Med. Biol. 199:239–279.

WEIEL, J. and HAPNER, K.D. (1976) Barley proteinase inhibitors: a possible role in grasshopper control?. Phytochemistry 15: 1885–1887.

WESELAKE, R.J., MACGREGOR, A.W. and HILL, R.D. (1983a) An endogenous α–amylase inhibitor in barley kernels. Plant Physiol. 72:809–812.

WESELAKE, R.J., MACGREGOR, A.W. and HILL, R.D. (1985) Endogenous α–amylase inhibitor in various cereals. Cereal Chem. 62:120–123.

WESELAKE, R.J., MACGREGOR, A.W., HILL, R.D. and DUCKWORTH, H.W. (1983b) Purification and characteristics of an endogenous α–amylase inhibitor from barley kernels. Plant Physiol. 73:1008–1012.

WIECZOREK, M., OTLEWSKI, J., COOK, J., PARKS, K., LELUK, J., WILIMOWSKA-PELC, A., POLANOWSKI, A., WILUSZ, T. and LASKOWSKI, M. (1985) The squash family of serine proteinase inhibitors. Amino acid sequences and association equilibrium constants of inhibitors from squash, summer squash, zucchini, and cucumber seeds. Biochem. Biophys. Res. Commun. 126:646–652.

WILLIAMSON, M., FORDE, J., BUXTON, B. and KREIS, M. (1987) Nucleotide sequence of barley chymotrypsin inhibitor-2 (CI-2) and its expression in normal and high-lysine barley. Eur. J. Biochem. (in press).

WILSON, K.A. (1980) The release of proteinase inhibitors from legume seeds during germination. Phytochemistry 19:2517–2519.

WILSON, K.A. and CHEN, J.C. (1983) Amino acid sequence of mung bean trypsin inhibitors and its modified form appearing during germination. Plant Physiol. 71:341–349.

WILSON, K.A. and LASKOWSKI, M.Jr. (1975) The partial amino acid sequence of trypsin inhibitor II from garden bean, Phaseolus vulgaris, with location of the trypsin and elastase-reactive sites. J. Biol. Chem. 250:42 61–4267.

WILUSZ, T., WIECZOREK, M., POLANOWSKI, A., DENTON, A., COOK, J. and LASKOWSKI, M. (1983) Amino-acid sequence of two trypsin isoinhibitors, ITDI and ITDIII from squash seed (Cucurbita maxima). Hoppe-Seyler's Z. Physiol. Chem. 364:93–95.

YAMAMOTO, M., HARA, S. and IKENAKA, T. (1983) Amino acid sequences of two trypsin inhibitors from winged bean seeds (Psophocarpus tetragonolobus (L) DC.). J. Biochem. 94:849–863.

YOSHIKAWA, M., IWASAKI, T., FUJII, M. and OOGAKI, M. (1976) Isolation and some properties of a subtilisin inhibitor from barley. J. Biochem. 79:765–773.

YOSHIKAWA, M., KIYOHARA, T., IWASAKI, T., ISHII, Y. and KIMURA, N. (1979a) Amino acid sequences of proteinase inhibitors II and II' from azuki beans. Agric. Biol. Chem. 43:787–796.

YOSHIKAWA, M., KIYOHARA, T., IWASAKI, T. and YOSHIDA, I. (1979b) Modification of proteinase inhibitor II in azuki beans during germination. Agric. Biol. Chem. 43:1989–1990.

YOSHIKAWA, M. and OGURA, S. (1978) Reactive sites of a trypsin and chymotrypsin inhibitor, proteinase inhibitor II, from azuki beans (Phaseolus angularis). Agric. Biol. Chem. 42:1753–1759.

ZAWISTOWSKA, U. and BUSHUK, W. (1986) Electrophoretic characterization of low-molecular weight wheat protein of variable solubility. J. Sci. Food Agric. 37:409–417.

ZHANG, Y.S., LOS, S., TAN, F.L., CHI, C.W., XU, L.X. and ZHANG, A.L. (1982) Complete amino acid sequence of mung bean trypsin inhibitor. Sci. Sinica 25:268–277.

NEWS AND VIEWS
SECTION

Oxford Surveys of Plant Molecular & Cell Biology Vol.4 (1987) 337–342

AMINO ACID HOMOLOGIES SUGGEST FUNCTIONS
FOR PATHOGENESIS–RELATED PROTEINS

W. S. Pierpoint and P. R. Shewry

*AFRC Institute of Arable Crops Research,
Rothamsted Experimental Station,
Harpenden, Herts. AL5 2JQ, UK*

The usefulness of computer–aided comparisons of the amino acid sequences of proteins is well appreciated: these comparisons recognize and quantify structural similarities; they indicate evolutionary relationships; and they suggest possible functions that were not suspected. A recent and unexpected example of this usefulness is in recognizing the structural similarities between three proteins from three very diverse plant origins, namely virus–infected tobacco leaves, the sweet arils of the fruit of a West African shrub, and the seeds of cultivated maize (Richardson et al., 1987). The revealed similarities invite speculation in a number of directions; thus they suggest the existence of a wide–spread class of plant proteins with a characteristic tightly–coiled structure stabilized by eight disulphide bonds, and that these themselves may be part of a larger superfamily of proteins.

The protein from virus–infected leaves, variously referred to as R or S (Pierpoint, 1986) is one of the Pathogenesis–Related (PR–) proteins. PR–proteins (Antoniw et al., 1980) have been recognized for the last 17 years as a debatable and enigmatic group of proteins produced in relatively large amounts in tobacco leaves which are responding hypersensitively to infection with a virus or other pathogen (see van Loon, 1985 and Fraser this volume). The four originally recognized on PAGE analysis of extracts of such leaves have been extended to at least 13 (Pierpoint, 1986) and more recently reports (Hogue and Asselin, 1987) suggest that there are some ten more which are too basic to be seen in standard Ornstein–Davis gels. The proteins have enough properties in common to be considered as a group; thus they are produced in leaves responding hypersensitively to a pathogen, and accumulate in the inter–cellular fluid; they are soluble at pH 3, relatively resistant to proteolysis, and many of them are also induced by specific chemicals including ethylene and aspirin (see van Loon, 1985). However, they clearly belong to a number of different sub–groups of specific structural and immuno-logical relationships (Table 1), and this indeed is the basis of a widely accepted nomenclature (Antoniw et al., 1980). Proteins related immunogenically to some of the tobacco proteins can be detected in a wide range of plants, including monocotyledons, when these are challenged with appropriate pathogens or the chemical inducers (van

Table 1. Some of the PR-proteins of TMV-infected
tobacco (N. tabaccum cv. xanthi-nc) leaves

Formal Name	Provisional Name	M_r (from SDS-PAGE)	Properties
Ia	–	~14,500	}
Ib	–	~14,500	} Acidic proteins with structural and immunological similarities
Ic	–	~14,500	}
IIa	II	~40,000	}
IIb	N	~40,000	} Charge isomers on electrophoresis
–	O	~40,000	}
–	P	~28,000	
–	Q	~28,000	
–	R or R'	~15,000	Some N-acetylglucosamine: possibly 2 subunits
–	S or R	~23,000	Thaumatin-like protein

M_r values averaged from cited literature:

Some PR-protein-SDS complexes are believed to migrate anomalously

Loon, 1985; White et al., 1987). But the aspect that has generated most interest and debate, is their possible relationship to the processes that restrict pathogen spread in a hypersensitive infection and which may render the leaf, as well as neighbouring leaves, more resistant to a subsequent infection. Although sceptical views have been expressed (Fraser, 1982), it is possible to cite examples where the appearance of PR-proteins goes hand-in-hand with some measure of induced-resistance; thus leaves in which PR-proteins have been induced by ethylene or aspirin produce fewer, smaller lesions on subsequent inoculation with tobacco mosaic virus (see van Loon, 1985, see also Fraser, this volume).

A great embarrassment to those who argue for a protective function for PR-proteins is the lack of ideas as to how they might function. Purified or semi-purified preparations have been injected into leaves and introduced into protoplast suspensions, but have not produced any consistent convincing protection against viruses. This may, of course, simply reflect on the state of the proteins as isolated, and the difficulties of introducing proteins into intact cells. The proteins, or some of them, have been examined for easy-to-test enzyme activities (van Loon, 1982), as enzyme inhibitors, and as agglutinating agents for particular micro-organisms (Pierpoint, 1986), but again without convincing results. Such direct attempts to find a role for PR-proteins have been disappointing, although recent and unpublished evidence suggests that a PR-protein that occurs in the intercellular fluid of beans infected with the fungus *Colletotrichum*, co-purifies with a chitinase activity (Metraux, Streit and Staub, 1987) and chitinases are recognized as powerful anti-fungal agents (Schlumbaum et al., 1986). It is likely that a better view of functional roles for PR-proteins will come from comparing their amino acid sequences, once these are known, with those of other proteins in primary structure databases.

Bol and his colleagues at Leiden, recently identified in a library of c-DNA from tobacco plants, some 32 clones that correspond to virus-induced RNAs (Hooft van Huijsduijnen et al., 1986; Cornelissen et al., 1986). They were classified into six clusters, the DNA inserts of which would hybridize only with those of the same cluster. One cluster contained a single cloned DNA insert which on analysis was seen to contain an open reading frame corresponding to a protein of some 226 amino acids flanked by both 5' and 3' non-coding regions. When trimmed of a leader sequence, this protein has a M_r of about 21,596 and corresponds closely to the PR-protein R which had been previously isolated from the intercellular fluid of TMV-infected tobacco leaves (Pierpoint et al., 1987) and which attracted attention because of its content of 16 cysteine residues per molecule. More surprisingly the protein has a high homology ($\approx$65% identical residues) with thaumatin, the intensely sweet-tasting protein present in arils of the West African shrub *Thaumatococcus daniellii*. This homology includes conservation of all 16 cysteine residues.

Thaumatin is an un-glycosylated protein, or more properly one of a family of about five closely related proteins (M_rs about 22,000) and is some 30-100,000 times as sweet as sucrose on a molar basis (van der Wel, 1983). It has always seemed an oddity among proteins. Its sweet properties were known and utilized by ancient Africans, brought to European attention by a British surgeon, W. F. Daniell, in 1839, and rediscovered by Inglett over a hundred years later in the course of a search for non-calorific sweeteners of plant origin. Now marketed by Tate and Lyle it has commercial importance as a sweetener or taste enhancer in both human and animal foods. Its molecular structure is known in some detail as a result of chemical and X-ray studies (see van der Wel, 1986; Kinghorn and Soejarto, 1986). The relevant gene from *Thaumatococcus* can be transferred to, and expressed in eukaryotic organisms. However, its relationship to PR-R, although interesting, did not go far to suggest a function for PR-R. For the function of thaumatin was presumed to be concerned with its attractiveness to animals, and PR-R is apparently not sweet: indeed the two lysine and the tyrosine residues on which the sweetness of thaumatin is thought to depend (Edens and van der Wel, 1985) are not conserved in protein R (Pierpoint et al., 1987). It is true that thaumatin has been claimed to have a latent proteolytic activity, but this is only revealed when its tertiary structure is disrupted by reduction, and its biological significance is questionable and uncertain.

The maize protein (MAI) is an inhibitor of bovine trypsin and of α-amylase from the beetle *Tribolium castaneum*, a pest of stored grains. MAI has an M_r of about 22,000 and again contains 16 cysteine residues (Richardson et al., 1987). Such inhibitors of digestive enzymes are widespread in the plant kingdom (Richardson, 1977), especially in storage organs where they may deter both phytophagous insects and fungi: an inhibitor purified from wheat, for instance, has a deleterious effect on the larval development of *Callosobruchus*, a pest of stored legumes, but not on the wheat-adapted larvae of *Tribolium* (Gatehouse et al., 1986). Other inhibitors are produced in response to wound damage or infection, and may contribute to a wide spectrum defense mechanism.

Comparison of the amino acid sequence of MAI with those of other inhibitors of plant origin (Richardson et al., 1987) shows regions of homology (8-30 residues) with an α-amylase inhibitor (I-2) from finger millet (*Eleusine coracana*) (Campos and Richardson, 1984) and a probable amylase and protease inhibitor (PAPI) from barley aleurone (Mundy and Rogers, 1986). Limited homology with Kunitz-type trypsin inhibitors is also present, but the biological significance of this is open to doubt.

There is, however, no uncertainty about the homologies of MAI with thaumatin and PR-R, which have 52% and 57% identical residues. These homologies suggest not only a common evolutionary origin of the three proteins, but also that thaumatin and PR-R are, or have been derived from, inhibitors of α-amylase and/or trypsin. While this may not seem a particularly relevant or useful role for a PR-protein induced by virus infection, it does fit in with a view that is emerging of the hypersensitive response. This response is clearly non-specific in that it is triggered by a wide range of pathogens or chemical treatments, and it seems probable that it is a 'blunderbus' response conferring a wide spectrum resistance to bacteria, fungi and virus (and perhaps also to insect predators!). It is a response that protects not only against the triggering pathogen, but also against any secondary pathogen to which the lesion and neighbouring tissue is further exposed. When viewed in this light the α-amylase/trypsin inhibitory activity is clearly relevant.

Although the homologies reported by Richardson and co-workers should start a resurgence of interest in identifying inhibitory functions for thaumatin and PR-R, we should not be surprised if these attempts are unsuccessful. The sweet taste of thaumatin may well have been selected to the detriment of its inhibitory activity, while PR-R could be an inactive descendant of a previously active protein.

Thaumatin, MAI and PR-R clearly constitute a closely related family of proteins, with similar M_rs and amino acid sequences and with 16 conserved cysteine residues. To emphasize these common characteristics, we will refer to them as "hexadecathionins". In thaumatin the cysteine residues are all involved in intra-chain disulphide bonds, and their conservation indicates that the three-dimensional structures of the three proteins also have conserved elements. This conservation is, however, not inconsistent with divergence in function: conserved cysteine residues are also present in another seed protein superfamily that includes cereal inhibitors of α-amylase and trypsin (not related to those discussed here, see also Garcia-Olmeda this volume), 2S storage globulins of some dicotyledonous species and specific structural domains of some cereal prolamins (Kreis et al., 1985).

The two inhibitors, PAPI and the α-amylase inhibitor (I-2) of finger millet, are less closely related to MAI having lower M_rs (9694 and 9333 respectively) and less cysteine residues (8 and 7 respectively). However, seven of these cysteines are conserved in both inhibitors, and three are also conserved in MAI, PR-R and thaumatin. These five proteins can, therefore, be considered to be members of a single protein superfamily as defined by Dayhoff (1978), and as two separate families within this (Fig. 1). Previous experience with other families of plant inhibitors suggests that further members will not be long in arriving!

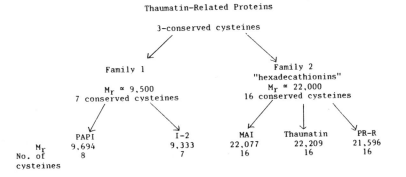

FIGURE 1. The Superfamily of Thaumatin-Related Proteins.

REFERENCES

ANTONIW, J.F., RITTER, C.E., PIERPOINT, W.S. and VAN LOON, L.C. (1980) J. Gen. Virol. 47: 79–87.

CAMPOS, F.A.C. and RICHARDSON, M. (1984) FEBS Lett. 167: 221–225.

CORNELLISSEN, B.J.C., HOOFT van HUIJSDUIJNEN, R.A.M. and BOL, J.F. (1986) Nature 321: 531–532.

DAYHOFF, M.O. (1978) Atlas of Protein Sequence and Structure 5, Suppl. 3.

EDENS, L. and VAN DER WEL, H. (1985) Trends in Biotech. 3: 61–64.

FRASER, R.S.S. (1982) J. Gen. Virol. 58: 305–313.

GATEHOUSE, A.M.R., FENTON, K.A., JEPSON, I. and PAVEY, D.J. (1986) J. Sci. Food Agric. 37: 727–734.

HOGUE, R. and ASSELIN, A. (1987) Can. J. Bot. 65: 476–481.

HOOFT van HUIJSDUIJNEN, R.A.M., VAN LOON, L.C. and BOL, J.F. (1986) EMBO J. 5: 2057–2061.

KINGHORN, A.D. and SOEJARTO, D.D. (1986) Crit. Rev. in Plant Sci. 4: 79–120.

KREIS, M., SHEWRY, P.R., FORDE, B.G., FORDE, J. and MIFLIN, B.J. (1985) Oxford Surveys of Plant Molecular and Cell Biology 2: 253–317.

METRAUX, J.P., STREIT, L. and STAUB, Th. (1987) Physiol. Mol. Plant Pathol., in press.

MUNDY, J. and ROGERS, J.C. (1986) Planta 169: 51–63.

PIERPOINT, W.S. (1986) Phytochemistry 25: 1595–1601.

PIERPOINT, W.S., TATHAM, A.S. and PAPPIN, D.J.C. (1987)   Physiol. Mol. Plant Pathol., 31: 291–298.

RICHARDSON, M. (1977) Phytochem. 16: 159–169.

RICHARDSON, M., VALDES-RODRIGUEZ, S. and BLANCO-LABRA, A. (1987) Nature 327: 432–434.

SCHLUMBAUM, A., MAUCH, F., VÖGELI, U. and BOLLER, T. (1986) Nature 324: 365–7.

VAN DER WEL, H. (1983) Chemistry and Industry 1: 19–22.

VAN DER WEL, H. (1986) In: Hudson, B.J.F. (ed.) Developments in Food Proteins. Elsevier: New York and London.

VAN LOON, L.C. (1985) Plant Molecular Biol. 4: 111–116.

VAN LOON, L.C. (1982) In: Wood, R.K.S. (ed.)   Active Defense Mechanisms in Plants. Plenum Press: New York.

WHITE, R.F., RYBICHI, E.P., von WECHMAR, M.B., DEKKER, J.L. and ANTONIW, J.F. (1987) J. Gen. Virol. 68: in press.

Oxford Surveys of Plant Molecular & Cell Biology Vol.4 (1987) 343-346

RELATIONSHIP BETWEEN DNA SYNTHESIS AND
CYTODIFFERENTIATION TO TRACHEARY ELEMENTS

Munetaka Sugiyama and Atsushi Komamine

*Biological Institute, Faculty of Science,
Tohoku University, Sendai, 980, Japan*

The differentiation of tracheary elements from quiescent parenchymatous cells in various tissues and organs can be induced *in
vitro*. The relationship between such differentiation and the cycle of
cell division has been the focus of considerable interest (recently
reviewed by Dodds, 1981; Fukuda and Komamine, 1985). During
studies of this relationship, it became apparent that the role of DNA
synthesis in the differentiation needed to be elucidated.

In early studies, the effects of inhibitors of DNA synthesis on the
differentiation of tracheary elements was examined in an attempt to
determine whether or not cell division was essential for the
differentiation. In cultured stem segments of *Coleus blumei*, both
5-fluorodeoxyuridine and mitomycin C, two potent inhibitors of DNA
synthesis, blocked the differentiation of tracheary elements (Fosket,
1968; 1970). Moreover, the blockage by fluorodeoxyuridine was
overcome by exogenously supplied thymidine, which might have been
expected from our knowledge of the mechanism of inhibition by
fluorodeoxyuridine of DNA synthesis. The differentiation of tracheary
elements in the root cortical parenchyma of *Pisum sativum* was also
blocked by fluorodeoxyuridine (Shininger, 1975). Phillips (1980)
reported that fluorodeoxyuridine inhibited the formation of tracheary
elements in cultured tuber explants of *Helianthus tuberosus*, although
the data were not published. These results suggested that inhibition
of DNA replication in the S phase, normally followed by mitosis and
cytokinesis, brought about an inhibition of the differentiation of
tracheary elements, and have been interpreted as evidence of a
requirement for DNA replication prior to the differentiation. On the
other hand, the differentiation of tracheary elements in the presence
of 10^{-5} M fluorodeoxyuridine was reported with cultured pith
explants of *Lactuca sativa* as the experimental plant materials
(Turgeon, 1975). Total inhibition of DNA synthesis, however, was not
strictly demonstrated by this study, as pointed out by Shininger
(1979).

Inhibitors of DNA synthesis prevented the differentiation of
tracheary elements in three different experimental systems, as

described in the previous paragraph. Furthermore, in many other cases, experimentally induced formation of tracheary elements was observed to follow cell proliferation (e.g. Clutter, 1960; Dalessandro and Roberts, 1971). Thus, it appeared reasonable that DNA replication, accompanied by cell division, was generally a pre- requisite for the differentiation of tracheary elements from quiescent parenchymatous cells. However, it must be emphasized that DNA synthesis has, thus far, been confused with the replication of DNA in the S phase, if we review all the data critically. Inhibition of DNA synthesis, caused by fluorodeoxyuridine and mediated by thymidylate deficiency, includes inhibition of synthesis of mitochondrial and chloroplast DNA, gene amplification, and repair DNA synthesis, as well as the replication of genomic DNA.

The need for a reconsideration of the relationship between DNA synthesis and cytodifferentiation was evident from results of experiments with an isolated cell system from *Zinnia*. Kohlenbach and Schmidt (1975) observed the differentiation of tracheary elements in single cells isolated from the mesophyll of *Zinnia elegans*. Based on this report, Fukuda and Komamine (1980a) established a new and reproducible system. They clarified, through serial observations, microdensitometry and autoradiography, that the majority of tracheary elements differentiated directly from cells in the G_1 phase, i.e., that the differentiation required neither DNA replication nor cytokinesis (Fukuda and Komamine, 1980b; 1981a). Moreover, they demonstrated in the same *Zinnia* system that various inhibitors of DNA synthesis, such as fluorodeoxyuridine, fluorouracil, mitomycin C, arabinosyl cytosine and aphidicolin, all caused a nearly complete inhibition of the differentiation (Fukuda and Komamine, 1981b). These chemicals act at different sites in the biosynthetic pathway of DNA and, therefore, it is unlikely that the inhibition of the differentiation was the result of a possible side effect. The results obtained in the *Zinnia* system can be explained by the hypothesis that some minor amount of DNA synthesis, separate from DNA replication in the S phase, is essential for the differentiation of tracheary elements (Fukuda and Komamine, 1981b; 1985). This hypothesis may be applied not only to *Zinnia* system but also to the other systems mentioned above. Differentiation of tracheary elements in tuber explants of *Helianthus*, which was considered to require the replication of genomic DNA, was not inhibited when the progression into the S phase was blocked by γ -irradiation (Phillips, 1981), a result which supports the involvement of some minor amount of DNA synthesis during cytodifferentiation in this system also.

Aphidicolin, which prevented the differentiation of tracheary elements in the *Zinnia* system, is an inhibitor specific for DNA polymerase-α (Oguro et al., 1979), an enzyme which is localized in the nucleus (Bensch et al., 1982). Therefore, it is likely that some minor amount of DNA synthesis, which is essential for cytodiffer- entiation, is catalyzed by DNA polymerase-α in the nucleus. However, this possibility may represent an oversimplification, in view of the following results (unpublished). We have found recently that, at a very early stage (12 h) in culture, the differentiation was inhibited by fluorodeoxyuridine but not by aphidicolin, while, at a later stage, both drugs inhibited the differentiation. The result suggests the involvement of DNA synthesis catalyzed by the DNA polymerase(s) other than polymerase-α . After isolation of *Zinnia* cells, aphidicolin-insensitive synthesis of chloroplast DNA occurred rapidly and very actively. Moreover, a current autoradiographic study has demonstrated a low level of incorporation of [³H]thymidine into nuclei, which is resistant to aphidicolin and is distinguishable from the replication of DNA in the S phase. In *Zinnia* cells, synthesis of chloroplast DNA and/or this additional DNA synthesis revealed by the low level of incorporation of [³H]thymidine may be required for the differentiation of tracheary elements. Since the incorporated [³H]thymidine is dispersed all over nuclei, this incorporation does not seem to indicate amplification of ribosomal genes, as has been observed in differentiating xylem cells in roots of *Allium cepa* (Avanzi et al., 1973; Durante et al., 1977). It is possible that the low level of incorporation of [³II]thymidine reflects some kind of repair-type DNA synthesis, because inhibitors of ADP-ribosyltransferase, which is associated with DNA excision repair (Durkacz et al., 1980), prevented the differentiation of tracheary elements in tuber explants of *Helianthus* (Hawkins and Phillips, 1983), in root explants of *Pisum* (Phillips and Hawkins, 1985) and in mesophyll cells of *Zinnia* (Sugiyama and Komamine, 1987). Up to now, aphidicolin-sensitive synthesis of any DNA other than genomic DNA during normal replication has not been found in cultures of *Zinnia* mesophyll cells. If it is truly the repair-type synthesis of DNA that is a prerequisite for the differentiation of tracheary elements, failure in finding aphidicolin-sensitive synthesis of some minor DNA is not without explanation, since both DNA polymerases α and β are believed to be involved in excision repair of DNA and inhibition of DNA polymerase-α alone does not always cause distinct reduction in repair DNA synthesis (e.g. Morita et al., 1982).

We have not discussed xylogenesis *in situ* and the differentiation of tracheary elements in cultured dedifferentiated callus, since these processes may be different from the experimentally induced differentiation from quiescent parenchyma which is accompanied by a reprogramming of each cell. The significance of DNA synthesis in tracheary differentiation presents a fascinating problem, and it is still unknown even what DNA must be synthesized to permit the differentiation of tracheary elements. The nature of this DNA synthesis which is essential for cytodifferentiation remains to be elucidated.

REFERENCES

AVANZI, S., MAGGINI, F. and INNOCENTI, A. M. (1973) Protoplasma. 76:197–210.
BENSCH, K. G., TANAKA, S., HU, S. -Z., WANG, T. and KORN, D. (1982) J. Biol. Chem. 257:8391–8396.
CLUTTER, M. E. (1960) Science. 132:548–549.

DALESSANDRO, G. and ROBERTS, L. W. (1971) Amer. J. Bot. 58:378- 385.
DODDS, J. H. (1981) In: Barnett, J. R. (eds.) Xylem Cell Development, Castle House Publ. Ltd., Tunbridge Wells. pp.153–167.
DURANTE, M., CREMONINI, R., BRUNORI, A. AVANZI, S. and INNOCENTI, A. M. (1977) Protoplasma. 93:289–303.
DURKACZ, B. W., OMIDIJI, O., GRAY, D. A. and SHALL, S. (1980) Nature. 283:593–596.
FOSKET, D. E. (1968) Proc. Natl. Acad. Sci. USA. 59:1089–1096.
FOSKET, D. E. (1970) Plant Physiol. 46:64–68.
FUKUDA, H. and KOMAMINE, A. (1980a) Plant Physiol. 65:57–60.
FUKUDA, H. and KOMAMINE, A. (1980b) Plant Phisiol. 65:61–64.
FUKUDA, H. and KOMAMINE, A. (1981a) Physiol. Plant. 52:423–430.
FUKUDA, H. and KOMAMINE, A. (1981b) Plant Cell Physiol. 22:41–49.
FUKUDA, H. and KOMAMINE, A. (1985) In: Vasil, I. K. (eds.) Cell Culture and Somatic Cell Genetics of Plants, Vol. 2, Academic Press, New York. pp.149–212.
HAWKINS, S. W. and PHILLIPS, R. (1983) Plant Sci. Lett. 32:221–224.
KOHLENBACH, H. W. and SCHMIDT, B. (1975) Z. Pflanzenphysiol. 75:369–374.
MORITA, T., TSUTSUI, Y., NISHIYAMA, Y., NAKAMURA, H. and YOSHIDA, S. (1982) Int. J. Radiat. Biol. 42:471–480.
OGURO, M., SUZUKI-HORI, C., NAGANO, H., MANO, Y. and IKEGAMI, S. (1979) Eur. J. Biochem. 97:603–607.
PHILLIPS, R. (1980) Int. Rev. Cytol. Suppl. 11A:55–70.
PHILLIPS, R. (1981) Planta. 153:262–266.
PHILLIPS, R. and HAWKINS, S. W. (1985) J. Exp. Bot. 36:119–128.
SHININGER, T. L. (1975) Dev. Biol. 45:137–150.
SHININGER, T. L. (1979) Ann. Rev. Plant Physiol. 30:313–337.
SUGIYAMA, M. and KOMAMINE, A. (1987) Plant Cell Physiol. 28:541–544.
TURGEON, R. (1975) Nature (London). 257:806–808.

Oxford Surveys of Plant Molecular & Cell Biology Vol.4 (1987) 347–357

RECENT ADVANCES IN PLANT ELECTROPORATION

H. Jones, [1]M.J. Tempelaar and M.G.K. Jones

AFRC Institute of Arable Crops Research,
Biochemistry Department, Rothamsted Experimental Station,
Harpenden, Herts. AL5 2JQ. U.K.
and [1]Department of Genetics, University of Groningen,
P.O. Box 14, 9751 AA, Haren, The Netherlands.

There is currently considerable interest in techniques of direct transfer of genes into plant protoplasts and in particular, in the use of electroporation. This is because direct introduction of molecular constructs by this relatively simple approach allows both rapid transient expression of introduced genes, to study tissue specific and developmental regulation of expression, and can result in stable integration of genes. The latter aspect is of particular interest in experiments to transform cereals, which are not at present amenable to the use of *Agrobacterium* as a gene vector.

Like its allied technique of electrofusion, relevant aspects of which have been discussed in a previous article in this series (Tempelaar and Jones, 1985a), electroporation is based on the reversible dielectric breakdown of membranes when they are exposed to brief (μs–ms) electric fields. The advantages of electrofusion and electroporation over other techniques lies in their simplicity, greater degree of control and reproducibility, the avoidance of toxic chemicals such as PEG, and high fusion and transformation frequencies.

The technique of electroporation has been applied successfully to transform protoplasts of both dicotyledonous and monocotyledonous species, as indicated in Table 1. It is evident from Table 1, which summarizes the types of pulses and conditions employed, that a bewildering number of experimental conditions have been used successfully. These include varied pulse types (square pulses, and exponentially decaying "RC" pulses from capacitor discharges), field strengths, pulse durations and numbers, and different ionic conditions, temperatures, DNA concentrations and presence or absence of carrier DNA. It is evident in many instances that conditions have been arrived at empirically, with little consideration of the underlying biophysical principles. In order to reconcile the different conditions employed, and arrive at general conclusions for the best conditions to use for electroporation, it is necessary to discuss the underlying principles of electroporation. Where relevant,

TABLE 1. Transient and stable expression of genes introduced into higher plant cells by electroporation.

(a) Transient

Species	Plasmid	Pulse type	Field strength (V/cm)	Pulse duration (ms)	Reference
Tobacco (sc). carrot (sc). maize (sc)	pNOSCAT pCaMVCAT	Exponential	875 (1)	54	Fromm et al. (1985)
Rice (sc/m). Wheat (sc), sorghum (sc)	pCopia CAT		*10000 (20.sc) 7500 (10,m)	0.0625	Ou-Lee et al. (1986)
Tobacco (sc)	pNCAT4 pCaMVCAT pEND4K	Exponential	300 (1)	6	Okada et al. (1986a)
Maize (sc)	pAdh+CAT	Exponential	625 (1)	50	Howard et al. (1987)

(b) Viral RNA/particle transfer

Species	RNA	Pulse type	Field strength (V/cm)	Pulse duration (ms)	Reference
Tobacco (sc). Vinca rosea (sc)	TMV-RNA CMV-RNA CMV/TMV particles	Exponential	300 (1)	6	Okada et al. (1986b)
Tobacco (m)	TMV-RNA CMV-RNA	Square	5000-10000 (8)	0.09	Nishiguchi et al. (1986)
Tobacco (m)	TMV-RNA	Square	550-800 (10)	0.05	Hibi et al. (1986)
Tobacco (m)	TMV particles	Square	670 (1)	10	Nishiguchi et al. (1987)

(c) Stable

Species	Plasmid	Pulse type	Field strength (V/cm)	Pulse duration (ms)	*Transformation Frequency (%)	Reference
Tobacco (m)	pABD1	Exponential	15000 (3)	0.01	2.1	†Shillito et al. (1985)
Maize (sc)	pCaMVNEO	Exponential	500 (1)	2	1.0	Fromm et al. (1986)
Carrot (sc)	pTiC58	Square	500-3300 (6)	0.06	1.6	Langridge et al. (1985)
Tobacco (m)	pMON200	Exponential	2000 (1)	0.25	0.1	Riggs and Bates (1986)

* Transformation frequencies are expressed per number of colonies recovered
† Shillito et al. (1985) used a combination of electroporation. PEG, and heat shock.
‡ Used a non-contact Baekon system.
Number in brackets refer to the number of pulses applied.
SC – suspension culture protoplasts; m – mesophyll prctoplasts.

parallels are drawn between mechanisms common to both electrofusion and electroporation.

Induced Membrane Events

Exposure of protoplasts to brief electric fields results in ion displacement and the generation of a transmembrane potential (Fig. 1). The magnitude of this transmembrane potential (V) is proportional to the field strength (E) and the protoplast radius (r) and is described by the following equation (Zimmermann, 1982):

$$V = 1.5 \cdot E \cdot r \cdot \cos \theta \qquad \text{Equation 1}$$

where 1.5 is a shape factor, and θ are those membrane sites on the surface of the protoplasts at a certain angle with respect to the field lines (Fig. 1). Charge separation across the plasmalemma results in an electrostatic force leading to membrane compression and pore formation. If the cells are first brought into close membrane contact by dielectrophoresis, the application of the field pulse results in protoplast fusion (Tempelaar and Jones, 1985a,b). Pore formation (or fusion) only occurs if the membrane voltage exceeds a certain threshold or critical voltage, Vc (see below). Pores will first form at those membrane sites positioned on the 'equator' of the protoplasts (Fig. 1). Larger protoplasts will be exposed to a greater transmembrane voltage than smaller protoplasts in response to a given field strength. If E is increased so that V > Vc, a corresponding increase in the number and size of pores occurs leading eventually to membrane rupture and protoplast lysis when V >> Vc.

There is evidence for interaction between the applied field, E, and the transmembrane potential normally present in plant cells (inside about 100 mv negative with respect to outside the plasmamembrane), such that the first pores to form occur at the equator of cells opposite the anode, since E and the transmembrane potential are additive. With increasing applied field, Vc is also reached on the opposite pole, and pores will then form on both equatorial faces of the cell (Mehrle et al. 1985).

Parameters for Electroporation and DNA Uptake

The biophysical concepts outlined above are implicit in the practical results illustrated in Figs. 2A,B. These show data that relate protoplast viability to expression of the marker gene chloramphenicol acetyl transferase (CAT, driven by CaMV 35S promoter), for protoplasts of *Solanum brevidens*, in response to increasing field strength (electroporation by capacitor discharge – 50 μF; 3 pulses; 20 mM KCl with mannitol as osmoticum, pH 7.2 – 1.7 kΩ). The field strength indicated in Fig. 2B is the initial field strength, Eo, which decays exponentially with time to lower values

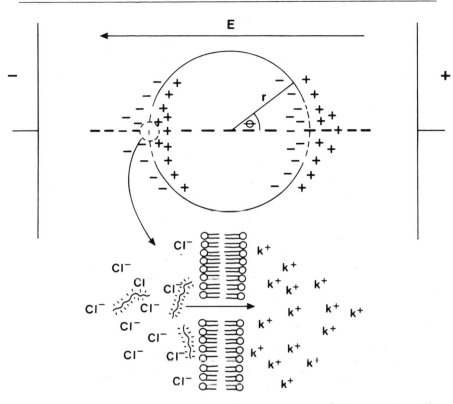

FIGURE 1. Schematic diagram of a cell exposed to an externally applied electric field, E. For details see text.

according to the equation,

$$E = Eo \cdot e^{-t/RC}$$

where R is the electrode chamber resistance and C is the capacitance. R can be measured using an LCR meter (eg. AVO B183 LCR meter; Thorn-EMI Ltd., Dover, Kent U.K.). By convention the time taken for E to decay to 1/e of Eo is defined as the time constant, τ = R.C. In Fig. 2B τ was calculated to be 85ms. The observation of peak activity in Fig. 2B at 225V corresponds to approximately 50% protoplast death (Fig. 2A) and confirms numerous other observations to this effect (e.g. Fromm et al. 1985). Similar results are obtained when protoplasts of commercial potato varieties are electroporated. These results demonstrate transient expression of CAT activity, but it should be noted that optimum conditions for transient expression and stable integration of genes are not necessarily identical. The reason is that longer-term growth of protoplasts is not needed to study transient expression, whereas for stable integration it is desirable to maintain good viability. Excessive damage to protoplasts also reduces protoplast density and releases metabolites that inhibit subsequent division and growth. Therefore, somewhat milder electro-

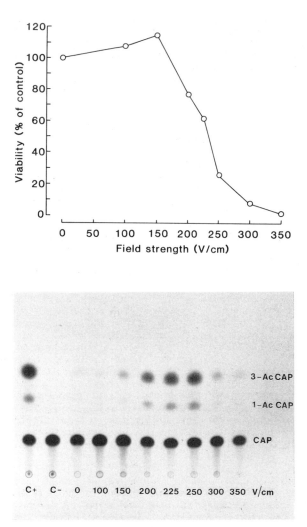

FIGURE 2. Effect of increasing field strength on viability (A) and transient expression of pCaMVCAT gene (B) in Solanum brevidens mesophyll protoplasts. On the left is the positive and on the right is the negative control. CAP represents chloramphenicol; 1-AcCAP and 3-AcCAP represent the monoacetylated forms of chloroamphenicol. Capacitor discharge (50 µF); 3 pulses; pulse duration of 85 ms; medium used was 20 mM KCl + mannitol as osmoticum, pH 7.2; electrode chamber resistance was 1.7 kΩ.

poration conditions (lower applied voltages or smaller capacitors) and modified post-transfection culture conditions are recommended.

For both the processes of electrofusion and electroporation the external field strength must be sufficient to induce a transmembrane voltage above the breakdown voltage (i.e. V > Vc). However, Vc is not an invariable constant and has been shown in previous studies to depend on pulse duration (Zimmermann and Benz, 1980; Zimmermann, 1982). Information on the pulse length dependence of Vc in protoplasts would be useful for both fusion and electroporation. One way to investigate this relationship is by the analysis of protoplast fusion–voltage curves performed at different pulse durations, similar to those described by Tempelaar and Jones, 1985b. A set of such data for dihaploid *Solanum* protoplasts is shown in Fig. 3 (from van der Steege et al. in preparation). In order for fusion to occur a certain threshold field strength (voltage) has to be exceeded. This critical field strength is identified as Vc, and depends on pulse duration. When values for Vc were calculated from equation 1 and plotted against pulse duration the relationship in Fig. 4 was obtained. This shows Vc to decrease exponentially from around 3 to 0.4 V (protoplast transmembrane voltage) on increasing the pulse duration from 50 to 1000 µs. This result is quantitatively similar to biophysical measurements performed on bilayer, mammalian, and plant cell membranes (Zimmermann, 1982). Fusion–voltage curves produced at different pulse durations can therefore yield useful information regarding the dependence of Vc on pulse duration for particular protoplast systems. A knowledge of Vc combined with the use of equation 1 can be used in predicting the field requirements. Thus, when *S. tuberosum* mesophyll protoplasts were electroporated using field strengths of 150–300 V cm^{-1} and a pulse duration of 85 ms (50 µF capacitor), transient expression of CAT was obtained. However, when the RC pulse duration was reduced to 85 µs (50 nF capacitor) the optimal field strength was 3000 V cm^{-1} for CAT activity. This dependence of Vc on pulse duration also explains the varied field conditions used by different workers in Table 1. The fact that the underlying mechanisms of electrofusion and electroporation are similar is confirmed in recent studies which show close agreement between electrofusion and electroporation responses (Hibi et al. 1986; Riggs and Bates, 1986).

The pulse length range in which Vc can shift is species specific (Zimmermann, 1982). It may also depend on the type of protoplasts used. This may explain differences observed in the field requirements of mesophyll when compared to suspension culture protoplasts. Thus, when bread wheat (*T. aestivum*) suspension culture protoplasts were electroporated in our laboratory, CAT expression was obtained using a capacitor discharge pulse of field strength 400 V cm^{-1} and 155 ms duration. Compare this result to those obtained using mesophyll derived protoplast shown in Fig. 2. Differences between the field requirements of suspension culture compared to mesophyll protoplasts have also been documented in electrofusion studies (Tempelaar and Jones, 1985b). Changes in Vc in this way must be due to differences in membrane properties such as composition, thickness, dielectric constant, capacitance, elasticity, and surface charge. The internal cytoplasmic and vacuolar conductivity may also be important.

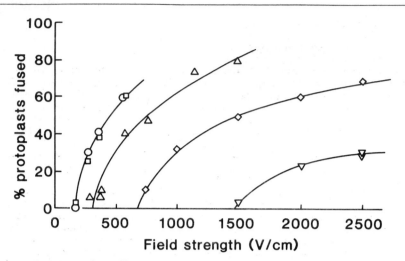

FIGURE 3. Effect of field strength on % dihaploid Solanum protoplasts fused for different pulse durations. Pulse durations used were 50 μs (▼), 100 μs (♦), 200 μs (▲), 500 μs (O) and 1000 μs (■).

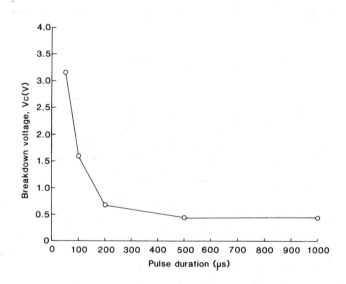

FIGURE 4. Relationship between the breakdown voltage, Vc, of dihaploid Solanum protoplasts and pulse duration, derived from fusion curves such as those in Fig. 3. Vc was calculated using equation 1. For details see text.

Types of Electroporation Pulses

As shown in Table 1 transient and stable expression may be obtained using either square or capacitor discharge pulses. With square pulses the applied field is constant over the whole duration of the pulse. With capacitor discharge it drops exponentially. With the latter the field strength is above the breakdown voltage of the membrane for only a relatively short while until it decays down to values below Vc where there is no further pore formation. The 'tail' of the pulse represents a gradually decreasing current for ion and DNA flow through the transmembrane pores. DNA may also penetrate the cell interior by diffusion. This is indicated by studies which show that expression may occur when RNA is introduced to a protoplast suspension a few minutes after the application of the field pulse (Okada et al. 1986b). The same mechanism operates with square pulses, however, the duration has to be relatively short because of the sustained transmembrane voltage above the breakdown voltage of the membrane. When using square pulses, the usual strategy is therefore to use short pulse durations and many pulses (e.g. Langridge et al. 1985, Table 1). The time interval between the pulses, and temperature in particular, may be important for membrane resealing and thus on the ability of the protoplast membrane to build up a breakdown potential between the pulses. These aspects have not yet been studied in any sufficient detail with plant protoplasts. With capacitor discharge pulses suitable values for RC must be chosen of adequate duration (depending on the field strength used) to induce pore formation but without excessive heating as to cause protoplast death.

Pore Parameters, DNA Uptake and Integration

The detailed physico-chemical mechanisms of pore formation and DNA uptake are unknown. The fact that Ca^{2+} ions can increase uptake (Fromm et al. 1985) suggests that DNA binding (by electrostatic bridging) to the plasmalemma may play a role in uptake. However, experiments in our laboratory using *S. brevidens* protoplasts failed to detect any stimulatory effect of Ca^{2+} ions on DNA uptake. This suggests that the role of Ca^{2+} ions may be species (protoplast) specific and/or occurs only by using certain pulse media. It has been shown that under suitable field conditions even TMV particles may be taken up by protoplasts (Table 1; Nishiguchi et al. 1987; Okada et al. 1986b). These particles are 18 x 300 nm in size and much larger than the suggested pore size of 3-5 nm under electroporation conditions (Benz and Zimmermann, 1981). If the mechanisms of DNA permeation is the same as for TMV particle uptake, it should be possible for quite large fragments of DNA to pass through the membrane. This may be important in any 'shotgun' strategy of DNA transformation.

Once the DNA is introduced by electroporation into the cytoplasm, it may be acted upon by endonucleases and some may be broken down before it is expressed. The use of carrier DNA in electroporation is helpful in saturating endonuclease activity and hence may increase the chances of transient and stable DNA expression. However, there are indications that in stable transformation, carrier DNA cannot be replaced by plasmid DNA, implying that the carrier may be acting as a homologous intermediate for recombination of the plasmid DNA with the genomic DNA (Shillito et al. 1985). Related to this is a recent study which shows that highly species specific repeated DNA sequences may promote the insertion of genes at sites of homology when used as a carrier or introduced into the plasmid vector itself (Sala et al. 1987). This is an interesting observation and opens the possibility of increasing transformation frequencies as well as directing DNA to specific sites on the chromosome. However, it implies that carrier DNA should not perhaps be used if stable integration of a gene is required, since carrier DNA itself may become integrated into the genome and act as a mutagen.

Another avenue for increasing transformation frequencies is through the transformation of protoplasts derived from synchronised cultures in the M phase of the cell cycle (Okada et al. 1986a). Frequencies of stable transformants may also be enhanced by using linear as opposed to close circular/supercoiled DNA (e.g. Shillito et al. 1985).

Because of the host-range limitation of *Agrobacterium*-mediated DNA transformation, electroporation appears particularly promising for the transformation of agriculturally important monocotyledonous as well as dicotyledonous crop plants. Recent advances in the regeneration of whole plants from rice protoplasts (e.g. Fujimura et al. 1985; Toriyama et al. 1986; Abdullah et al. 1986) and the development of new marker genes such as GUS (Jefferson et al. 1987) and luciferase (Ow et al. 1986) consolidate this view. Biophysical understanding as outlined above, the use of synchronised cell suspension protoplasts, and the use of species specific repetitive DNA sequences as carrier or in the plasmid vector itself, will contribute to increasing gene transformation and expression as well as directing DNA integration to particular loci in the host chromosomes. This should make electroporation a universal and versatile tool for both fundamental studies, as well as in more applied aspects involved in the genetic manipulation of crop plants.

Acknowledgements

We thank the Biotechnology Action Programme of the EEC (Contract numbers BAP.0101.UK(H)MGKJ & MJT) for financial support.

REFERENCES

ABDULLAH, R., COCKING, E.C. and THOMPSON, J.A. (1986) Bio/Tech. 4: 1087-1090.

BENZ, R. and ZIMMERMANN, U. (1981) Planta 152: 314–318.

FROMM, M., TAYLER, L.P. and WALBOT, V. (1985) Proc. Natl. Acad. Sci. USA. 82: 5824–5828.

FROMM, M., TAYLER, L.P. and WALBOT, V. (1986) Nature 319:791–793.

FUJIMURA, T., SAKURAI, M. AKAGI, H., NEGISHI, T. and HIROSE, A. (1985) Plant Tissue Culture Lett. 2: 74–75.

HOWARD, E.A., WALKER, J.C., DENNIS, E.S. and PEACOCK, W.J. (1987) Planta 170: 535–540.

HIBI, T., KANO, H., SUGIURA, M., KAZAMI, T. and KIMURA, S. (1986) J. Gen. Virol. 67: 2037–2042.

JEFFERSON, R.A., KLASS, M., WOLFF, N. and HIRSH, D. (1978) J. Mol. Biol. 193: 41–46.

LANGRIDGE, W.H.R., LI, B.J. and SZALAY, A.A. (1985) Plant Cell Rep. 4: 355–359.

MEHRLE, W., ZILMMERMANN, U. and HAMPP, R. (1985) FEBS 185: 89–94.

NISHIGUCHI, M., LANGRIDGE, W.H.R., SZALAY, A.A. and ZAITLIN, M. (1986) Plant Cell Rep. 5: 57–60.

NISHIGUCHI, M., SAKO, T. and MOTOYOSHI, F. (1987) Plant Cell Rep. 6: 90–93.

OKADA, K., TAKEBE, I. and NAGATA, T. (1986a) Mol. Gen. Genet. 205: 398–403.

OKADA, K., NAGATA, T. and TAKEBE, I. (1986b) Plant Cell Physiol. 27: 619–626.

OU-LEE, T-M., TURGEON, R. and WU, R. (1986) Proc. Natl. Acad. Sci. USA 83: 6815–6819.

OW, D.W., WOOD, K.V., DELUCA, M., DE WET, J.R., HELINSKI, D.R. and HOWELL, S.H. (1986) Science 234: 856–859.

RIGGS, C.D. and BATES, G.W. (1986) Proc. Natl. Acad. Sci. USA 83: 5602–5606.

SALA, F. et al. Biotechnology Action Programme. Meeting. Louvain-la-Neuve, March 23–26, 1987.

SHILLITO, R.D., SAUL, M.W., PASZKOWSKI, J., MÜLLER, M. and POTRYKUS, I. (1985) Bio/Tech. 3: 1099–1103.

TEMPELAAR, M.J. and JONES, M.G.K. (1985a) Oxford Surveys of Plant Molecular and Cell Biology 2: 347–351.

TEMPELAAR, M.J. and JONES, M.G.K. (1985b) Planta 165: 205–216.

TORIYAMA, K. et al. (1986) Theor. Appl. Genet. 73: 16–19.

ZIMMERMANN, U. and BENZ, R. (1980) J. Membr. Biol. 53: 33–43.

ZIMMERMANN, U. (1982) Biochim. Biophys. Acta 694: 227–277.

Oxford Surveys of Plant Cell & Molecular Biology, Vol.4 (1987) 359–365

THE USE OF PLANT PROTEINS IN IMMUNOTOXIN CONSTRUCTIONS

Jane Gould and Peter T. Richardson

*Department of Biological Sciences,
University of Warwick, Coventry.*

Many plant extracts contain proteins which are highly toxic to intact cells or to cell–free extracts. The inhibitory proteins, known as ribosome inactivating proteins (RIPs), are found in different plant tissues in varying concentrations ranging from a few μg to several hundred mg per 100g of plant tissue (Barbieri and Stirpe, 1982). These proteins can be subdivided into two groups; type I or single chain RIPs which, being unable to enter intact cells, are only capable of intoxicating cell free systems and type II RIPs or cytotoxic lectins which are highly toxic to intact cells.

Type I RIPs.

Type I RIPs are single chain proteins which are frequently, but not always, glycosylated. Many plants contain single chain RIPs and it is possible that they may be ubiquitously distributed throughout the plant kingdom. In experiments designed to find novel plant toxins almost all crude extracts tested were shown to exert some inhibitory influence on the protein synthetic activity of a cell free system (Barbieri & Stirpe, 1982). This effect was not correlated with high toxicity of the extracts for intact cells. Proteins with similar physiochemical and biological properties to the A chains of type II RIPs (see below) and which were responsible for ribosomal inactivation could be purified from the crude extracts. Table I gives examples of some type I RIPs characterised to date.

Type II RIPs.

Type II RIPs, for example ricin, abrin and modeccin (see Table II), consist of two dissimilar peptide chains joined covalently by a single disulphide bond. The A chain is similar to the type I RIPs described above and inactivates 60S ribosomal subunits, apparently by the same mechanism. The B chain is a lectin which binds to cell

Table 1. Examples of type I RIPs.

Plant Source	Inhibitor	M_r
Phytolacca americana (pokeweed) seeds or leaves	pokeweed antiviral protein (PAP)	30,000
Triticum aestivum (wheat) seeds	tritin	30,000
Gelonium multiflorum seeds	gelonin	30,000
Momordica charantia (bittergourd) seeds	momordin	31,000
Saponaria officinalis (soapwort) seeds	saporin	29,500
Dianthus carophyllus (carnation) leaves	dianthin	30,000
Zea mays seeds	corn inhibitor	23,000
Secale cereale seeds	rye inhibitor	30,000

Data taken from Barbieri & Stirpe (1982) and Lord & Roberts (1987).

---

surface receptors containing terminal galactose residues, facilitating toxin binding to cells and subsequently, release of A chain from an intracellular vesicle into the cytoplasm. Type II RIPs are thus highly toxic to intact cells as well as cell free systems and as such are structurally and functionally analogous to the bacterial cytotoxins – diphtheria toxin, shigella toxin and pseudomonas exotoxin A. With the possible exception of shigella toxin, these bacterial toxins inhibit protein synthesis by a different mechanism to that of cytotoxic lectins, the toxic fragment of the bacterial proteins catalysing ADP-ribosylation of EF2 (Olsnes and Pihl, 1982).

Table 2. Examples of type II RIPs.

Plant Source	Inhibitor	M_r
Ricinus communis (castor bean) seeds	ricin	65,000
	A chain	32,000
	B chain	34,000
Abrus precatorius (jequirity bean) seeds	abrin	65,000
	A chain	30,000
	B chain	36,000
Viscum album (mistletoe) leaves	viscumin	60,000
	A chain	29,000
	B chain	32,000
Adenia digitata roots	modeccin	63,000
	A chain	28,000
	B chain	31,000
Adenia volkensii roots	volkensin	62,000
	A chain	29,000
	B chain	36,000
Robinia pseudacacia	robin	
Phaseolus vulgaris (common bean)	phasin	

Data taken from Lord & Roberts (1987).

Mechanism of Action.

Type I RIPs.

Single chain RIPs inhibit protein synthesis by inactivating the 60S subunit of eukaryote ribosomes. They have no effect on prokaryote ribosomes. The mechanism by which this inactivation occurs is not yet known. Pokeweed antiviral protein renders the 60S subunit incapable of interacting with EF2 causing inhibition of EF2–dependent GTPase activity and arrest of the elongation reaction (reviewed by Barbieri

& Stirpe, 1982). It is presumed that all RIPs act on ribosomes by the same mechanism. In all cases studied they act in the absence of any cofactors and inactivate a molar excess of ribosomes, suggesting a catalytic action. Type I RIPs are relatively non-toxic to intact cells.   This is presumably because, unlike type II RIPs, they are unable to bind to and enter cells and cannot therefore readily reach their site of action. Toxicity of type I RIPs can be greatly enhanced by coupling to a cell binding protein (eg antibody or lectin) or by incorporating the RIP into liposomes or erythrocyte ghosts which can be fused to intact cells. Type I RIPs have been coupled to a variety of targeting molecules for use in cancer therapy (Thorpe 1984; Olsnes & Pihl, 1982).

Type II RIPs.

The A chain of type II RIPs functions enzymically in the absence of any cofactors in an analogous manner to the single chain RIPs (reviewed by Olsnes & Pihl, 1982). Most work has been done with abrin and ricin A chains where it has been shown that a single A chain molecule is capable of inactivating about 1500 ribosomes per minute. Modeccin also catalytically inactivates 60S ribosomal subunits although somewhat less efficiently than abrin and ricin. The exact nature of the reaction catalysed by A chain is not clear; A chain apparently acts on the ribosome at a site identical with, or close to, the binding site of EF2 such that its affinity for EF2 is reduced (Olsnes & Pihl, 1982). More recently it has been suggested that ricin modifies 28S rRNA rendering it susceptible to hydrolysis (Endo et al. 1987). The ability of type II RIPs to intoxicate intact cells is attributable to the cell-binding property of the B chain. Internalisation of ricin has been extensively studied (van Deurs et al., 1986) and it seems clear that endocytosis is the major mechanism by which ricin enters a cell. Once inside the cell the A chain is transferred from an intracellular compartment into the cytoplasm where it exerts its toxic effect.

Therapeutic Potential of RIPs.

There is considerable interest in the use of RIPs as cytotoxic agents in, for example, cancer therapy or regulation of the immune response (reviewed by Thorpe, 1984). Native RIPs are differentially toxic to various cells, peritoneal macrophages being particularly sensitive (Barbieri & Stirpe, 1982).   Mommordin and Pokeweed antiviral protein have strong immunosuppressive activity in mice but, unlike other immunosuppressive agents, only appear to prevent the immune response against new antigens, without modifying immunity already established. This property could be exploited both experimentally and therapeutically. Most interest at present focuses on the use of RIPs as antitumour agents after coupling to carriers capable of directing them to specific cell targets. The remainder of this article deals with the use of ricin in such conjugates and the work being done in this laboratory to improve the efficiency and specificity of these conjugates.

Immunotoxins

At the beginning of this century Paul Ehrlich recognised the potential of antibodies as carriers of pharmacological agents; conventional drugs, radioisotopes, toxins or response-modifiers conjugated to antibodies should be directed to cells carrying specific antigens recognised by the antibodies. Since tumour cells carry antigens which are largely or wholly absent from normal cells, and with the development of hybridoma technology, antibodies capable of specifically recognising tumour cells can now be produced. It was originally hoped that such monoclonal antibodies might themselves be capable of selectively eliminating cancer cells *in vivo*. However, in general, they have shown very poor toxicity (Baubert, 1984). As a result, monoclonal antibodies have been conjugated to other molecules in an attempt to enhance their toxicity. The most widely used such conjugates, immunotoxins (ITs), contain protein toxins such as ricin or diphtheria toxin. The properties of ricin based ITs and their potential as pharmacological agents has been discussed in some detail in recent reviews (Vitteta & Uhr, 1985; Lord et al., 1985, Krolich, 1985). Two classes of ricin-based ITs have been studied; those in which whole ricin is coupled to the antibody and those where only the A chain subunit is used. Both types of ITs have given disappointing results due to the inherent properties of ricin or its subunits.

A chain based ITs exhibit a high degree of targeting specificity which is due to the antibody, but toxicity of such conjugates is very variable (Thorpe, 1985). This lack of toxicity is attributed to the absence of the B chain which is required for efficient translocation of A chain into the cytoplasm (Vitteta, 1986). A second problem with such A chain ITs is related to the difficulty of purifying the A chain subunit from any B chain. Even traces of B chain contamination facilitate indiscriminate binding of the A chain to cell surface galactose residues and subsequent intoxication of non-target cells. The availability of active recombinant A chain, purified from *E. coli* (O'Hare et al., 1987) should overcome this problem by eliminating the requirement for chemical purification of the A chain.

In the case of ITs containing whole ricin, the presence of the B chain can lead to predictable problems. In contrast to A chain based conjugates, whole ricin ITs are very toxic to their target cells, but the B chain renders them relatively non-specific. The opportunistic binding of B chain to cell surface galactose residues overrides the antibody specificity making such conjugates unsafe for use *in vivo*. However, whole ricin ITs have been used successfully *in vitro*, where excess lactose or galactose can be used to out-compete B chain galactose binding, for example neoplastic bone marrow has been treated with whole ricin ITs in the presence of excess galactose and returned to patients free of tumour cells (Thorpe et al., 1984). Since neither type of IT described above is ideal for *in vivo* use, attempts have been made to modify the B chain such that its non-specific galactose binding function is eliminated, whilst it still retains the ability to translocate A chain into the cytoplasm. Chemical modification of the B chain readily abolishes its sugar binding activity, but also appears to affect its ability to translocate A chain thus ITs constructed with chemically modified B chain have reduced toxicity (Vitteta, 1986).

The two B chain functions are thought to reside on different domains of the molecule (Vitteta, 1986) and it should therefore be possible to modify the sugar binding property without affecting A chain translocation. The availability of ricin cDNA clones (Lamb et al., 1985) has enabled us to approach this problem by producing ricin, ricin A chain and ricin B chain in various heterologous expression systems. Ricin B chain, expressed cytoplasmically or in the secretory system of yeast, has been shown to be active (Richardson et al., 1987 in prep.). Other workers (Chang et al., 1987) have expressed B chain in animal cells, albeit in small amounts. We can now express, in yeast, B chain constructs modified by site-directed mutagenesis and determine whether the sugar-binding and A chain translocation functions have been affected. A prerequisite for the modification of B chain by such methods is delineation of the "active site", so that it is known which residues should be modified to abolish sugar binding. This information is now available from the recently published 2.8Å Xray crystallographic structure of ricin (Montfort et al., 1987). The eventual aim of this work is to produce genetically modified B chain from heterologous expression systems which in combination with recombinant A chain would allow the construction of immunotoxins whose specificity is solely determined by the targeting antibody.

Summary and Future Prospects.

It has been shown that native ricin A and B chains can be successfully expressed in heterologous systems (O'Hare et al., 1987; Chang et al., 1987; Richardson et al., 1987). Using such an approach, combined with site-directed mutagenesis to modify the B chain, it should soon be possible to produce an altered ricin molecule. An IT constructed from such a molecule should be highly specific, binding only those cells carrying the antigen recognised by the antibody part of the IT, and potently toxic due to the inclusion of the modified B chain, which can potentiate A chain translocation into the cytosol. It is envisaged that other modifications of ricin may further increase the usefulness of ricin based ITs. For example, removal of the sugar groups from the A and B chains should decrease the rate of clearance of ITs from the blood stream (Foxwell et al., 1987). This could be achieved by modifying the cDNA clone such that potential glycosylation sites are removed. The patients' immune systems may also reduce the effectiveness of ITs and work is in progress to minimise the size of foreign proteins used in IT construction, thus reducing their antigenicity. This would also enhance penetration of ITs into solid tumours. The use of plant toxins in cancer therapy is currently attracting much interest. Clinical trials are already underway with ricin A chain ITs and it is envisaged that more useful drugs may emerge from this work within the next five years.

REFERENCES

BAUBERT, R.B., Cancer Drug Deliv. (1984) 1 p. 251-258

BARBIERI, L. & STIRPE, F., Cancer Surveys (1982) 1 , No. 3 pp 489-520

CHANG, M., RUSSELL, D.W., UHR, J.W. & VITTETA, E.S., PNAS (1987 in press)

EHRLICH, P., in Collected Studies on Immunology (ed. R.B. Herbeman & Martinus Nijhoff) (1983) Boston p. 159

ENDO, Y., MITSUI, K., MOTIZUKI, M & TSURUGI, K. (1987) J. Biol. Chem. 262, 5908-5912

FROHNE, D. and PFONDER, H.J. (1983) In, A Colourful Atlas of Poisonous Plants. Wolfe Publishing Ltd.

FOXWELL, B.M.J., BLAKEY, D.C., BROWN, A.N.F., DONOVAN, T.A. & THORPE, P. E. (1987) Biochem. Biophys. Acta 923, 59-65

KROLICH, K.A. ASM News, (1985) 51, 569-573

LAMB, F.I., ROBERTS, L.M. & LORD, J.M. Eur. J. Biochem (1985) 148, 265-270

LORD, J.M., ROBERTS, L.M., THORPE, P.E. & VITTETA, E.S. Trends Biotech. (1985) 3, 175-179

LORD, J.M., & ROBERTS, L.M. Microbiological Sciences (1987 in press)

MONTFORT, W., VILLAFRANCA, J.E., MONZINGO, A.F., ERNST, S.R., KATZIN, B., RUTENBER, E., XUONG, N.H., HAMLIN, R., ROBERTUS, J.D. J. Biol. Chem. (1987) 262, 5398-5403

O'HARE, M., ROBERTS, L.M., THORPE, P.E., WATSON, G.J., PRIOR, B. & LORD, J.M. F.E.B.S. Letters (1987) 216, No. 1:73-78

OLSNES, S. & PIHL, A., Cancer Surveys (1982) 1, No 3 pp 467-487

RICHARDSON, P.T., ROBERTS, L.M., GOULD, J.H., SMITH, A.B. & LORD, J.M., Proc. Biochem. Soc. (1987 in press)

THORPE, P.E., ROSS, W.C.J., BROWN, A.N.F., MEYERS, C., CUMBER, A.J., FOXWELL, B.M.J. & FORRESTER, J.A. Eur. J. Biochem (1984) 140, 63-71

THORPE, P.E. in Monoclonal Antibodies 1984 – Biological and Clinical Applications (1985) (ed. A. Pinchara, G. Doria, F. Dammacco & A. Borgelliri) Editrice Kurtise, Rome pp 475-512

VAN DEURS, B., TONNESSEN, T.I., PETERSEN, O.W., SANDVIG, K., & OLSNES, S. J. Cell Biol (1986) 102, 37-47

VITTETA, E.S. & UHR, J.W. Cell (1985) 41, 653-654

VITTETA, E.S. J. Immunol. (1986) 136, 1880-1887

OXFORD SURVEYS OF PLANT MOLECULAR AND CELL BIOLOGY

Volume 3: 1986

Contents

Oxford Surveys of Plant Molecular and Cell Biology

Oxford Surveys of Plant Molecular and Cell Biology may be obtained from your local bookseller or on direct subscription through the Journals Subscription Department at Oxford University Press.

Once recorded as a direct subscriber, you will be notified some six to eight weeks prior to the publication of each volume (you may cancel at any time), and **Oxford Surveys of Plant Molecular and Cell Biology** will be delivered directly to you, immediately on publication.

Simply fill in the reservation form overleaf and return it to the Journals Subscription Department, Oxford University Press, Walton Street, Oxford OX2 6DP, UK. Back volumes may also be ordered on this form.

We regret, it is not possible to despatch any volumes until payment has been received.

ORDER FORM

I have purchased Volume 4 of **Oxford Surveys of Plant Molecular and Cell Biology** and I wish to be recorded as a direct subscriber to future volumes. I understand that I will receive regular notification of the price of future volumes before they are despatched and that I may cancel my subscription at any time.

Name ....

Address ....

....

Postcode .... Country ....

Please supply additional volume of **Oxford Surveys of Plant Molecular and Cell Biology** as below.

☐ Volume 3, 1986 £24.00

☐ I enclose my remittance.

☐ Please debit my Access/American Express/Diners/Visa Account*

* *Delete as appropriate.*

Card Number ☐☐☐☐☐☐☐☐☐☐☐☐☐☐☐☐

Expiry Date .... Signature ....
If address registered with card company differs from above, please give details.

Please send me details of:

☐ Oxford Reviews of Reproductive Biology

☐ Oxford Surveys in Eukaryotic Genes

☐ Oxford Surveys in Evolutionary Biology

Please note that prices given are correct at the time of printing but may be changed at the Publishers discretion and without notice.

Discounts are available for members of the ISPMB